# SIMPLIFIED BASIC ALGEBRA AND DIFFERENTIAL CALCULUS

The Ultimate Guide to Mastering Basic Algebra and Differential Calculus

Kingsley Augustine

Printed by Amazon KDP

# Table of Contents

# PREFACE

**Simplified Basic Algebra and Differential Calculus** is a book that contains topics under algebra and differential calculus. This book serves as a useful companion for students in high schools, colleges and universities. It is a valuable textbook for students who want to write entrance test or examination into colleges and universities. This book consists of step-by-step explanation of topics presented in a way that is easy for students to understand. It contains very many worked examples and many self-assessment exercises to ensure that students get a mastery of each topic covered. The answers to the exercises are provided at the end of the book.

What makes this book a unique mathematical asset, is its detailed step by step approach in explaining the topics covered in this branch of mathematics. Instead of solving questions by going straight to the point, leaving you confused and frustrated, this textbook teaches you in simple English, explaining each step taken at a time. Thus, allowing anyone, regardless of their experience in algebra and differential calculus, to understand each topic with ease, and hence make mathematics more interesting.

I give all thanks and Glory to God Almighty, for giving me the grace to write this book. I also wish to express my deep appreciation to my wife Mrs. Mercy Augustine for her patience, understanding and encouragement when I was writing this book. I also thank my children, Dora, Merit and Elvis for their moral support.


Kingsley Augustine.

# CHAPTER 1
## BASIC ARITHMETIC OPERATIONS

In order to fully understand the content of this book, it is important to be familiar with basic arithmetic operations. The following topics should remind us of these basics.

**Addition and Subtraction of Integers (Directed Numbers)**

a.  -11 + 7 = -4
Since the two numbers have different signs, one way of solving this problem is to subtract the smaller number from the larger number. Then take the sign of the higher number in the question as the sign of the answer. In this example, 7 is the smaller number while 11 is the larger number. Hence, subtract 7 from 11 to get 4. The sign of the higher number (i.e. 11) is minus, so the answer 4 will carry a minus sign. This will finally give us -4

b.  -6 + 15 = 9
Since their signs are different, subtract the smaller number from the larger number. Finally, take the sign of the higher number as the sign of the answer. In this example, 6 subtracted from 15 is 9. The sign of the higher number (i.e. 15) is plus, so the answer 9 will carry a plus sign. This will finally give us +9 or 9 as the answer as shown above.

c.  3 - 10 = -7
Their signs are different. So, subtract 3 from 10. This gives 7. The sign of the larger number (i.e. 10) is minus, so the answer 7 will carry a minus sign. This will finally give us –7.

d.  -4 - 8= -12
In this case, their signs are the same. So, add the two numbers, then give the sign that they have, to the answer. In this example, 4 added to 8 gives 12. Their sign is minus, so the answer 12 will carry a minus sign. This will finally give us -12.

e.  5 - (-6) = 5 + 6 = 11
Two negative signs that are closed to each other multiply to become positive.

f.  -6 - (-2) = -6 + 2 = -4.   (This is similar to (a) above where the larger number has a negative sign.

g.  -14 - 17 = -31
Since their signs are the same, add the two numbers to get the answer, and remember to give the sign that they have to the answer. In this example, 14 added to 17 gives 31. Their sign is minus, so the answer 31 will carry a minus sign. This will finally give us -31.

## Multiplication and Division of Integers (Directed Numbers)
a.  -2 x 3 = -6

Carry out multiplication and division as the usual way of multiplying and dividing numbers. However, when one of the numbers is negative, the answer will be negative. When the two numbers are negative, the answer will be positive.

b. $5 \times -4 = -20$

c. $-2 \times -6 = 12$

d. $-7 \times -3 = 21$

e. $-10 \div 2 = \dfrac{-10}{2} = -5$

f. $18 \div -6 = \dfrac{18}{-6} = -3$

g. $-48 \div -12 = \dfrac{-48}{-12} = 4$

h. $-24 \div -6 = \dfrac{-24}{-8} = 3$

## Precedence (Rule of PEMDAS or BODMAS)

Precedence is the order which arithmetic operations are followed during simplifications. The order of operation is easily remembered by using the acronym below:

**PEMDAS**: Parenthesis, Exponent, Multiplication, Division, Addition, Subtraction.

This means that parenthesis is done first before exponents (powers and roots), then multiplication or division, and addition or subtraction. Multiplication and division are at the same level, hence deal with anyone that comes first from left to right. Similarly, addition and subtraction are at the same level, hence deal with anyone that comes first from left to right. Another acronym that can be used for precedence is:

**BODMAS**: Bracket, Order, Division, Multiplication, Addition, Subtraction.

This means that bracket is done first before order (powers and roots), then multiplication or division, and addition or subtraction.

Note that the 'O' in BODMAS does not mean 'Of' as popularly used in some books. However, the operation, 'Of', which can be used in some arithmetic processes, should be treated as multiplication.

Between the two acronyms i.e. PEMDAS and BODMAS, I prefer the use of BODMAS because I think it is easier to remember.

**Examples**

1. $8 \div 2 \times 5$

Here, the rule of BODMAS has to be applied. We have to carry out the division first before the multiplication. This is done as follows:

$8 \div 2 \times 5 = (8 \div 2) \times 5$     (The brackets show the operation that will be done first)

$$= 4 \times 5 = 20$$

2. $15 \times 2 \div 5$

Since division and multiplication are at the same level, we simplify anyone that comes first. We can work from left to right and carry out the multiplication first before division. This is done as follows:

$$15 \times 2 \div 5 = (15 \times 2) \div 5$$
$$= 30 \div 5$$
$$= 6$$

3. $6 + 24 \div 3 \times (5 - 3) \text{ of } \dfrac{3}{4}$

By applying the rule of BODMAS, the part in the bracket will be handled first. This gives:

$$6 + 24 \div 3 \times (5 - 3) \text{ of } \frac{3}{4} = 6 + 24 \div 3 \times 2 \text{ of } \frac{3}{4}$$

The next step will be division. 24 divided by 3 gives 8. This is written as follows:

$$6 + 8 \times 2 \text{ of } \frac{3}{4}$$

The 'Of' is regarded as multiplication. This now simplifies to give

$$6 + 8 \times 2 \times \frac{3}{4} = 6 + 16 \times \frac{3}{4}$$
$$= 6 + 4 \times 3 \qquad \text{(Note that 4 divides into 16 to give 4)}$$
$$= 6 + 12$$
$$= 18$$

4. $4 \times 5 - 9 \div [3 + 2^3 - (3 \times 2)]$

The inner bracket simplifies to give 6. This is given below.

$$4 \times 5 - 9 \div (3 + 2^3 - 6)$$

The exponent in the bracket gives 8 ( i.e. $2^3 = 8$) as shown below.

$$4 \times 5 - 9 \div (3 + 8 - 6)$$

The addition and subtraction in the bracket simplifies to give $3 + 8 - 6 = 5$. This gives:

$$4 \times 5 - 9 \div 5$$

The multiplication and division simplify to give:

$$20 - \frac{9}{5} = \frac{100 - 9}{5}$$
$$= \frac{91}{5} = 18\frac{1}{5}$$

5. $4 + 6 - 3^2 \times (2 \times \sqrt{16})7 + 1$

The root in the bracket simplifies to give 4 as shown below:

$$4 + 6 - 3^2 \times (2 \times 4)7 + 1$$

The term in the bracket gives 8 as follows:

$$4 + 6 - 3^2 \times (8)7 + 1$$

The exponent or order or power simplifies to give 9 as shown below:

$$4 + 6 - 9 \times (8)7 + 1$$

The term outside the bracket multiplies the term in the bracket to give:

$$4 + 6 - 9 \times 56 + 1$$

We simplify the multiplication operation to give:

4 + 6 − 504 + 1

Starting from left to right we carry out the addition and subtraction as follows:

10 − 504 + 1 = − 494 + 1

= − 493

## Exercise 1

1. Evaluate the following:

a. −20 + 5
b. −32 + 21
c. 2 − 7
d. −3 − 9
e. 8 − (−11)
f. −11 − (−17)
g. −56 − 26

2. Find the values of the following:

a. −5 x 2
b. 2 x −11
c. −4 x −9
d. −6 x −5
e. −24 ÷ 6
f. 42 ÷ −7
g. −72 ÷ − 9
h. −48 ÷ −12

3. Apply precedence in evaluating the following:

a. 22 ÷ 11 x 6
b. 8 x 15 ÷ 5
c. $13 + 30 \div 15 \times (16 - 8) \text{ of } \frac{2}{3}$
d. $12 \times \frac{5}{24} - 9 \div [1 + 3^4 - (2 \times 5)]$
e. $2 + 11 - 4^3 \times (3 \times \sqrt{4})5 - 3$

# CHAPTER 2
## SIMPLIFICATION, FACTORIZATION AND SUBSTITUTION IN ALGEBRA

The use of letters to represent numbers is known as algebra. In algebra, we imagine the letters to stand for numbers in order for operations to be easy.

**Examples**
1. What is y kilograms in grams

Solution
    Since 1kg = 1000g
    and 2kg = 2 x 1000g
    Then ykg = y x 1000g
        = 1000yg

Note that y x 1000 = 1000 x y = 1000y. In algebra, when a number is multiplied by a letter, the number is usually written before the letter even if the letter comes first in the multiplication. Also, when two or more letters are multiplied together, the letters are usually written in alphabetical order. For example $x$ x d x f is written as df$x$ (in alphabetical order).

Also note that m x n is written as mn, m÷ n is written as $\dfrac{m}{n}$, m + n remains m + n, while m − n remains m − n.

2. A boy was 13 years old $y$ years ago. How old is he now? How old will he be in $c$ years time?

Solution
It is easy to evaluate this question if the letters are imagined to be numbers. For example, if the boy was 13 years old 4 years ago, then he will be 17 years old now (i.e. 13 + 4 = 17).
Therefore, the boy's present age is obtained by addition. So, his present age is (13 + $y$) years.
Similarly, in $c$ years time, his age will be obtained by addition. So, in $c$ years time, the boy's age will be his present age + $c$. This gives:
(13 + $y$ + $c$)years      (Note that his present age is 13 + y)

3. A man is $x$ years old. How old was he p years ago? How old will he be in 5 years time?

Solution
His present age is $x$ years. His age p years ago will be obtained by subtracting p from his present age.
So, p years ago, the man's age was ($x$ − p) years
Also, in 5 years time, his age will be:
($x$ + 5)years

4(a) What is $a in cents?

(b) Express $x$ minutes in hours.

Solution

(a) Since $1 = 100 cents

and $2 = 2 x 100 cents

Then, $a = a x 100 cents

= 100a cents

(b) Since 60 minutes = 1 hour

1 minute = $\dfrac{1}{60}$ x 1 = $\dfrac{1}{60}$ hours

2 minutes = $\dfrac{1}{60}$ x 2 hours

10 minutes = $\dfrac{1}{60}$ x 10 hours

Then, $x$ minutes = $\dfrac{1}{60}$ x $x$ hours

∴ $x$ minutes = $\dfrac{x}{60}$ hours

5. The perimeter of a rectangle is 50cm and the breadth is $y$cm. Find the area of the rectangle in terms of $y$.

Solution

Perimeter of a rectangle = 2($l$ + b), where $l$ = length and b = breadth.

50 = 2($l$ + y)          (since breadth = y)

$\dfrac{50}{2} = \dfrac{2(l+y)}{2}$          (When you divide both sides by 2 in order to eliminate the 2)

∴ 25 = $l$ + y

∴ 25 – y = $l$          (Note that when y is taken to the other side of the equation, its sign will change. Hence +y becomes –y).

∴ length, $l$ = 25 – y

Area of rectangle = $l$ x b

∴ Area = (25 – y) x y

= 25y – y$^2$

Note that y x y = y$^2$. Also, when a number or letter multiplies an expression in a bracket, the letter or number is used to multiply each term in the bracket.

6. One side of a rectangle is ($2x$ + 6)cm. If the perimeter of the rectangle is ($6x$ + 8)cm, find the area of the rectangle in terms of $x$.

Solution

Let $l$ = ($2x$ + 6)

∴ perimeter = 2($l$ + b)

$6x$ + 8 = 2($2x$ + 6 + b)

Divide both sides by 2. This gives:

$$\frac{6x + 8}{2} = \frac{2(2x + 6 + b)}{2}$$

$$\frac{6x}{2} + \frac{8}{2} = 2x + 6 + b \qquad \text{(2 has canceled out)}$$

$$3x + 4 = 2x + 6 + b$$

Collecting terms in $x$ and constant terms on the same side of the equation gives b as follows:

$$3x + 4 - 2x - 6 = b$$

$$3x - 2x + 4 - 6 = b$$

$$x - 2 = b$$

$\therefore$ Breadth, b = $x - 2$

The area of the rectangle is given by:

$$\text{Area} = l \times b$$

$$= (2x + 6) \times (x - 2)$$

$$= (2x + 6)(x - 2)$$

$$\text{Area} = 2x^2 - 4x + 6x - 12$$

$$\text{Area} = 2x^2 + 2x - 12$$

Note that in expanding the bracket above, each term in the first bracket is used to multiply each term in the second bracket. Ensure you carry the sing of each term along with the term when carrying out the expansion.

## Expansion of Brackets

The expression a$(x + y)$ means a x $(x + y)$, and it is expanded as follows:

a$(x + y)$ = a$x$ + a$y$

The expression (a + b)$(x + y)$ means (a + b) x $(x + y)$, and it is expanded as follows:

(a + b)$(x + y)$ = a$x$ + a$y$ + b$x$ + b$y$. Each term in the first bracket is used to multiply each term in the second bracket.

**Examples**

1.  Expand 2$(x - 3y)$

Solution

2$(x - 3y)$ = 2$(x)$ + 2(-3y)

$\qquad$ = 2$x$ - 6y

2.  Expand and simplify the expression, 2$x$ - [3 - (a + b)]

Solution

2$x$ - [3 - (a + b)] = 2$x$ - (3 - a - b)

$\qquad$ = 2$x$ - 3 + a + b

Note that the inner bracket was expanded first before the outer bracket. Also, a negative sign outside a bracket is taken as -1 outside the bracket. For example

-(a+b) = -1(a + b) = (-1 x a) + (-1 x b) = -a - b as expressed above.

3. Remove bracket and simplify a + 2(a − b) − (a + 3b)

Solution
a + 2(a − b) − (a + 3b) = a + 2a − 2b − a − 3b
Collecting like terms together gives:
    a + 2a − a − 2b − 3b
  = 2a − 5b
Note that terms that have the same letters are like terms. Numbers without letters are also like terms.

4. Simplify $2x(x − 3) − 2(x^2 + 2x)$

Solution
  $2x(x − 3) − 2(x^2 + 2x) = 2x(x) + 2x(-3) − 2(x^2) − 2(+2x)$
                        $= 2x^2 − 6x − 2x^2 − 4x$
Collecting like terms together gives:
    $2x^2 − 2x^2 − 6x − 4x$
  $= −10x$    (since $2x^2 − 2x^2 = 0$)

5. Expand (a − b)(a − c)

Solution
  (a − b)(a − c) = a x a + (a x −c) + (−b x a) + (−b x −c)
               $= a^2 − ac − ab + bc$    (This cannot simplify further)
Take note of how the negative sign was used in the expansion.

6. Expand and simplify $(2x + 5y)(3x − 2y)$

Solution
$(2x + 5y)(3x − 2y) = (2x$ x $3x) + (2x$ x $−2y) + (5y$ x $3x) + (5y$ x $−2y)$
                    $= 6x^2 − 4xy + 15xy − 10y^2$
                    $= 6x^2 + 11xy − 10y^2$
-Note that −4xy and +15xy are like terms which simplify to +11xy. Note that x x x = $x^2$. Similarly, $x^2$ x x = $x^3$. This is done by adding their powers. Note that x has a power of 1 which is not usually written.

7. Expand and simplify $(3a − 3)^2$

Solution
$(3a − 3)^2 = (3a − 3)(3a − 3)$
      = (3a x 3a) + (3a x −3) + (−3 x 3a) + (−3 x −3)
      $= 9a^2 − 9a − 9a + 9$
      $= 9a^2 − 18a + 9$

Note : that $- x - = +$ (i.e. minus multiplied by minus gives plus), $- x + = -$, $+ x - = -$. For example $-2a \times -a = +2a^2$, and $2 \times -3 = -6$

8. Find the coefficient of ab in the expansion and simplification of: $(2a - 3b)^2 - (a - b)^2$

<u>Solution</u>

$(2a-3b)^2 - (a - b)^2 = (2a - 3b)(2a - 3b) - [(a-b)(a-b)]$

$\qquad\qquad\qquad = 4a^2 - 6ab - 6ab + 9b^2 - (a^2 - ab - ab + b^2)$

$\qquad\qquad\qquad = 4a^2 - 6ab - 6ab + 9b^2 - a^2 + ab + ab - b^2$

Collecting like terms together gives

$\qquad 4a^2 - a^2 - 6ab - 6ab + ab + ab + 9b^2 - b^2$

$\qquad = 3a^2 - 10ab + 8b^2$

$\therefore$ The coefficient of ab $= -10$

Note that coefficient is a number multiplying a particular variable or letter/letters

9. Simplify and find the coefficient of $d^2$ in $3d(2d + 3) - (3d + 1)(2d + 1)$

<u>Solution</u>

$3d(2d + 3) - (3d + 1)(2d + 1) = 3d(2d + 3) - [(3d + 1)(2d + 1)]$

Note that the outer bracket is introduced due to the negative sign multiplying the bracket

$\qquad = 6d^2 + 9d - (6d^2 + 3d + 2d + 1)$

$\qquad = 6d^2 + 9d - 6d^2 - 3d - 2d - 1$

$\qquad = 6d^2 - 6d^2 + 9d - 3d - 2d - 1$

$\qquad = 0d^2 + 4d - 1$

$\qquad = 4d - 1$

$\therefore$ The coefficient of $d^2 = 0$

## Lowest Common Multiple (L.C.M) of Algebraic Terms

The LCM of a given set of numbers is the first multiple of the largest number which can divide all the other numbers without a remainder .

For letters, the LCM is obtained by multiplying all the letter together. However, when a particular letter is repeated in the various terms, the LCM is the letter that has the highest power.

**Examples**

Find the LCM of the following:

1. $x, y, z$
2. $2ab, 3ax, ac$
3. $5ax^2, 3a^2y, 3xy$
4. $6a$ and $3b$
5. $(x - y)$ and $(2x + y)$
6. $(a - 2b)$ and $(4a - 8b)$

<u>Solutions</u>

1. The LCM of $x$, y, z = $xyz$   (Their multiplication)

2. 2ab, 3a$x$, ac. Here, first find the LCM of 2 and 3. The larger number is 3. The multiples of 3 are 3, 6, 9, 12, ... e.t.c.  Out of these multiples the first one that can divide 2 and 3 without a reminder is 6. So the LCM of 2 and 3 is 6.
For the letters, the highest power of a is 1, so the LCM of a in all the terms is a. Similarly, the LCM of b, c, and $x$ are b, c, and $x$, since they all have power of 1. We now multiply all these LCMs to obtain the overall LCM. Therefore the LCM of 2ab, 3a$x$ and ac is 6abc$x$.

3. 5a$x^2$, 3a$^2$y, 3$xy$. For the numbers, the largest is 5. The multiples of 5 are 5, 10, 15, 20 ... e.t.c. The first multiple that can divide 3, 3 and 5 is 15. Hence, the LCM of 5, 3, 3 is 15.
For the letters, a in its highest power is a$^2$, for $x$ is $x^2$ and for y is y. So the LCM of 5a$x^2$, 3a$^2$y and 3$xy$ is 15a$^2x^2$y.

4. 6a and 3b. For the numbers, the larger value is 6. The multiples of 6 are 6, 12, 18, ... e.t.c. The first multiple that can divide 6 and 3 is 6. The letters are single in the two terms. So the LCM of 6a and 3b is 6ab.

5. $(x - y)$ and $(2x + y)$. Here, the terms in the bracket are not the same and they cannot be factorized to look alike, so the LCM is obtained by multiplying them together.
$\therefore$  The LCM of $(x - y)$ and $(2x + y)$ is $(x - y)(2x + y)$

6. (a − 2b) and (4a − 8b). If we look at the second expression carefully, we will observe that it can be factorized to have an expression like the first one. If we factorize the second expression it will give 4(a − 2b). Hence, we can say that we are looking for the LCM of (a − 2b) and 4(a − 2b). Since (a − 2b) is common, we simply pick one as an LCM and then we take 4 as an LCM since it is the only number present. When we multiply 4 and (a − 2b) (i.e. one of the common expression) it gives 4(a − 2b).
$\therefore$  The LCM of (a − 2b) and (4a − 8b) is 4(a − 2b).

## Addition and Subtraction of Algebraic Fractions

**Examples**
1. Simplify the following:

(a) $\dfrac{3x}{8} + \dfrac{x}{8}$

(b) $\dfrac{1}{5x} + \dfrac{3}{4x}$

(c) $\dfrac{2a}{3b} - \dfrac{3}{4x}$

(d) $\dfrac{3x+2}{5a} + \dfrac{2b-1}{10b}$

(e) $\dfrac{a+3b}{a} - \dfrac{2a-b}{4b} - 3$

Solutions

(a) $\dfrac{3x}{8} + \dfrac{x}{8}$

The LCM of 8 and 8 is 8. So, divide 8 (i.e. LCM) by each of the denominators and multiply the answer by the corresponding numerator. This gives:

$$\dfrac{3x}{8} + \dfrac{x}{8} = \dfrac{(1 \times 3x) + (1 \times x)}{8} = \dfrac{3x+x)}{8} = \dfrac{4x}{8}$$

$$= \dfrac{x}{2} \quad \text{(in its lowest term)}$$

(b). $\dfrac{1}{5x} + \dfrac{3}{4x}$

The LCM of $5x$ and $4x$ is $20x$. Divide $20x$ by each denominator and multiply the value obtained by the corresponding numerator. This gives:

$$\dfrac{1}{5x} + \dfrac{3}{4x} = \dfrac{(4 \times 1) + (5 \times 3)}{20x} = \dfrac{4+15}{20x} = \dfrac{19}{20x}$$

Note that $\dfrac{20x}{5x} = 4$, while $\dfrac{20x}{4x} = 5$. These values are used to multiply their corresponding numerators as shown in the solution above.

(c) $\dfrac{2a}{3b} - \dfrac{3}{4x} = \dfrac{(4x \times 2a) - (3b \times 3)}{12bx} = \dfrac{8ax - 9b}{12bx}$

Note that the answer above cannot be simplified further. Also, $\dfrac{12bx}{3b} = 4x$, while $\dfrac{12bx}{4x} = 3b$, as b and $x$ cancels out respectively.

(d) $\dfrac{3x+2}{5a} + \dfrac{2b-1}{10b} = \dfrac{2b(3x+2) + a(2b-1)}{10ab} = \dfrac{6bx + 4b + 2ab - a}{10ab}$  (This cannot be simplified further)

(e) $\dfrac{a+3b}{a} - \dfrac{2a-b}{4b} - \dfrac{3}{1}$  (The 3 is expressed as $\dfrac{3}{1}$)

This gives: $\dfrac{a+3b}{a} - \dfrac{2a-b}{4b} - \dfrac{3}{1} = \dfrac{4b(a+3b) - a(2a-b) - 4ab(3)}{4ab}$

$$= \dfrac{4ab + 12b^2 - 2a^2 + ab - 12ab}{4ab}$$

$$= \dfrac{4ab + ab - 12ab + 12b^2 - 2a^2}{4ab}$$

$$= \dfrac{-7ab + 12b^2 - 2a^2}{4ab}$$

$$= \frac{12b^2 - 7ab - 2a^2}{4ab}$$

2. Express each of the following as a single fraction in its simplest form.

(a) $\dfrac{3x + 2}{3} - \dfrac{x - 1}{4} - \dfrac{5}{12}$

(b) $\dfrac{5}{a + 4} - \dfrac{2}{a - 2}$

(c) $\dfrac{a - 3}{a + 2} - \dfrac{a - 2}{a - 5}$

Solutions

(a) $\dfrac{3x + 2}{3} - \dfrac{x - 1}{4} - \dfrac{5}{12} = \dfrac{4(3x+2) - 3(x - 1) - (1 \times 5)}{12}$

$$= \frac{12x + 8 - 3x + 3 - 5}{12}$$

$$= \frac{12x - 3x + 8 + 3 - 5}{12}$$

$$= \frac{9x + 6}{12}$$

$$= \frac{3(3x + 2)}{12} \quad \text{(After factorizing the numerator)}$$

$$= \frac{3x + 2}{4} \quad \text{(After equal division by 3)}$$

(b) $\dfrac{5}{a + 4} - \dfrac{2}{a - 2}$

The LCM of a + 4 and a − 2 is (a + 4)(a − 2)

Also, if we use this LCM to divide the first denominator, it gives: $\dfrac{(a + 4)(a - 2)}{a + 4} = a - 2$ (Since a + 4 will cancel out).

Similarly, the LCM divided by the second denominator gives: $\dfrac{(a + 4)(a - 2)}{a - 2} = a + 4$ (Since a − 2 will cancel out)

$$\therefore \frac{5}{a + 4} - \frac{2}{a - 2} = \frac{(a-2)5 - (a+4)2}{(a + 4)(a - 2)} = \frac{5a - 10 - (2a + 8)}{(a + 4)(a - 2)}$$

$$= \frac{5a - 10 - 2a - 8}{(a + 4)(a - 2)}$$

$$= \frac{3a - 18}{(a + 4)(a - 2)}$$

$$= \frac{3(a-6)}{(a + 4)(a - 2)}$$

(c) $\dfrac{a - 3}{a + 2} - \dfrac{a - 2}{a - 5}$

The LCM of a + 2 and a − 5 is (a + 2)(a − 5)

$$\therefore \quad \frac{a - 3}{a + 2} - \frac{a - 2}{a - 5} = \frac{(a-3)(a-5) - (a-2)(a+2)}{(a + 2)(a - 5)}$$

$$= \frac{a^2 - 5a - 3a + 15 - (a^2 + 2a - 2a - 4)}{(a + 2)(a - 5)}$$

$$= \frac{a^2 - 8a + 15 - (a^2 - 4)}{(a + 2)(a - 5)} \quad \text{(Note that +2a − 2a = 0)}$$

$$= \frac{a^2 - 8a + 15 - a^2 + 4)}{(a + 2)(a - 5)}$$

$$= \frac{a^2 - a^2 - 8a + 15 + 4}{(a + 2)(a - 5)}$$

$$= \frac{-8a + 19}{(a + 2)(a - 5)}$$

## Highest Common Factor (H.C.F) of Algebraic Terms

The HCF of a given set of numbers is the number which is just lower than or equal to the smallest number in the given set, and which all the numbers in the given set can divide without a remainder.

For letters, the HCF of a particular letter is obtained by taking the letter with the lowest power in the given set.

**Examples**

Find the HCF of the following:

1. 4, 6, 8
2. 3, 9, 15
3. $8a^2b$, $20ab^2$
4. $a^2b^4c$, $a^3b^2c^3$
5. $36xy^3z^5$, $54xy^5z^3$, $72xy^4z^5$

Solutions

1. The H.C.F of 4, 6, 8 is 2. The smallest number in the set is 4. So, a number that is just lower than or equal to 4 which 4, 6, and 8 can divide without a remainder is 2.

So, their H.C.F is 2.

2. The H.C.F of 3, 9, and 15 is 3. This is because the lowest number is 3, and all the numbers can divide 3 without remainder.

3. For the terms, $8a^2b$ and $20ab^2$, the H.C.F of 8 and 20 is 4. The lower number is 8, and a number which is just lower than 8 or equal to 8, and which 8 and 20 can divide without remainder is 4. Concerning the letters, between $a^2$ and a, we take "a", since it has the lower power. Similarly, between b and $b^2$ we take b since it has the lower power. Therefore, combining all these answers together gives the H.C.F of $8a^2b$ and $20ab^2$ as 4ab.

4. $a^2b^4c$, $a^3b^2c^3$
For the letter a, we take $a^2$ since it has the smaller power. For the letter b, we take $b^2$, while for the letter c, we take c since it has the lower power.
Therefore, the H.C.F of $a^2b^4c$ and $a^3b^2c^3$ is $a^2b^2c$.

5. $36xy^3z$, $54xy^5z^3$, $72xy^4z^5$
For the numbers, 36, 54 and 72, the smallest is 36. All the numbers can divide 2, 3, 6, 9 and 18. Note that these factors must not be greater than 36 which is the smallest in the set. So, among 2, 3, 6, 9 and 18, the highest is 18. So, 18 is the number which is just lower than or equal to 36, and which all the numbers (i.e. 36, 54, 72) can divide without remainder.
For the letters, the lowest power of $x$ is 1 (i.e. $x$), the lowest power of y is 3 (i.e. $y^3$), while the lowest power of z is 3 (i.e. $z^3$). So, the H.C.F of the letters is $xy^3z^3$
∴ The HCF of $36xy^3z^5$, $54xy^5z^3$, and $72xy^4z^5$ = $18xy^3z^3$.

## Factorization of Algebraic Expression

An algebraic expression can be expressed as a product of its factors. When factorizing expressions we simply take the HCF of the terms in the expression and put it outside a bracket. Then we divide each individual term in the expression by the HCF in order to obtain the new terms that will be inside the bracket.

**Examples**
1. Factories 3a – 12b

Solution
The HCF of 3 and 12 is 3. For the letters, since the letters are not repeated, then their HCF which is 1 will not be used here. When there is no common letter or number to a set of terms, then their HCF is taken as 1.
So, factorize the expression by taking the HCF and then introducing a bracket. Then, divide each term of the expression by the HCF. This is done as follows:

$3a - 12b = 3(\dfrac{3a}{3} - \dfrac{12b}{3})$    (The HCF of 3a and 12b is 3, which is used outside the bracket, and also used to divide each term in the expression)

= 3(a − 4b)    (After simplifying the division above)

∴  3a − 12b = 3(a − 4b)

(Note that the expansion of the factorized expression will give the original expression)

2.  Factorize 15y + 10.

Solution

15y + 10

The HCF of 15y and 10 is 5. The letter is not considered since it appears in only one of the terms

∴  $15y + 10 = 5(\dfrac{15y}{5} + \dfrac{10}{5})$

$\quad\quad\quad = 5(3y + 2)$    (This is obtained after the division of 15 by 5, and 10 by 5)

∴  15y + 10 = 5(3y + 2)

3.  Factorize: −6m + 15mn

Solution

−6m + 15mn

The HCF of −6m and 15m is −3m. Always take the sign of the first term.

Note that n is not present in both terms, so it is not considered in the HCF.

∴  $-6m + 15mn = -3m(\dfrac{-6m}{-3m} + \dfrac{15mn}{-3m})$

$\quad\quad\quad = -3m(2 - 5n)$

(Note that −6m divided by −3m is 2, while 15mn divided by −3m is −5n)

4.  Factorize  $-4a^5 + 2a^3b - 10a^2b^2$

Solution

$-4a^5 + 2a^3b - 10a^2b$

The HCF of $-4a^5$, $2a^3b$ and $10a^2b$ is $-2a^2$. Note that b is not present in all the three terms, so it is not considered in the HCF. Since $-2a^2$ is the HCF of the terms, it also means that $-2a^2$ is what is common to the three terms in the expression.

∴  $-4a^5 + 2a^3b - 10a^2b^2 = -2a^2(\dfrac{-4a^5}{-2a^2} + \dfrac{2a^3b}{-2a^2} - \dfrac{10a^2b^2}{-2a^2})$

$\quad\quad\quad = -2a^2(2a^3 - ab + 5b^2)$

Note that $\dfrac{a^5}{a^2} = a^{5-2} = a^3$   (simply subtract the powers). Also note that a negative sign outside a bracket changes the sign of every term in the bracket.

5.  Factorize $3m^3 - 2m^2 + m$

Solution

$3m^3 - 2m^2 + m$

The terms $m^3$, $m^2$ and m have m in common. This means that m is their HCF. 3 and 2 have nothing but 1 in common. So, their HCF is not considered here.

$$\therefore \quad 3m^3 - 2m^2 + m = m\left(\frac{3m^3}{m} - \frac{2m^2}{m} + \frac{m}{m}\right)$$
$$= m(3m^2 - 2m + 1)$$

## Factorization by Grouping

In factorization by grouping four terms are usually given in the expression. The first two of the terms will be factorized before the other two terms are factorized. After this, the common term obtained is used to carry out the final factorization of the terms.

**Examples**

1.  Factorize $ax + ay + 3bx + 3by$

Solution

The terms $ax$ and $ay$ have 'a' in common. This also means that their HCF is 'a'. The terms $3bx$ and $3by$ have $3b$ in common. This also means that their HCF is $3b$. We now factorize each pair as follows:

$ax + ay + 3bx + 3by = a(x + y) + 3b(x + y)$

The final step is to take the factors outside each of the bracket (i.e. a and 3b) and enclose them in a bracket. Then, one of the two equal bracket terms is also taken along. This gives:

$a(x + y) + 3b(x + y) = (a + 3b)(x + y)$

This also means that $(x + y)$ is common to the terms above. So, it can be factorized as follows:

$$a(x + y) + 3b(x + y) = (x + y)\left[\frac{a(x + y)}{(x + y)} + \frac{3b(x + y)}{(x + y)}\right]$$

$(x + y)$ cancels out to give: $(x + y)(a + 3b)$ as obtained above.

Hence, $ax + ay + 3bx + 3by = (x + y)(a + 3b)$

2.  Factorize $2ab - 5a + 2b - 5$

Solution

'a' is common to the first two terms. The next two terms have nothing in common, so we can take 1 to be the common term between them.

$\therefore \quad 2ab - 5a + 2b - 5 = a(2b - 5) + 1(2b - 5)$

Taking one of the bracket (since they are equal) along with the factors outside each of the bracket gives:

$a(2b - 5) + 1(2b - 5) = (2b - 5)(a + 1)$

3.  Factorize $2ax - 2ay - 3bx + 3by$

Solution

2a is common to the first two terms, while $-3b$ is common to the last two terms. Note that the sign of the first of the two terms must be taken. For example, $-ab + 2ax$ has $-a$ as the common term. Also, when a negative term is used as a common term, the inner sign of the bracket

23

formed will change. For example, $-2a - 3ax$ will give $-a(2 + 3x)$. Notice the change of sign when the bracket is formed, especially as $-3ax$ becomes $+3x$ in the bracket.

$$\therefore \quad 2ax - 2ay - 3bx + 3by = 2a(x - y) - 3b(x - y)$$
$$= (x - y)(2a - 3b)$$

Notice the change in sign of $+3by$ to $-y$ after factorizing.

4. Factorize $15 - xy + 5y - 3x$

Solution

Since the first two terms have nothing in common, and the last two terms have nothing in common, then we regroup the expression. To regroup the expression means to rearrange the terms such that the first two terms will have a common factor and the last two terms will have a common factor. This gives:

$15 - xy + 5y - 3x = 15 + 5y - xy - 3x$

$\therefore$ The expression to be factorized is:

$15 + 5y - xy - 3x$

The first two terms have 5 as a factor while the last two terms have $-x$ as a factor. The expression is now factorized as follows:

$$15 + 5y - xy - 3x = 5(3 + y) - x(y + 3)$$
$$= (3 + y)(5 - x)$$

Note that $(3 + y)$ is also equal to $(y + 3)$.

## Difference of Two Squares

$$(a + b)(a - b) = a^2 - ab + ab - b^2$$
$$= a^2 - b^2 \quad (\text{since } -ab + ab = 0)$$
$$\therefore \quad a^2 - b^2 = (a + b)(a - b)$$

This can be used to factorize the difference of the squares of two quantities.

**Examples**

1. Factorize $81 - w^2$

Solution

$81 - w^2$

We can express 81 as a square in order to make the expression a difference of two squares. This gives $81 - w^2 = 9^2 - w^2$

$$= (9 + w)(9 - w) \quad [\text{This is similar to } a^2 - b^2 = (a + b)(a - b) \text{ as explained above}]$$

2. Factorize $16y^2 - 4z^2$

Solution

$$16y^2 - 4z^2 = 4^2y^2 - 2^2z^2 \quad (\text{Since } 16 = 4^2 \text{ and } 4 = 2^2)$$
$$= (4y)^2 - (2z)^2 \quad (\text{This is a difference of two squares})$$

$$= (4y + 2z)(4y - 2z)$$

3.  Factorize $5 - 5m^2$

Solution

The common factor is 5. This is used to factorize the expression as follows:

$$5 - 5m^2 = 5(1 - m^2)$$

But, $1 - m^2$ is a difference of two squares because it can be expressed as

$$1^2 - m^2 \quad \text{(Since } 1^2 = 1)$$

$$\therefore \ 5(1 - m^2) = 5(1^2 - m^2)$$

We now factorize the terms in the bracket (i.e. difference of two squares) as follows:

$$5(1^2 - m^2) = 5(1 + m)(1 - m)$$

$$\therefore \ 5 - 5m^2 = 5(1 + m)(1 - m)$$

4.  Factorize $25x^2 - 9$

Solution

$$\begin{aligned} 25x^2 - 9 &= 5^2 x^2 - 3^2 \\ &= (5x)^2 - 3^2 \\ &= (5x + 3)(5x - 3) \end{aligned}$$

5.  Find the value of $62^2 - 38^2$ without using calculator.

Solution

$$\begin{aligned} 62^2 - 38^2 &= (62 + 38)(62 - 38) \\ &= 100 \times 24 = 2400 \end{aligned}$$

Notice that what was obtained from one bracket was used to multiply what was obtained from the other bracket.

6.  Find the value of $84^2 - 16$ without using calculator.

Solution

$84^2 - 16 = 84^2 - 4^2$   (Both terms must be expressed in squares).

$$\therefore \quad 84^2 - 4^2 = (84 + 4)(84 - 4)$$

$$= 88 \times 80$$

$$= 7040$$

## Factorization of Simple Quadratic Expression

A quadratic expression is an expression in which 2 is the highest power of the unknown. For example, $2x^2 - 5x + 8$ is a quadratic expression.

**Examples**

1.  Factorize $x^2 + 8x + 12$

Solution

Since the co–efficient of $x^2$ is 1 (i.e. $1x^2$ also means $x^2$), then we can apply a short method as follows:

$$x^2 + 8x + 12 = (x \quad )(x \quad )$$

Find two numbers such that their product is +12 and their sum is +8. These two numbers must be factors of 12 since their product has to give 12. An easy way to find the factors of a number is represented below:

```
2 | 12
2 | 6
3 | 3
  | 1
```

∴ The factors are 12 and 1, 6 and 2, 3 and 4. These are the numbers (or multiplication of numbers) on the right side and left side of the line above, along with a number that must be used to multiply them in order to obtain 12.

∴ Among the three pairs of factors of 12 above, the factors whose product is +12, and sum is +8 are +6 and +2. Note that the correct sign must be taken along with the factors. These two numbers are now used to complete the two brackets above. This gives:

$$x^2 + 8x + 12 = (x + 6)(x + 2)$$

2.  Factorize $c^2 - 8c - 20$

Solution

The co–efficient of $c^2$ is 1. So let's apply the direct method.

$$c^2 - 8c - 20 = (c \quad )(c \quad )$$

Find two numbers such that their product is −20 (i.e. the last term) and their sum is −8 (i.e. the middle term). These two numbers must be factors of 20. We can obtain the factors as follows:

```
2 | 20
2 | 10
5 | 5
  | 1
```

∴ The factors are 20 and 1 (since 20 x 1 = 20). The second factors are 10 and 2 (since 10 x 2 = 20). The third factors are 5 and 4 (since 5 x 4 = 20). Note that 4 is obtained from 2 x 2. Among these factors, it is −10 and +2 that will give a product of −20 and a sum of −8. Note that you have to try different signs with the factors in order to get the right sign. The two numbers are now used to factorize the expression by putting them in the bracket above. This gives:

$$c^2 - 8c - 20 = (c - 10)(c + 2)$$

3. Factorize $m^2 - 11m + 24$

<u>Solution</u>

$$m^2 - 11m + 24 = (m \quad )(m \quad )$$

Two numbers whose product is +24 and sum is –11 are –8 and –3. This is because –8 x (–3) = +24,

and –8 + (–3) = –11. These two numbers are now used to complete the brackets above.

$$\therefore \quad m^2 - 11m + 24 = (m - 8)(m - 3)$$

Note that when these brackets are expanded, the original quadratic expression will be obtained.

4. Factorize $2a^2 - 10a + 12$

<u>Solution</u>

The co–efficient of $a^2$ is 2. It is not 1, so the direct method used above cannot be applied here. Hence the first step in this case is to multiply the first and last terms (i.e. $2a^2$ and +12). This gives: $2a^2$ x 12 = $24a^2$

We now find two numbers as coefficients of 'a' such that their product is $24a^2$ and their sum is –10a (i.e. the middle term in the original expression). These two numbers must be factors of $24a^2$. They are –6a and –4a. We now substitute these two terms for –10a in the original expression. This gives:

$$2a^2 - 10a + 12 = 2a^2 - 6a - 4a + 12$$

We now factorize by grouping. Between the first two terms 2a is common while between the last two terms –4a is common. This now factorizes as follows:

$$2a^2 - 10a + 12 = 2a^2 - 6a - 4a + 12$$
$$= 2a(a - 3) - 4(a - 3)$$
$$= (a - 3)(2a - 4)$$

## Substitution

**Examples**

1. Evaluate the following if $x$ = 2, and y = 3
(a) $2x - 3y$
(b) $x + 2(3x - y)$

<u>Solutions</u>

(a) $2x - 3y$

Substitute 2 for $x$ and 3 for y. Note that $2x$ means 2 x $x$. Similarly, 3y means 3 x y

$\therefore \quad 2x - 3y$

$\quad = (2 \text{ x } 2) - (3 \text{ x } 3)$

$$= 4 - 9$$
$$= -5$$

(b) $x + 2(3x - y)$  (Substitute 2 for $x$ and 3 for y)
$$= 2 + 2[(3 \times 2) - 3]$$
$$= 2 + 2(6 - 3)$$
$$= 2 + 2(3)$$
$$= 2 + 6 = 8$$

Note that the term in the bracket is first evaluated. Also, a term outside a bracket is multiplying the contents of the bracket.

2. Evaluate the following given that a = −1, b = 3 and c = −5
(a) $2(a - 2c) - 4b$
(b) $(b - a)(c + 4a)$
(c) $\dfrac{a + bc}{b - ac}$

Solutions
(a) $2(a - 2c) - 4b$
$$= 2[-1 - (2(-5))] - (4 \times 3) \quad \text{(Substitute the values of the letters as given in the question)}$$
$$= 2[-1 - (-10)] - 12$$
$$= 2(-1 + 10) - 12$$
$$= 2(9) - 12$$
$$= 18 - 12$$
$$= 6$$

b. $(b - a)(c + 4a)$
$$= [3 - (-1)][-5 + (4(-1)]$$
$$= (3 + 1)[-5 + (-4)]$$
$$= 4(-5 - 4)$$
$$= 4 \times (-9)$$
$$= -36$$

Note that it is advisable to enclose each multiplication in a bracket. For example 4b is expressed as (4 x 3), and 4a is expressed as [4 x (−1)]. Always evaluate the contents of inner brackets before the outer brackets and then outside the brackets.

(c) $\dfrac{a + bc}{b - ac}$

$$= \frac{-1 + [3 \times (-5)]}{3 - [(-1) \times (-5)]}$$

$$= \frac{-1 + (-15)}{3 - (5)}$$

$$= \frac{-1 - 15}{3 - 5}$$

$$= \frac{-16}{-2}$$

= 8    (The negative sign cancels out)

Note that the numerator and denominator have to be reduced to a single value before division is possible.

3. Given that p = 3, and q = –1, find the value of $x$ when $x = p^2q - q^2p$.

<u>Solution</u>

$x = p^2q - q^2p$

$x = [3^2 \times (-1)] - [(-1)^2 \times 3]$    (Simply substitute the given values of p and q)

$= [9 \times (-1)] - (1 \times 3)$

$= -9 - 3 = -12$

Note that the terms having powers must be evaluated first before further evaluations.

4. Evaluate: $ab\sqrt{c^2 + b^2}$, given that a = 2, b = –3, c = 4

<u>Solution</u>

$ab\sqrt{c^2 + b^2}$

$= [2 \times (-3)] \times \sqrt{4^2 + (-3)^2}$

$= (-6) \times \sqrt{16 + 9}$

$= -6 \times \sqrt{25}$

$= -6 \times 5 = -30$

5. If x = –7 and y = 3, calculate the values of:

(a) $\left(\frac{x + y}{x - y}\right)^2$

(b) $2x^2y + y^2x$

<u>Solutions</u>

(a) $\left(\frac{x + y}{x - y}\right)^2$

$= \left(\frac{-7 + 3}{-7 - 3}\right)^2$

$= \left(\frac{-4}{-10}\right)^2$

$= \left(\frac{2}{5}\right)^2$    (After simplifying the term in the bracket)

$$= \frac{2^2}{5^2}$$

$$= \frac{4}{25}$$

Note that the contents of the bracket must be evaluated to a single digit before the use of the power. The power is used for both the numerator and denominator.

(b) $2x^2y + y^2x$
$= [2(-7)^2(3)] + [(3^2)(-7)]$
$= (2 \times 49 \times 3) + [9 \times (-7)]$
$= 294 + (-63)$
$= 294 - 63 = 231$

## Exercise 2

1. What is y kilometers in meters?
2. A man was 38 years old m years ago. How old is he now? How old will he be in p years time?
3. A man is b years old. How old was he c years ago? How old will he be in 12 years time?
4.(a) What is $y in cents?
(b) Express n seconds in minutes. What is n seconds in hours?
5. The perimeter of a rectangle is 44cm and the breadth is $x$cm. Find the area of the rectangle in terms of $x$.
6. One side of a rectangle is $(x + 10)$cm. If the perimeter of the rectangle is $(4x + 18)$cm, find the area of the rectangle in terms of $x$.
7. Expand $5(2x - y)$
8. Expand and simplify the expression, $11x - [5 - 3(x + y)]$
9. Remove bracket and simplify $3a + (a - 2b) - (2a + b)$
10. Simplify $3x(7x + 2) - (10x^2 - 9x)$
11. Expand $(2a + 3b)(a - 5b)$
12. Expand and simplify $(3x - 2y)(5x - y)$
13. Expand and simplify $(4a - 5)^2$
14. Find the coefficient of $xy$ in the expansion and simplification of: $(3x + 2y)^2 - (2x - 3y)^2$
15. Simplify and find the coefficient of $y^2$ in $5y(y - 2) - (2y - 3)(y - 4)$
16. Find the LCM of the following:
a. $l$, m, n
b. 6cd, 2ed, ec
c. $2a^3x$, $5ay^2$, $6x^2y$
d. 20m and 10n
e. $(x + y)$ and $(3x - y)$
17. Simplify the following:
a. $\frac{2x}{5} + \frac{x}{5}$

b. $\dfrac{4}{9x} + \dfrac{1}{3x^2}$

c. $\dfrac{a}{10b} - \dfrac{7}{15x}$

d. $\dfrac{2x+5}{8a} + \dfrac{b-3}{5b}$

e. $\dfrac{5a-b}{3} - \dfrac{a-5b}{5a^2} - 1$

18. Express each of the following as a single fraction in its simplest form.

a. $\dfrac{x-4}{10} - \dfrac{3x-2}{5} + \dfrac{4}{15}$

b. $\dfrac{2}{3a+7} - \dfrac{1}{6a+14}$

c. $\dfrac{2a-1}{3a-1} - \dfrac{3a+4}{a-2}$

19. Find the HCF of the following:
a. 5, 20, 30
b. 6, 14, 24
c. $12a^3b^2$, $30a^2b^3$
d. $ab^2c^3$, $ab^5c^2$
e. $24x^2yz^3$, $72x^2y^2z^4$, $36x^3yz^2$
20. Factorize $4m - 8n$
21. Factorize $12y - 8$.
22. Factorize: $-20y - 50yz$
23. Factorize $5m^4 - 10m^2 - m^3$
24. Factorize $2x + 6y + 5mx + 15my$
25. Factorize $3ab + 7a - 3b - 7$
26. Factorize $5bx - 5by + nx - ny$
27. Factorize $15 - xy + 3y - 5x$
28. Factorize $100 - m^2$
29. Factorize $25x^2 - 9y^2$
30. Factorize $7 - 7b^2$
31. Factorize $36p^2 - 4$
32. Find the value of $63^2 - 37^2$ without using calculator.
33. Find the value of $92^2 - 64$ without using calculator.
34. Factorize $x^2 + 6x + 9$
35. Factorize $b^2 - b - 20$
36. Factorize $n^2 - 14n + 48$
37. Factorize $5a^2 - 11a - 12$

38. Evaluate the following if $x = -3$, and $y = -2$

a. $x^2y - y^2x$

b. $(y^2 - x^2)(y - x)$

39. Evaluate the following given that $a = -1$, $b = 3$ and $c = -5$

a. $2(4a - c) + 7b$

b. $(3a - b)(b - 2c)$

c. $\dfrac{ab - bc}{a - bc}$

40. Given that $p = 1$, and $q = -3$, find the value of $x$ when $x = pq - q^2q$.

41. If $x = -4$ and $y = 2$, calculate the values of:

a. $\left(\dfrac{x + 2y}{x - 6y}\right)^2$

b. $2y^2 - x^2y$

# CHAPTER 3
# LAWS OF INDICES

The following are the laws of indices. They are true for all values of a, b and $x \neq 0$

Law 1: $x^a \times x^b = x^{a+b}$

Law 2: $x^a \div x^b = x^{a-b}$

Law 3: $x^0 = 1$

Law 4: $x^{-a} = \dfrac{1}{x^a}$    or    $bx^{-a} = \dfrac{b}{x^a}$    or    $\left(\dfrac{b}{x}\right)^{-a} = \left(\dfrac{x}{b}\right)^a$

**Examples**
Simplify the following:

1. $10^5 \times 10^4$

2. $m^8 \div m^5$

3. $\dfrac{a^{-8}}{a^3}$

4. $5x^2 \times 4x^0 \times 2x^{-6}$

5. $y^{-5} \div b^0$

<u>Solution</u>

1. $10^5 \times 10^4 = 10^{5+4} = 10^9$

2. $m^8 \div m^5 = m^{8-5} = m^3$

3. $\dfrac{a^{-8}}{a^3} = a^{-8-3} = a^{-11} = \dfrac{1}{a^{11}}$

4. $5x^2 \times 4x^0 \times 2x^{-6} = (5 \times 4 \times 2)x^{2+0+(-6)} = 40x^{2-6} = 40x^{-4} = \dfrac{40}{x^4}$

5. $y^{-5} \div b^0 = y^{-5} \div 1 = y^{-5} = \dfrac{1}{y^5}$

## Product of Indices
In applying product of indices, the following are true:

$(x^a)^b = x^{ab}$

Similarly, $(x^a y^b)^c = x^{ac} y^{bc}$ and $\left(\dfrac{x}{y}\right)^a = \dfrac{x^a}{y^a}$

**Examples**
Simplify the following:
1. $(h^4)^{-5}$
2. $(2^{-3})^2$
3. $(-c^3)^2$
4. $(-4u^2v)^3$

Solution
1. $(h^4)^{-5} = h^{4\times(-5)} = h^{-20} = \dfrac{1}{h^{20}}$

2. $(2^{-3})^2 = 2^{-3\times2} = 2^{-6} = \dfrac{1}{2^6} = \dfrac{1}{64}$

3. $(-c^3)^2 = -c^{3\times2} = -c^6 = c^6$ (A negative number that is raised to an even number power will give a positive value).

4. $(-4u^2v)^3 = -4^{1\times3} \, u^{2\times3} \, v^{1\times3} = -4^3 u^6 v^3 = -64u^6v^3$

## Fractional Indices
In applying fractional indices, the following are true:

$$x^{1/a} = \sqrt[a]{x} \quad \text{and} \quad x^{a/b} = \sqrt[b]{x^a} \quad \text{or} \quad x^{a/b} = (\sqrt[b]{x})^a$$

In all cases, $x \neq 0$

**Examples**
Simplify the following:
1. $27^{\frac{1}{3}}$
2. $9^{-\frac{1}{2}}$
3. $(25a^2)^{\frac{1}{2}}$
4. $\sqrt{1\dfrac{9}{16}}$
5. $\left(\dfrac{16}{54}\right)^{-\frac{2}{3}}$

Solutions
1. $27^{\frac{1}{3}} = \sqrt[3]{27} = 3$

2. $9^{-\frac{1}{2}} = \dfrac{1}{9^{\frac{1}{2}}} = \dfrac{1}{\sqrt{9}} = \dfrac{1}{3}$  (Note that $\sqrt[2]{9}$ is $\sqrt{9}$ since 2 is not usually written with the square root sign).

3. $(25a^2)^{\frac{1}{2}} = 25^{\frac{1}{2}} a^{(2 \times \frac{1}{2})} = 25^{\frac{1}{2}} a^1 = (\sqrt{25}) \times a = 5a$

4. $\sqrt{1\frac{9}{16}} = \sqrt{\frac{25}{16}} = \frac{5}{4}$

5. $\left(\frac{16}{54}\right)^{-\frac{2}{3}} = \left(\frac{8}{27}\right)^{-\frac{2}{3}}$ (When the fraction is expressed in its lowest term)

$\left(\frac{8}{27}\right)^{-\frac{2}{3}} = \left(\frac{27}{8}\right)^{\frac{2}{3}} = \frac{27^{2/3}}{8^{2/3}} = \frac{(\sqrt[3]{27})^2}{(\sqrt[3]{8})^2} = \frac{3^2}{2^2} = \frac{9}{4}$

Note that by taking the inverse of the term in the bracket, the negative power becomes positive.

## Equations in Indices

**Examples**

Solve the following equations:

1. $4^{x-1} = 64$
2. $n^{-\frac{2}{3}} = 9$
3. $2a^{-3} = -16$
4. $9^x = 27$
5. $5x = 40x^{-\frac{1}{2}}$

Solutions

1. $4^{x-1} = 64$

This is solved by expressing both sides of the equation in the same base and then equating the powers. This gives:

   $4^{x-1} = 4^3$   (Note that $64 = 4^3$)

Equating the powers gives:

   $x - 1 = 3$

   $x = 3 + 1$

   $x = 4$

2. $n^{-\frac{2}{3}} = 9$

In this case, the unknown letter is the base. To solve this, make the power of n to 1 by multiplying this power by its inverse and using the same sign of the power. The other side of the equation should also be raised to the same power. This gives:

   $(n^{-\frac{2}{3}})^{-3/2} = 9^{-3/2}$

   $n^{(-2/3 \times -3/2)} = \frac{1}{9^{3/2}}$   (The inverse of $\frac{2}{3}$ is $\frac{3}{2}$. The sign of the original power is also used with the

inverse)

   $n^1 = \frac{1}{(\sqrt{9})^3} = \frac{1}{3^3} = \frac{1}{27}$   (Note that $-\frac{2}{3} \times -\frac{3}{2} = 1$. This gives $n^1$)

   $n = \frac{1}{27}$

3. $2a^{-3} = -16$

Divide both sides by 2.

$$a^{-3} = \frac{-16}{2}$$

$$a^{-3} = -8$$

Now make the power of 'a' to be 1 by multiplying this power by its inverse. This gives:

$$(a^{-3})^{-\frac{1}{3}} = (-8)^{-\frac{1}{3}}$$

$$a^1 = \frac{1}{(-8)^{\frac{1}{3}}} = \frac{1}{\sqrt[3]{-8}} = \frac{1}{-2}$$

$$a = -\frac{1}{2}$$

4. $9^x = 27$

Expressing both sides of the equation in the same base gives:

$$(3^2)^x = 3^3$$

$$3^{2x} = 3^3$$

Equating the powers gives

$$2x = 3$$

$$x = \frac{3}{2}$$

5. $5x = 40x^{-\frac{1}{2}}$

Divide both sides by 5.

$$x = \frac{40x^{-\frac{1}{2}}}{5}$$

$$x = 8x^{-\frac{1}{2}} \quad \text{(Since } 40 \div 5 \text{ gives 8)}$$

Divide both sides by $x^{-\frac{1}{2}}$

$$\frac{x}{x^{-\frac{1}{2}}} = \frac{8x^{-\frac{1}{2}}}{x^{-\frac{1}{2}}}$$

Cancelling out $x^{-\frac{1}{2}}$ on the right hand side gives,

$x^{1-(-\frac{1}{2})} = 8$ (Note that $x$ can be expressed as $x^1$. Also, from the law of indices, $x \div x^{-\frac{1}{2}} = x^{1-(-\frac{1}{2})}$)

$$\therefore x^{1+\frac{1}{2}} = 8$$

$$x^{3/2} = 8$$

Make the power of $x$ to be 1 by multiplying it by its inverse. Also raise the power of 8 to the same inverse. This gives:

$$(x^{3/2})^{\frac{2}{3}} = 8^{\frac{2}{3}}$$

$$x^1 = 8^{\frac{2}{3}}$$

$$x = (\sqrt[3]{8})^2 = (2)^2$$

$$x = 4$$

## Exercise 3

1. Simplify the following:
   a. $-3(te^3)^4$
   b. $(4ab^3)^3$
   c. $\dfrac{(-a)^2 \times a^7}{(-a)^5}$
   d. $(-g^4)^5$
   e. $\dfrac{(m^2)^3}{m^4 \times (-m)}$

2. Simplify the following:
   a. $(3a)^{-1}$
   b. $(a^2)^{-\frac{1}{2}}$
   c. $(49x^3)^{\frac{1}{2}}$
   d. $(27x^{3/2})^{\frac{2}{3}}$

3. Solve the following equations:
   a. $x^{-\frac{1}{2}} = 5$
   b. $a^{-2} = 9$
   c. $9^{x-2} = 27$
   d. $\dfrac{4^{2x-1}}{16^2} = 64$

# CHAPTER 4
## LINEAR EQUATIONS AND CHANGE OF SUBJECT OF FORMULAE

### Linear Equations

Linear equations can sometimes be expressed with brackets, with fractions or both. When solving linear equations with fractions, clear the fractions by multiplying each term of the equation by the L.C.M of the denominators of the equation.

**Examples**

1. Solve the following equations:
a. $2x + 4(3 - x) = 11$
b. $6(a - 3) - 2(5a - 8) = -4$
c. $a - 5(2 + a) - (3a - 4) = 2(2a - 1) - 7$

Solutions

a. $2x + 4(3 - x) = 11$

Expanding the bracket gives:

$\quad 2x + 12 - 4x = 11$

Collect like terms. This means to collect terms in $x$ on one side of the equation and constant terms on the other side of the equation.

$\quad 12 - 11 = 4x - 2x$ (When a number crosses the equality sign, its sign will change)

$\quad 1 = 2x$

Divide both sides by 2

$\quad \therefore \ x = \dfrac{1}{2}$

b. $6(a - 3) - 2(5a - 8) = -4$

Expanding the brackets gives:

$6a - 18 - 10a + 16 = -4$ (Note that $-2 \times -8 = +16$)

Collect like terms in 'a' on one side of the equation.

$6a - 10a = -4 + 18 - 16$

$-4a = -2$

Divide both sides by $-4$

$\dfrac{-4a}{-4} = \dfrac{-2}{-4}$

$a = \dfrac{-2}{-4}$

$a = \dfrac{1}{2}$

c. $a - 5(2 + a) - (3a - 4) = 2(2a - 1) - 7$

Expanding the brackets gives:

$a - 10 - 5a - 3a + 4 = 4a - 2 - 7$ (Note that $-5 \times (+a) = -5a$, and $- \times -4 = -1 \times -4 = +4$)

38

Collect like terms on one side of the equation.

$7 + 2 + 4 - 10 = 5a - a + 3a + 4a$

$3 = 11a$

Divide both sides by 11

$$\frac{3}{11} = \frac{11a}{11}$$

$$a = \frac{3}{11}$$

2. Solve the following equations:

a. $\frac{3}{4}x - \frac{1}{3}(x - 2) = \frac{5}{6} - (2x - 1)$

b. $\frac{1}{6}(5x - 2) - \frac{2}{3}(4 - x) = 1$

c. $\frac{5}{2m - 3} - \frac{3}{4} = \frac{1}{6} + 7$

Solutions

a. $\frac{3}{4}x - \frac{1}{3}(x - 2) = \frac{5}{6} - (2x - 1)$

In order to clear the fractions, multiply each term in the equation by 12 which is the L.C.M of the denominators, i.e. 4, 3 and 6. This gives:

$12(\frac{3x}{4}) - 12 \times \frac{1}{3}(x - 2) = 12(\frac{5}{6}) - 12(2x - 1)$

Cancelling out by using the 12 to divide the various denominators gives:

$3(3x) - 4(x - 2) = 2(5) - 12(2x - 1)$

$9x - 4x + 8 = 10 - 24x + 12$

Collect like terms:

$9x - 4x + 24x = 10 + 12 - 8$

$29x = 14$

$$x = \frac{14}{29}$$

b. $\frac{1}{6}(5x - 2) - \frac{2}{3}(4 - x) = 1$

In order to clear the fractions, multiply each term in the equation by 6 which is the L.C.M of the denominators, i.e. 6 and 3. This gives:

$6 \times \frac{1}{6}(5x - 2) - 6 \times \frac{2}{3}(4 - x) = 6 \times 1$

Cancelling out by using the 6 to divide the various denominators gives:

$(5x - 2) - 2 \times 2(4 - x) = 6$

$$5x - 2 - 4(4 - x) = 6$$
$$5x - 2 - 14 + 4x = 6$$
$$5x + 4x = 6 + 2 + 14$$
$$9x = 22$$
$$\therefore \; x = \frac{22}{9}$$

c. $\dfrac{5}{2m - 3} - \dfrac{3}{4} = \dfrac{1}{6} + 7$

Multiply each term in the equation by 12(2m − 3) which is the L.C.M of the denominators, i.e. (2m − 3), 4 and 6. This gives:

$$12(2m - 3)\frac{5}{2m - 3} - 12(2m - 3)\frac{3}{4} = 12(2m - 3)\frac{1}{6} + 12(2m - 3) \times 7$$

Cancelling out gives:

$$12(5) - 3(2m - 3)3 = 2(2m - 3) + 12(2m - 3)7$$
$$60 - 9(2m - 3) = 4m - 6 + 84(2m - 3)$$
$$60 - 18m + 27 = 4m - 6 + 168m - 252$$
$$60 + 27 + 252 + 6 = 4m + 168m + 18m$$
$$345 = 190m$$
$$m = \frac{345}{190}$$

$$\therefore \; m = \frac{69}{38} \qquad \text{(After equal division by 5)}$$

### Change of Subject of Formulae

If m = b + c, then m is the subject of the formula. If it is rearranged to give b = m − c, then b is now the new subject of the formula.

In changing the subject of a formula, simply solve the equation for the letter which is to become the new subject.

**Examples**

1. Make h the subject of the formula: $s = \dfrac{wd}{h}\left(h - \dfrac{d}{2}\right)$

Solution

$$s = \frac{wd}{h}\left(h - \frac{d}{2}\right)$$

Expanding the bracket gives:

$$s = \frac{wd}{h}(h) - \frac{wd}{h}\left(\frac{d}{2}\right)$$

Canceling out the h gives:

$$s = wd - \frac{wd^2}{2h}$$

To clear fractions, multiply throughout by 2h (LCM)

$$2h(s) = 2h(wd) - 2h(\frac{wd^2}{2h})$$

2hs = 2hwd − wd²    (Note that the 2h at the end on the right side has cancelled out).

Collect terms in h on one side.

wd² = 2hwd − 2hs

Factorizing the right hand side gives:

wd² = h(2wd − 2s)

Divide both sides by (2wd − 2s)

$$\frac{wd^2}{2wd-2s} = \frac{h(2wd-2s)}{2wd-2s}$$

Cancelling out the 2wd − 2s on the right hand side gives:

$$h = \frac{wd^2}{2wd-2s}$$

2.  Given that  $I = \dfrac{E}{\sqrt{R^2 + W^2 L^2}}$ , express R in terms of I, E, W and L.

<u>Solution</u>

$$I = \frac{E}{\sqrt{R^2 + W^2 L^2}}$$

Cross multiply to obtain:

$$I\sqrt{R^2 + W^2 L^2} = E$$

Square both sides to remove the square root sign.

$$(I\sqrt{R^2 + W^2 L^2})^2 = E^2$$

$I^2(R^2 + W^2L^2) = E^2$    (Note that the square also affect I in the bracket)

$I^2R^2 + I^2W^2L^2 = E^2$

$I^2R^2 = E^2 - I^2W^2L^2$

Divide both sides by $I^2$

$$R^2 = \frac{E^2 - W^2 I^2 L^2}{I^2}$$

This can also be simplified as follows:

$$R^2 = \frac{E^2}{I^2} - \frac{I^2 W^2 L^2}{I^2}$$

Canceling out $I^2$ on the right side gives:

$$R^2 = \frac{E^2}{I^2} - W^2L^2$$

Take the square root of both sides in order to remove the square on $R^2$.

$$\therefore \quad R = \sqrt{\frac{E^2}{I^2} - W^2 L^2}$$

Note that when a term/terms in a square root sign are squared, it gives only the terms in the root sign. For example, $(\sqrt{m^2 b^2})^2 = m^2 b^2$. Take note of the absence of the root sign

3. Make $x$ the subject of the formula $R = \sqrt{\dfrac{ax - P}{Q + bx}}$

Solution

$$R = \sqrt{\frac{ax - P}{Q + bx}}$$

Square both sides to remove the square root sign
$$R^2 = \frac{ax - P}{Q + bx}$$
By cross multiplication it gives:
$R^2(Q + bx) = ax - P$
$R^2 Q + R^2 bx = ax - P$
Collecting terms in $x$ gives
$ax - R^2 bx = R^2 Q + P$
Factorizing the left hand side gives:
$x(a - R^2 b) = R^2 Q + P$
Divide both sides by $(a - R^2 b)$
$$\frac{x(a - R^2 b)}{(a - R^2 b)} = \frac{R^2 Q + P}{a - R^2 b}$$

$$x = \frac{R^2 Q + P}{a - R^2 b}$$

Exercise 4
1. Solve the following equations:
a. $5x + 2(3 - x) = 10$
b. $2(2a - 3) - 5(4a - 1) = -6$
c. $b - 4(1 + b) - (5b - 1) = -(b - 3) - 2$

2. Solve the following equations:
a. $\dfrac{1}{4}x - \dfrac{2}{3}(x - 1) = \dfrac{3}{4} - (5x - 2)$

b. $\dfrac{1}{6}(5x-2) - \dfrac{5}{12}(3-2x) = 1$

c. $\dfrac{2}{2n-3} - \dfrac{3}{5} = 2\dfrac{1}{2}$

3. Make p the subject of the formula: $tp = md(p - \dfrac{d}{3})$

4. Given that $V = \dfrac{P}{\sqrt{E^2 + I^2 C^2}}$, express C in terms of I, V, E and P.

5. Make $x$ the subject of the formula $R = \sqrt{\dfrac{bx - S}{T + ax}}$

6. Make m the subject of the formula: $Amq = md(\dfrac{q}{5} - \dfrac{dA}{5})$

7. Given that $L = \dfrac{TC^2}{Y^3 - B}$, express Y in terms of L, T, B and C.

8. Make $x$ the subject of the formula $\sqrt{\dfrac{Q^2}{x}} = \sqrt{\dfrac{M}{2x^3}}$

# CHAPTER 5
## LINEAR EQUATIONS FROM WORD PROBLEMS

**Examples**

1. The sum of 6 and one-third of a certain number is one more than twice the number. Find the number.

<u>Solution</u>

Let the number be $x$

One-third of the number = $\frac{1}{3}x$

The sum of 6 and one-third of the number = $6 + \frac{1}{3}x$.

Twice the number = $2x$

$6 + \frac{1}{3}x$ is one more than $2x$. This means that:

$6 + \frac{x}{3} - 2x = 1$

Multiply each term by 3 which is the LCM of all the denominators. Note that all the whole numbers have 1 as their denominator. This multiplication by the LCM of their denominators is done in order to clear the fraction. Each term is now multiplied by 3 as follows:

$(6 \times 3) + 3(\frac{x}{3}) - (2x \times 3) = 1 \times 3$

$18 + x - 6x = 3$   (Note that in $3(\frac{x}{3})$, the 3 cancels out to give $x$)

Collect terms in $x$ on one side and the other terms on the other side of the equation. This gives:

$18 - 3 = 6x - x$   (Note the change in the signs of the terms that cross the equality sign).

$15 = 5x$

Divide both sides of the equation by 5 (i.e. the co-efficient of $x$ which is the unknown).

$\frac{15}{5} = \frac{5x}{5}$

$\therefore \quad 3 = x$   (The 5 cancels out)

$\therefore \quad x = 3$


2. When 10 is subtracted from the product of 6 and a certain number, and the result is divided by 4, the answer is equal to the number. What is the number?

<u>Solution</u>

Let the number be a

The product of 6 and the number = 6a

When 10 is subtracted from this product it gives: 6a − 10

This result divided by 4 is $\frac{6a-10}{4}$

The answer is equal to the number means that:

$\frac{6a-10}{4} = a$

Cross multiply to solve this equation. This gives:

$6a - 10 = 4a$

collect like terms in a on the left hand side

$6a - 4a = 10$

$2a = 10$

Divide both sides by 2

$$\frac{2a}{2} = \frac{10}{2}$$

$\therefore a = 5$

3. A man is three times as old as his son. Five years ago, the sum of their ages was 38. Find the age of the man and his son.

Solution

Let the son's age be $x$

$\therefore$ The man's age = $3x$ (i.e. three times his son's age)

Five years ago, the son's age was: $x - 5$

Similarly, five years ago the man's age was: $3x - 5$

The sum of their age five years ago, was:

$x - 5 + 3x - 5$

Since this sum is 38, we express it as follows:

$x - 5 + 3x - 5 = 38$

$x + 3x = 38 + 5 + 5$

$4x = 48$

$\therefore \qquad x = \dfrac{48}{4}$

$x = 12$

$\therefore$ The son's age = 12 years

The man's age = $3x$ = 3 x 12 = 36years

4. A girl is 5 years old and her mother is 32 years old. In how many years time will the mother be twice as old as the girl?

Solution

Let the number of years time be y.

At that time, the girl's age will be = 5 + y

At that time the mother's age will be = 32 + y

Since the mother's age will be twice the girl's age, it means that:

$\qquad 2(5 + y) = 32 + y$ \qquad (Note that 2(5 + y) is twice the girl's age)

$\therefore \qquad 10 + 2y = 32 + y$

$\qquad 2y - y = 32 - 10$ \qquad (After collecting like terms)

$\qquad y = 22$

$\therefore$ In 22 years time, the mother will be twice as old as the girl.

5. A cyclist takes 5hours to travel from a town A to a town B. When coming back, the journey took him 1hour less because he increased his speed by 15km/h. What was his speed when he

was going from A to B.

<u>Solution</u>
Let his speed from A to B be $x$

$$\text{Speed} = \frac{\text{Distance}}{\text{Time}}$$

∴    Distance = speed x time

        = $x$ x 5 = $5x$    (When going from A to B)

When coming from B to A (i.e. the return journey), his speed was $x + 15$ (since he increased His speed by 15km/h), while his time was: $5 - 1 = 4$    (Since he took 1 hour less)

∴    Distance = speed x time

        = $(x + 15)$ x 4

        = $4(x + 15)$

Since the distance from A to B is equal to the distance from B to A, then it follows that:

    $5x = 4(x + 15)$    (Distance A to B is $5x$, while distance B to A is $4(x + 15)$. Both are equal)

    $5x = 4x + 60$

    $5x - 4x = 60$

        $x = 60$

His speed when going from A to B was 60km/h


6.  A man walked for 3hours at 8km/h. He then cycled at 12km/h for a certain period of time. If his total distance was 54km, for how many hours did he cycle?

<u>Solution</u>

$$\text{Speed} = \frac{\text{Distance}}{\text{Time}}$$

When he walked, his total distance was:

    Distance = speed x time

        = 8 x 3

        = 24km

When he cycled, his total distance was:

    Distance = speed x time

Let the time taken to cycle be $x$

∴    Distance cycled = 12 x $x$

          = $12x$

His total distance is given by:

    $24 + 12x$

Since this total distance was 54km, then:

    $24 + 12x = 54$

        $12x = 54 - 24$

        $12x = 30$

∴        $x = \dfrac{30}{12}$

        $x = \dfrac{5}{2}$    (After equal division by 6)

$$x = 2\frac{1}{2}$$

∴    The man cycled for $2\frac{1}{2}$ hours

7. A sum of $324 is made up of $10 notes and $1 notes. If there are eight times as many $1 notes as there are $10 notes, find the number of each note.

Solution
Let there be $x$ $1 notes.
∴  The number of $10 notes = $8x$   (i.e. eight times as many $1 notes)
∴  Total amount of $1 notes = $x \times 1 = x$   (This means $$x$)
Similarly, total amount of $10 notes = $8x \times 10 = 80x$    (This means $$80x$)
Since this total gives the sum of $324, it means that:

   $x + 80x = 324$
   $81x = 324$      (Note that $x$ means $1x$)
∴      $x = \dfrac{324}{81}$
   $x = 4$
∴  Number of $1 = 4$
Number of $10 notes = $8x = 8 \times 4 = 32$

8.  Chloe has $30 and Peter has $186. If Chloe saves $5 a day and Peter spends $7 a day, after how many days will they have equal amounts?

Solution
Let the number of days for both of them to have equal amount of money be y.
If Chloe saves $5 a day, then after y days, she must have saved $5y (i.e. 5 x y)
If Peter spends $7 a day, then after y days he must have spent $7y (i.e. 7 x y)
So, total amount that Chloe will have = 30 + 5y     (i.e. her initial money plus her savings)
Total amount that Peter will have = 186 – 7y    (i.e. his initial money minus his spending)
Since they will both have equal amount of money, then:

   $30 + 5y = 186 - 7y$
∴    $5y + 7y = 186 - 30$
   $12y = 156$
∴      $y = \dfrac{156}{12}$
   $y = 13$
∴  They will have equal amounts of money after 13 days.

9. A total of $x$ litres of water is needed to fill 30 tanks with the same amount of water in each tank. If the size of each tank is 3 litres less, there will be enough water to fill 32 tanks. What is the value of $x$.

<u>Solution</u>

Amount of water in each of the 30 tank is: $\dfrac{x}{30}$

When the size of each of these 30 tanks is 3 litres less than its original size, then each tank will contain: $\dfrac{x}{30} - 3$

Since this amount of water can fill each of the 32 tanks, then the total amount of water in the 32 tanks is:

$$32(\dfrac{x}{30} - 3)$$

This total amount of water has not changed. It is also equal to $x$ litres (i.e. the original quantity of water). This means that:

$$32(\dfrac{x}{30} - 3) = x$$

$$\therefore \quad \dfrac{32x}{30} - 96 = x$$

Multiply each term by 30 in order to clear fraction. This gives:

$$30(\dfrac{32x}{30}) - 30(96) = 30(x)$$
$$32x - 2880 = 30x$$
$$32x - 30x = 2880$$
$$2x = 2880$$
$$\therefore \quad x = \dfrac{2880}{2}$$
$$x = 1440$$
$$x = 1440 \text{ litres}$$

10.  When a plane travels a certain journey at an average speed of 200km/h, it arrives an airport at 5am. When it travels at an average speed of 240km/h, it arrives at the airport at 3am. What is the length of the journey?

<u>Solution</u>

Let the duration of the journey when it travels at 200km/h be m hours.

∴ The duration when it travels at 240km/h will be:

    m – 2    (i.e. it arrives 2 hours earlier, from 5am to 3am. This means it takes lesser time)

But, average speed = $\dfrac{\text{Total distance}}{\text{Total time}}$

∴ At 200km/h:   $200 = \dfrac{\text{Total distance}}{m}$

∴ Total distance = 200 x m

              = 200m

Similarly, at 240km/h:   Total distance = 240(m – 2)

Since the distance is the same, then:

    200m = 240(m – 2)
    200m = 240m – 480
      480 = 240m – 200m
      480 = 40m

$$\therefore \quad m = \frac{480}{40}$$

$$m = 12$$

Recall that: Total distance = 200m or 240(m − 2)

∴   Total distance = 200 x 12 = 2400

∴   The length of the journey is 2400km


11.  A man walks to a village at 18km/h and returns at 12km/h. If the whole journey takes $6\frac{1}{4}$ hours, what is the total distance walked?


Solution

Let $x$ be the time taken when he walked at 18km/h

∴   The time taken for the return journey when he walked at 12km/h will be:

  $6\frac{1}{4} - x$   (since total time for the two journeys is $6\frac{1}{4}$)

Also, speed = $\dfrac{\text{Distance}}{\text{Time}}$

∴   At 18km/h:   $18 = \dfrac{\text{Distance}}{x}$

∴   Distance = 18$x$

Similarly,

At 12km/h:    $12 = \dfrac{\text{Distance}}{6\frac{1}{4} - x}$

∴   Distance = $12(6\frac{1}{4} - x)$

Since the to and fro distances are the same, then:

  $18x = 12(6\frac{1}{4} - x)$

  $18x = 12(\frac{25}{4} - x)$

  $18x = 12(\frac{25}{4}) - 12(x)$

  $18x = 3(25) - 12x$

  $18x + 12x = 75$

    $30x = 75$

$\therefore \qquad x = \dfrac{75}{30}$

      $x = 2\frac{1}{2}$ hours

∴    Distance = 18$x$ or 12(6¼ − $x$)

∴    Distance = 18 x 2½

      $= 18 \times \dfrac{5}{2} = 45$km

This distance is for one part of the journey

∴   Total distance for the to and fro journey is:

    45 x 2 = 90

∴   The man walked a total distance of 90km.

12. The result of taking 3 from a number and multiplying the answer by 4 is the same as taking 3 from five times the number. Find the number.

Solution
Let the number be c
3 taken from c means: $c - 3$
Multiplying this by 4 gives: $4(c - 3)$        (This is the first part of the statement)
Five times the number is 5c
3 taken from 5c is $5c - 3$      (This is the second part of the statement)
Since the first part = the second part, then:

$$4(c - 3 = 5c - 3$$
$$4c - 12 = 5c - 3$$
$$4c - 5c = 12 - 3$$
$$-c = 9$$
∴        $$-9 = c$$
∴        $$c = -9$$

Hence, the number is −9.

13. The sum of three consecutive numbers is 63. Find the numbers.

Solution
Let the first number be $x$.
∴    The second number will be $x + 1$, while the third number will be $x + 2$. For example, if the first number is 10, then the second number will be 11 which is (10 + 1), while the third number will be 12 which is (10 + 2).
So, the numbers are $x$, $x + 1$ and $x + 2$.
Since their sum is 63, then it follows that:

$$x + (x + 1) + (x + 2) = 63$$
$$3x + 3 = 63$$
$$3x = 63 - 3$$
$$3x = 60$$
∴        $$x = \frac{60}{3}$$
$$x = 20$$

∴ The numbers are 20, 21 and 22 from $x$, $x + 1$ and $x + 2$.

14. The sum of two numbers is 21. Five times the first number added to two times the second number is 66. Find the two numbers.

Solution
Let the first number be $x$.
∴ The second number is $21 - x$. Note that the value of the sum of two numbers minus the first number gives the second number. For example, if 8 + 6 = 14, then 14 − 8 = 6 or 14 − 6 = 8. This shows that

Sum − one of the numbers = the other number

∴ five times the first number is : 5 x $x$ = $5x$

2 times the second number is: 2 x (21 − $x$) = 2(21 − $x$)

When these two expressions are added together, it gives 66 as follows:

$$5x + 2(21 − x) = 66$$
$$5x + 42 − 2x = 66$$
$$5x − 2x = 66 − 42$$
$$3x = 24$$
$$∴ \quad x = \frac{24}{3}$$
$$x = 8$$

∴ The first number is 8 and the second number is 21 − 8 = 13, i.e. 8 and 13.


15.  2 is added to twice a certain number and the sum is doubled. The result is 10 less than 5 times the original number. Find the number.


Solution

Let the number be y.

Twice the number is 2 x y = 2y

2 added to twice the number = 2 + 2y

When this sum is doubled, it gives: 2 x (2 + 2y) = 2(2 + 2y)

From second sentence, 5 times the number = 5 x y = 5y

10 less than 5 times the number is: 5y − 10

Since the result from first sentence = the result from second sentence, then it follows that:

$$2(2 + 2y) = 5y − 10$$
$$4 + 4y = 5y − 10$$
$$4 + 10 = 5y − 4y$$
$$14 = y$$
$$∴ \quad y = 14$$

The number is 14.


16.  The sum of two numbers is 38. When 8 is added to twice one of the numbers, the result is 5 times the other number. Find the two numbers.


Solution

Let one of the numbers be n.

∴ The other number = 38 − n (Since their sum is 38)

Twice one of the numbers is: 2 x n = 2n

8 added to twice one of the numbers is: 8 + 2n

5 times the other number is: 5 x (38 − n) = 5(38 − n)

Since these two results are equal according to the question, then we equate them as follows:

$$8 + 2n = 5(38 − n)$$
$$8 + 2n = 190 − 5n$$
$$2n + 5n = 190 − 8$$

$$7n = 182$$
$$n = \frac{182}{7}$$
$$n = 26$$

∴   The two numbers are n = 26 and 38 − n = 38 − 26 = 12.

They are 26 and 12

17.  A woman's age and her son's age add up to 45 years. Five years ago, the woman was 6 times as old as her son. How old was the woman when the son was born?

Solution

Let the woman's age be $x$

∴ The son's age is = $45 - x$ (Since their sum is 45)

Five years ago, the woman's age = $x - 5$

Five years ago, the son's age was: $45 - x - 5 = 40 - x$

Since five years ago, the woman's age was 6 times that of her son, then this gives:

$$x - 5 = 6(40 - x)$$
$$x - 5 = 240 - 6x$$
$$x + 6x = 240 + 5$$
$$7x = 245$$
$$x = \frac{245}{7}$$
$$x = 35$$

∴ The woman's age is $x$ = 35years.

The son's age is $45 - x = 45 - 35 = 10$years.

∴   The woman's age when the son was born is given by the difference between their ages (i.e. how many years older is the woman than her son). This gives

$$35 - 10 = 25$$

∴   The woman was 25 years old when the son was born.

18.  The numerator of a fraction is 3. If 9 is added to the numerator and 4 is added to the denominator, the fraction is doubled. What is the fraction?

Solution

Let the denominator be y

∴ The fraction is $\frac{3}{y}$

6 added to the numerator = 3 + 6 = 9

4 added to the denominator = y + 4

∴ The new fraction is: $\frac{9}{y + 4}$

This new fraction is double of the initial fraction. This gives

$$\frac{9}{y + 4} = 2\left(\frac{3}{y}\right)$$
$$\frac{9}{y + 4} = \frac{6}{y}$$

Cross multiply to give:

$$9y = 6(y + 4)$$
$$9y = 6y + 24$$
$$9y - 6y = 24$$
$$3y = 24$$
$$\therefore \quad y = \frac{24}{3}$$
$$y = 8$$

$\therefore$ The fraction is: $\frac{3}{y}$, which is $\frac{3}{8}$

## Exercise 5

1. The sum of 9 and one-fourth of a certain number is five less than twice the number. Find the number.

2. When 2 is subtracted from the product of 5 and a certain number, and the result is divided by 3, the answer is equal to one and a half times the number. What is the number?

3. A man is six times as old as his daughter. Four years ago, the man was four times as old as his daughter. Find the age of the man and his daughter.

4. A boy is 16 years old and his mother is 42 years old. In how many years time will the mother be twice as old as the boy?

5. A cyclist takes 10hours to travel from a town A to a town B. When coming back, the journey took him 2hours more because he reduced his speed by 30km/h. What was his speed when he was coming from B to A.

6. A man runs for 2hours at 6km/h. He then cycled at 16km/h for a certain period of time. If his total distance was 72km, for how many hours did he cycle?

7. A sum of $186 is made up of $10 notes and $1 notes. If there are three times as many $10 notes as there are $1 notes, find the number of each note.

8. John has $20 and Rose has $200. If John saves $3 a day and Rose spends $9 a day, after how many days will they have equal amount of money?

9. A total of $x$ litres of wine is needed to fill 21 tanks with the same amount of wine in each tank. If the size of each tank is 55 litres more, the wine will fill only 10 tanks. What is the value of $x$.

10. When a plane travels a certain journey at an average speed of 180km/h, it arrives an airport at 2pm. When it travels at an average speed of 150km/h, it arrives at the airport at 4.30pm. What is the length of the journey?

11. A man walks to a village at 12km/h and returns at 15km/h. If the whole journey takes 4hours, what is the total distance walked?

12. The result of taking 5 from a number and multiplying the answer by 8 is the same as adding 8 to four times the number. Find the number.

13. The sum of three consecutive odd numbers is 45. Find the numbers.

14. The sum of two numbers is 29. Two times the first number added to six times the second number is 78. Find the two numbers.

15.  6 is added to thrice a certain number and the sum is halved. The result is 3 less than 2 times the original number. Find the number.

16.  The sum of two numbers is 40. When 10 is added to twice one of the numbers, the result is 4 times the other number. Find the two numbers.

17.  A woman's age and her son's age add up to 52 years. 8 years ago, the woman was 17 times as old as her son. How old was the woman when the son was born?

18.  The numerator of a fraction is 5. If 2.5 is added to the numerator and 8 is subtracted from the denominator, the fraction is halved. What is the fraction?

# CHAPTER 6
## SIMULTANEOUS LINEAR EQUATIONS

In simultaneous equations, we solve a pair of equations and determine the values of two variables. Apart from graphical method, the two methods of solving simultaneous equations are:
1. Substitution Method
2. Elimination Method

## Substitution Method

In this method, we make one of the variables a subject of formula from one of the equations, and substitute it into the other equation. It is easier to use this method when one of the variables has a coefficient of 1.

**Examples**

1.  Solve simultaneously the equations:   $x + y = 4$   and   $2x - y = 5$

<u>Solution</u>

Let us rewrite this equation and label them equation 1 and 2 for easy identification. This gives:

$x + y = 4$ .....................Equation (1)

$2x - y = 5$ .....................Equation (2)

A careful look at the two equations shows that $x$ and y in equation 1 have coefficients of 1, while y in equation 2, has a coefficient of −1. Any of these three variables can easily be made the subject of formula in their respective equation. So, let us make $x$ in equation 1 the subject of formula of the equation. This is done as follows:

$x + y = 4$ .....................Equation (1)

$x = 4 - y$

Let us represent this equation as equation 3 as follows:

$x = 4 - y$ .....................Equation (3)

Since $x = 4 - y$, let us substitute $4 - y$ for $x$ in equation 2. Note that we made $x$ the subject of formula from equation 1, so the substitution of the expression for $x$ should be done in the other equation (i.e. equation 2), and not in the same equation from which $x$ was made the subject of the formula. Hence we substitute $4 - y$ for $x$ in equation 2 as follows:

$2x - y = 5$ .....................Equation (2)

$2(4 - y) - y = 5$         (Note that $x$ has been replaced with $4 - y$ as stated above)

$8 - 2y - y = 5$

$-3y = 5 - 8$

$-3y = -3$

$y = \dfrac{-3}{-3}$         (When we divide both sides by −3)

$y = 1$

55

Now, substitute 1 for y in equation 3 in order to obtain $x$. Note that we can use any of the three equations, but equation 3 is the best to use since it directly gives us the value of $x$. This gives:

$x = 4 - y$ ........................Equation (3)

$x = 4 - 1$

$x = 3$

Therefore, $x = 3$ and $y = 1$

2.  Solve the simultaneous equations:  $3x + 2y = 10$ and $4x - y = 6$

Solution

Let us rewrite this equation and label them equation 1 and 2 for easy identification. This gives:

$3x + 2y = 10$ ....................Equation (1)

$4x - y = 6$ ....................Equation (2)

A careful look at the two equations shows that only y in equation 2 has a coefficients of −1. So, let us make y in equation 2 the subject of formula of the equation. This is done as follows:

$4x - y = 6$ ....................Equation (2)

Hence,   $4x - 6 = y$

Or,    $y = 4x - 6$        (Note that p = q also means that q = p)

Let us label this equation as equation 3 as shown below.

$y = 4x - 6$ ........................Equation (3)

Since y = 4x − 6, let us substitute $4x - 6$ for y in equation 1. Note that we made y the subject of formula from equation 2, so the substitution of the expression for y should be done in the other equation (i.e. equation 1), and not in the same equation from which y was made the subject of the formula. Hence we substitute $4x - 6$ for y in equation 1 as follows:

$3x + 2y = 10$ ....................Equation (1)

$3x + 2(4x - 6) = 10$        (Note that y has been replaced (substituted) with $4x - 6$)

$3x + 8x - 12 = 10$

$11x = 10 + 12$

$11x = 22$

$x = \dfrac{22}{11}$        (When we divide both sides by 11)

$x = 2$

Now, substitute 2 for $x$ in equation 3 in order to obtain y. Note that we can use any of the three equations, but equation 3 is the best to use since it directly gives us the value of y. This gives:

$y = 4x - 6$ ........................Equation (3)

$y = 4(2) - 6$

$y = 8 - 6$

$y = 2$

Therefore, $x = 2$ and $y = 2$

3.  Solve simultaneously, the equations $m - 6n = 1$ and $9m + n = -46$

<u>Solution</u>

$m - 6n = 1$ .....................Equation (1)

$9m + n = -46$ .....................Equation (2)

From equation 2:

$n = -46 - 9m$ .......................Equation (3)

Note that we can also easily make m the subject of the formula from equation 1. Since, n = -46 - 9m, let us substitute -46 - 9m for n in equation 1. This gives:

$m - 6n = 1$ .....................Equation (1)

$m - 6(-46 - 9m) = 1$

$m + 276 + 54m = 1$

$55m = 1 - 276$

$55m = -275$

$m = \dfrac{-275}{155}$

$m = -5$

Now, substitute -5 for m in equation 3 in order to find n. This gives:

$n = -46 - 9m$ .......................Equation (3)

$n = -46 - 9(-5)$

$n = -46 + 45$

$n = -1$

Therefore, n = -1 and m = -5

4. Solve simultaneously, the equations $3a - 5b = 25$ and $7a + 2b = 31$

<u>Solution</u>

$3a - 5b = 25$ .....................Equation (1)

$7a + 2b = 31$.....................Equation (2)

A careful look at the two equations shows that there is no variable with a coefficient of 1. However, we can still make any of the variable a subject of its equation, and substitute it into the other equation. Hence, from equation 1, let as make b the subject of the formula as follows:

$3a - 5b = 25$ .....................Equation (1)

$3a - 25 = 5b$

Or,     $5b = 3a - 25$     (Note that m = n also means that n = m)

Dividing both sides by 5 gives:

$b = \dfrac{3a - 25}{5}$ .......................Equation (3)

Since $b = \dfrac{3a - 25}{5}$, let us substitute $\dfrac{3a - 25}{5}$ for b in equation 2. This gives:

$7a + 2b = 31$.....................Equation (2)

$7a + 2(\dfrac{3a - 25}{5}) = 31$

$$7a + \frac{6a - 50}{5} = 31$$

Multiply each term by 5 in order to clear out the fractions. This gives:

$$5(7a) + 5(\frac{6a - 50}{5}) = 5(31)$$

$$35a + 6a - 50 = 155$$
$$41a = 155 + 50$$
$$41a = 205$$
$$a = \frac{205}{41}$$
$$a = 5$$

Now, substitute 5 for a in equation 3. This gives:

$$b = \frac{3a - 25}{5} \quad ......................\text{Equation (3)}$$

$$= \frac{3(5) - 25}{5}$$

$$= \frac{15 - 25}{5}$$

$$= \frac{-10}{5}$$

$$b = -2$$

Therefore, a = 5 and b = −2


5.  Solve the equations $8p - 3q = -5$  and $2p - 11q = -32$

<u>Solution</u>

$8p - 3q = -5$ .....................Equation (1)

$2p - 11q = -32$ ...................Equation (2)

A careful look at the two equations shows that there is no variable with a coefficient of 1. However, let us make p the subject of formula in equation 2. This is done as follows:

$2p - 11q = -32$ ...................Equation (2)

$2p = -32 + 11q$

Dividing both sides by 2 gives:

$$p = \frac{-32 + 11q}{2} \quad ......................\text{Equation (3)}$$

Substitute $\frac{-32 + 11q}{2}$ for p in equation 1. This gives:

$8p - 3q = -5$ .....................Equation (1)

$$8(\frac{-32 + 11q}{2}) - 3q = -5$$

$$4(-32 + 11q) - 3q = -5$$
$$-128 + 44q - 3q = -5$$
$$41q = -5 + 128$$
$$41q = 123$$
$$q = \frac{123}{41}$$
$$q = 3$$

Substitute 3 for q in equation 3 in order to find p. This gives:

$$p = \frac{-32 + 11q}{2} \quad \text{.....................Equation (3)}$$

$$= \frac{-32 + 11(3)}{2}$$

$$= \frac{-32 + 33}{2}$$

$$p = \frac{1}{2}$$

Hence, $p = \frac{1}{2}$ and q = 3

## Elimination Method

Elimination method is used in solving simultaneous equations, usually when there is no variable with a coefficient of 1. In elimination method, we get rid of (eliminate) one of the variables by making its coefficient the same in both equations.

**Examples**

1. Solve the equation: $4x - 5y = 5$ and $2x - 3y = 2$.

Solution

$$4x - 5y = 5 \quad \text{.....................Equation (1)}$$
$$2x - 3y = 2 \quad \text{.....................Equation (2)}$$

When using elimination method, any operation of multiplication and division can be used to manipulate any of the equations in order to make a particular variable to have equal coefficient.

A careful look at the two equations shows that we can make the coefficient of $x$ in equation (2) to become 4 so that the coefficient of $x$ in the two equations will be 4. In order to make the coefficient of $x$ to be 4 in equation (2), we multiply equation (2) by 2. This gives:

$$2(2x) - 2(3y) = 2(2) \quad \text{(Note that each term in equation (2) should be multiplied by 2)}$$
$$4x - 6y = 4 \quad \text{.....................Equation (3)} \quad \text{(Note that our result is labelled equation 3)}$$

Now, equation 1 and 3 have the same coefficient of $x$. This means that $x$ can now be eliminated by subtracting one equation from the other. Note that the coefficient of $x$ which is 4 in equation 1 and 3 is a positive value. Hence one equation will be subtracted from the other in order to eliminate $x$. However, when the coefficients of a variable have opposite signs (i.e.

positive and negative), we will have to add the two equations in order to eliminate that variable.

In this case, for us to eliminate $x$, let us subtract equation 3 from equation 1 (we can also subtract equation 1 from equation 3) as follows:

$4x - 5y - (4x - 6y) = 5 - 4$     (Take note of the use of bracket in the subtraction)

Note that when subtracting, you subtract the left hand side of both equations separately, and subtract the right hand side of both equations separately as shown above. This simplifies to give:

$4x - 5y - 4x + 6y = 1$     (Note the changes in sign of the terms in the bracket)

$4x - 4x - 5y + 6y = 1$

$y = 1$     (Note that $4x - 4x = 0$, and $-5y + 6y = y$)

It is important to note that the terms in equation 1 come first on both sides of the equation during subtraction since we are carrying out: Equation 1 minus Equation 3.

Now that we have obtained y = 1, let us substitute 1 for y in equation 2 in order to find $x$. Note that the substitution can be done in any of the three equations.

$2x - 3y = 2$ ......................Equation (2)

$2x - 3(1) = 2$     (Note that 1 has been substituted for y)

$2x - 3 = 2$

$2x = 2 + 3$

$2x = 5$

$x = \dfrac{5}{2}$

Hence, $x = \dfrac{5}{2}$ and y = 1


2.  Solve simultaneously, the equations $5x - 4y = -2$ and $4x + 5y = 23$.

Solution

$5x - 4y = -2$ ..................Equation (1)

$4x + 5y = 23$ ....................Equation (2)

An easy way of making the coefficient of $x$ to be the same in both equations is to multiply one equation by the coefficient of $x$ in the other equation. This means that we multiply equation (1) by 4 (i.e. the coefficient of $x$ in equation 2), and multiply equation (2) by 5 (i.e. the coefficient of $x$ in equation 1). This will make the coefficient of $x$ to be 20 in the two equations. Hence, let us multiply equation (1) by 4 and equation (2) by 5 in order to obtain equation (3) and equation (4) respectively as shown below:

$4(5x) - 4(4y) = 4(-2)$ ......................Equation (3)

$5(4x) + 5(5y) = 5(23)$ ......................Equation (4)

These two equations simplifies to give:

$20x - 16y = -8$ ..........................Equation (3)

$20x + 25y = 115$ ........................Equation (4)

Now, equation 3 and 4 have the same coefficient of $x$. This means that $x$ can now be eliminated by subtracting one equation from the other. Hence, let us subtract equation 3 from equation 4.

Note that any order can be followed, i.e. we can subtract equation 4 from equation 3. Therefore, equation 4 minus equation 3 gives:

$20x + 25y - (20x - 16y) = 115 - (-8)$   (Note the subtraction of left and right side separately)

$20x + 25y - 20x + 16y = 115 + 8$     (Note the changes in sign of the terms in the bracket)

$20x - 20x + 25y + 16y = 123$

$41y = 123$       (Note that $20x - 20x = 0$, and $25y + 16y = 41y$)

$$y = \frac{123}{41}$$

$$y = 3$$

Now that we have obtained y = 3, let us substitute 3 for y in equation 1 in order to find $x$. Note that the substitution can be done in any of the four equations.

$5x - 4y = -2$ ...................Equation (1)

$5x - 4(3) = -2$     (Note that 3 has been substituted for y)

$5x - 12 = -2$

$5x = 12 - 2$

$5x = 10$

$$x = \frac{10}{5}$$

$$x = 2$$

Hence, $x = 2$ and y = 3


3.  Solve simultaneously, the equations $11x + 3y = 4$ and $7x - 2y = -17$.

Solution

$11x + 3y = 4$ .....................Equation (1)

$7x - 2y = -17$ .....................Equation (2)

In our last two examples we made the coefficient of $x$ to be equal. In this example let us make the coefficient of y to be equal. An easy way of making the coefficient of y to be the same in both equations is to multiply equation (1) by 2 (i.e. the coefficient of y in equation 2), and multiply equation (2) by 3 (i.e. the coefficient of y in equation 1). This will make the coefficient of y to be 6 (we are not taking the signs into consideration) in the two equations. Hence, let us multiply equation (1) by 2 and equation (2) by 3 in order to obtain equation (3) and equation (4) respectively as shown below:

$2(11x) + 2(3y) = 2(4)$ .....................Equation (3)

$3(7x) - 3(2y) = 3(-17)$ .....................Equation (4)

These two equations simplifies to give:

$22x + 6y = 8$ ..........................Equation (3)

$21x - 6y = -51$ .......................Equation (4)

Now, equations 3 and 4 have the same coefficient of y. This means that $x$ can now be eliminated by adding the two equations together. Hence, let us add equation 3 with equation 4. Note that in example 1 and 2 above, we subtracted one equation from the other. This was because the signs of the coefficients of $x$ were the same. However, in this example (example 3), the signs of the coefficients of y (i.e. the variable to be eliminated) are not the same. Hence, we

61

have to add them together in order to eliminate y. This summarizes the rule that when the signs of the coefficient of the variable to be eliminated are the same, we subtract one equation from the other in order to eliminate that variable. However, when the coefficients of a variable to be eliminated have opposite signs (i.e. positive and negative), we will have to add the two equations in order to eliminate that variable.

Therefore, equation 4 plus equation 3 gives:

$22x + 6y + (21x - 6y) = 8 + (-51)$   (Take note of the addition of left and right side separately)

$22x + 6y + 21x - 6y = 8 - 51$

$22x + 21x + 6y - 6y = -43$

$43x = -43$

$x = \dfrac{-43}{43}$

$x = -1$

Now that we have obtained $x = -1$, let us substitute $-1$ for $x$ in equation 1 in order to find y.

$11x + 3y = 4$ .......................Equation (1)

$11(-1) + 3y = 4$

$-11 + 3y = 4$

$3y = 4 + 11$

$3y = 15$

$y = \dfrac{15}{3}$

$y = 3$

Hence, $x = -1$ and y = 3


4.   Solve simultaneously, the equations: 6m − 5n − 3 = 0 and 4m + 3n + 17 = 0

Solution

Each of the equations above can be rearranged by taking the constant term to the right hand side of the equations. This gives:

6m − 5n = 3 ........................Equation (1)

4m + 3n = −17 ......................Equation (2)

Let us make the coefficient of n to be equal in the two equations. An easy way of doing this is to multiply equation (1) by 3 (i.e. the coefficient of n in equation 2), and multiply equation (2) by 5 (i.e. the coefficient of n in equation (1). This will make the coefficient of n to be 15 in the two equations. Hence, equation (1) multiplied by 3 and equation (2) multiplied by 5 will give equation (3) and equation (4) respectively as shown below:

3(6m) − 3(5n) = 3(3) ........................Equation (3)

5(4m) + 5(3n) = 5(−17) ......................Equation (4)

These two equations simplify to give:

18m − 15n = 9 ..........................Equation (3)

20m + 15n = − 85 ......................Equation (4)

Now, equations 3 and 4 have the same coefficient of n. This means that n can now be eliminated by adding the two equations together. We have to add the two equations since the signs of the coefficients of n are different.

Another method of carrying out our operation (either addition or subtraction) is to arrange the two equations to be one above the other and then add or subtract the corresponding like terms. Hence, equation 3 plus equation 4 can be used to describe this method as shown below.

$$18m - 15n = 9 \dots\dots\dots\dots\text{Equation (3)}$$
$$20m + 15n = -85 \dots\dots\dots\text{Equation (4)}$$

Equation (3) + equation (4):    $38m = -76$

In the addition above, the followings were carried out in each column:

$18m + 20m = 38m$

$-15n + 15n = 0,$

and    $9 + (-85) = 9 - 85 = -76.$

Also, zero n was not written since n has been eliminated.

We now continue from above as follows:

$$38m = -76$$
$$m = \frac{-76}{38}$$
$$m = -2$$

Substitute −2 for m in equation (1) in order to obtain n. This gives:

$$6m - 5n = 3 \dots\dots\dots\dots\text{Equation (1)}$$
$$6(-2) - 5n = 3$$
$$-12 - 5n = 3$$
$$-5n = 3 + 12$$
$$-5n = 15$$
$$n = \frac{15}{-5}$$
$$n = -3$$

Therefore, m = −2 and n = −3

5.  Solve the simultaneous equations $5p - 8q = 16$ and $12q - p = 24$

Solution

The second equation above has been arranged differently from the first one. In the first equation the term in p is written first, but in the second equation the term in q has been written first. Therefore in order to avoid making mistake, it is advisable to have the same pattern of arrangement in the two equations.

Hence, let us now label the two equations and rearrange the second equation as shown below.

$$5p - 8q = 16 \dots\dots\dots\dots\text{Equation (1)}$$
$$- p + 12q = 24 \dots\dots\dots\dots\text{Equation (2)}$$

Let us make the coefficient of p to be equal in the two equations. In order to do this we simply multiply equation (2) by 5 so that the coefficient of p in the two equations will be 5 (we have ignored the signs). Hence, if we multiply equation (2) by 5, it will give us equation (3) as follows:

$$5(- p) + 5(12q) = 5(24) \dots\dots\dots\text{Equation (3)}$$

This simplifies to give:

$$-5p + 60q = 120 \dots\dots\dots\dots\text{Equation (3)}$$

Now, equations 1 and 3 have the same coefficient of p. This means that p can now be eliminated by adding the two equations together. We have to add the two equations since the signs of the coefficients of p are different.

Hence, equation (1) and (3) can be brought together and added to eliminate p as shown below.

$$5p - 8q = 16 \text{ ........................Equation (1)}$$

$$-5p + 60q = 120 \text{ .....................Equation (3)}$$

Equation (1) + equation (3): $\qquad 52q = 136$

Note that 5p + (−5p) = 0, −8q + 60q = 52q, and 16 + 120 = 136, as all obtained above when the corresponding like terms are added. Also, zero p was not written since it has been eliminated. We now continue from above as follows:

$$52q = 136$$
$$q = \frac{136}{52}$$
$$q = \frac{34}{13} \qquad \text{(After equal division by 4)}$$

Substitute $\frac{34}{13}$ for q in equation (2) in order to obtain p. This gives:

$$-p + 12q = 24 \text{ ......................Equation (2)}$$
$$-p + 12(\frac{34}{13}) = 24$$
$$-p + \frac{408}{13} = 24$$
$$-p = 24 - \frac{408}{13}$$
$$-p = \frac{312 - 408}{13}$$
$$-p = \frac{-96}{13}$$
$$p = \frac{96}{13} \qquad \text{(After dividing both sides by −1 in order to make p positive)}$$

Therefore, $p = \frac{96}{13}$, and $q = \frac{34}{13}$

6.  Solve simultaneously, the following equations:
$$2c + 5d = 0$$
$$3c - 2d = 19$$

Solution
$$2c + 5d = 0 \text{ ..................Equation (1)}$$
$$3c - 2d = 19 \text{ .................Equation (2)}$$

Let us make the coefficient of c to be equal in the two equations. In order to do this we multiply equation (1) by 3 (i.e. the coefficient of c in equation 2), and multiply equation (2) by 2 (i.e. the coefficient of c in equation 1). This will make the coefficient of c to be 6 in the two equations. Hence, equation (1) multiplied by 3 and equation (2) multiplied by 2 will give equation (3) and equation (4) respectively as shown below:

$3(2c) + 3(5d) = 3(0)$ ...................Equation (3)

$2(3c) - 2(2d) = 2(19)$ ..................Equation (4)

These two equations simplify to give:

$6c + 15d = 0$ .........................Equation (3)

$6c - 4d = 38$ .........................Equation (4)

We now eliminate c by subtracting equation (4) from equation (3). We have to subtract the two equations since the signs of the coefficients of c are the same.

Hence, equation 3 minus equation 4 is evaluated as shown below.

$6c + 15d = 0$ .........................Equation (3)

$6c - 4d = 38$ .........................Equation (4)

Equation (3) – equation (4):     $19d = -38$

Note that $6c - 6c = 0$, $15d - (-4d) = 19d$, and $0 - 38 = -38$, as all obtained above.

We now continue from above as follows:

$$19d = -38$$
$$d = \frac{-38}{19}$$
$$d = -2$$

Substitute –2 for d in equation (1) in order to obtain c. This gives:

$2c + 5d = 0$ ...................Equation (1)

$$2c + 5(-2) = 0$$
$$2c - 10 = 0$$
$$2c = 10$$
$$c = \frac{10}{2}$$
$$c = 5$$

Therefore, $c = 5$ and $d = -2$


7. Solve simultaneously, the following equations:

$$11x - 7y = -11$$
$$5x + 3y = 29$$

Solution

$11x - 7y = -11$ ......................Equation (1)

$5x + 3y = 29$ ......................Equation (2)

Let us make the coefficient of $x$ to be equal in the two equations. In order to do this, multiply equation (1) by 5 (i.e. the coefficient of $x$ in equation 2), and multiply equation (2) by 11 (i.e. the coefficient of $x$ in equation 1). This will give equation (3) and equation (4) respectively as shown below:

$5(11x) - 5(7y) = 5(-11)$ ......................Equation (3)

$11(5x) + 11(3y) = 11(29)$ ......................Equation (4)

These two equations simplifies to give:

$$55x - 35y = -55 \text{ ...........................Equation (3)}$$
$$55x + 33y = 319 \text{ ...........................Equation (4)}$$

We now eliminate $x$ by subtracting equation (4) from equation (3) (note that we can also subtract equation 3 from equation 4). We have to subtract the two equations since the signs of the coefficients of $x$ are the same.

Hence, equation 3 minus equation 4 is evaluated as shown below.

$$55x - 35y = -55 \text{ ...........................Equation (3)}$$
$$55x + 33y = 319 \text{ ...........................Equation (4)}$$

Equation (3) – equation (4):         $-68y = -374$

Note that $55x - 55x = 0$, $-35y - (+33y) = -68y$, and $-55 - 319 = -374$, as all obtained above.

Hence: $-68y = -374$     (From above)

$$y = \frac{-374}{-68}$$

$$y = \frac{11}{2} \qquad \text{(When expressed in its lowest term)}$$

$$y = 5\frac{1}{2}$$

Substitute $\frac{11}{2}$ for y in equation (2) (we can use any equation) in order to obtain $x$. This gives:

$$5x + 3y = 29 \text{ ......................Equation (2)}$$

$$5x + 3\left(\frac{11}{2}\right) = 29$$

$$5x + \frac{33}{2} = 29$$

$$5x = 29 - \frac{33}{2}$$

$$5x = \frac{58 - 33}{2}$$

$$5x = \frac{25}{2}$$

$$x = \frac{\frac{25}{2}}{5}$$

$$x = \frac{25}{2} \times \frac{1}{5}$$

$$= \frac{5}{2}$$

$$x = 2\frac{1}{2}$$

Therefore, $x = 2\frac{1}{2}$ and $y = 5\frac{1}{2}$

## Exercise 6

1. Solve simultaneously the equations:   $2x - y = -3$   and   $5x + y = -4$
2.  Solve the simultaneous equations:   $x + y = 5$ and $2x - y = 13$
3. Solve simultaneously, the equations $5m - n = 8$ and $3m - n = 4$
4. Solve simultaneously, the equations:  $a - 3b = -5$ and $4a - 3b = 25$
5. Solve the equations:  $p - q = 7$  and $p - 3q = 13$
6. Solve the equation:  $10x - 5y = -30$ and $20x + 3y = -34$
7. Solve simultaneously, the equations $4x - y = 6$ and $3x + y = 15$.
8. Solve simultaneously, the equations $x + 5y = 2$ and $9x - y = 12$.
9. Solve simultaneously, the equations: $11m - 3n - 11 = 3$ and $8m + 5n - 3 = 0$
10. Solve the simultaneous equations $2p - 15q = 1$ and $2q - 11p = 5$
11. Solve simultaneously, the following equations:

    $5c + 5d = 10$

    $c - d = 5$
12. Solve simultaneously, the following equations:

    $x - 4y = -20$

    $12x + 5y = 10$

# CHAPTER 7
## WORD PROBLEMS LEADING TO SIMULTAENOUS LINEAR EQUATIONS

**Examples**

1. Three spoons and five knives cost $46. Five spoons and ten knives cost $90. Find the cost of a spoon and a knife.

Solution

Let the cost of a spoon be $x$, and the cost of a knife be $y.

∴ From the first sentence, we have:

$3x + 5y = 46$ .................Equation (1)

Similarly, from the second sentence, we have:

$5x + 10y = 90$ .................Equation (2)

Bringing equation (1) and (2) together gives:

$3x + 5y = 46$ ...................Equation (1)

$5x + 10y = 90$ ....................Equation (2)

In order to make the coefficient of y to be the same in equation (1) and (2) (so as to solve by elimination method), lets simply divide equation (2) by 2. This gives:

$$\frac{5x + 10y = 90}{2}$$

This is simplified by dividing each term by 2 to give: $2.5x + 5y = 45$ .................... equation (3)

Note that the coefficient of y is made to be the same in order to eliminate y (elimination method).

Therefore bringing equations (1) and (3) together gives:

$3x + 5y = 46$ ......................Equation (1)

$2.5x + 5y = 45$ ....................Equation (3)

Equation (1) – equation (3): $0.5x = 1$

$$x = \frac{1}{0.5}$$

$$x = 2$$

Substitute 2 for $x$ in any of the equations. Let us use equation (1)

∴ $3x + 5y = 46$ .................Equation (1)

$3(2) + 5y = 46$

$6 + 5y = 46$

∴ $5y = 46 - 6$

$5y = 40$

∴ $y = \frac{40}{5}$

$y = 8$

∴ A spoon cost $2, while a knife cost $8.

2. One-quarter of a boy's score in English plus one-fifth of his score in Mathematics makes 31. Half of his score in mathematics added to two-third of his score in English gives 80. What was his score in each subject?

## Solution

Let his score in English be 'a' and his score in Mathematics be 'b'.

∴ From the first sentence, we have:

$$\frac{1}{4}a + \frac{1}{5}b = 31$$

Multiply each term by 20 (L.C.M of 4 and 5) in order to remove the fractions. This gives:

$$20\left(\frac{a}{4}\right) + 20\left(\frac{b}{5}\right) = 20 \times 31$$

∴   5a + 4b = 620 ................Equation (1)

From the second sentence of the question, we have

$$\frac{2}{3}a + \frac{1}{2}b = 80$$

Multiply each term by 6 (i.e. L.C.M of 3 and 2). This gives:

$$6\left(\frac{2a}{3}\right) + 6\left(\frac{b}{2}\right) = 6 \times 80$$

2(2a) + 3(b) = 480      (After 6 has been used to divide each of the denominator)

∴   4a + 3b = 480 .........................Equation (2)

Bringing equations (1) and (2) together in order to solve them simultaneously gives:

5a + 4b = 620 ........................Equation (1)

4a + 3b = 480 ........................Equation (2)

Multiply equation (1) by 3 and equation (2) by 4 in order to make the coefficients of b to be the same in the two equations. This gives:

3(5a + 4b = 620)

4(4a + 3b = 480)

These simplify to:

15a + 12b = 1860 .................... Equation (3)

16a + 12b = 1920 .................... Equation (4)

Equation (4) – Equation (3):     a         = 60

∴    a = 60

Substitute 60 for a in equation (2). This gives:

4a + 3b = 480 ..................Equation (2)

4(60) + 3b = 480

240 + 3b = 480

3b = 480 – 240

∴        3b = 240

$$b = \frac{240}{3}$$

∴        b = 80

∴   He scored 60 in English and 80 in Mathematics


3.  Divide 63 into two parts such that one part is two-fifth of the other.

Let the two parts be $x$ and y.

∴    $x$ + y = 63 ....................Equation (1)     (Note that the two parts sum up to 63)

$$\frac{2}{5}x = y$$ ......................Equation (2)  (Note that y has been made $\frac{2}{5}$ of $x$ as stated above)

Substitute $\frac{2x}{5}$ for y in equation (1)

$\qquad x + y = 63$ ....................Equation (1)

$\qquad x + \frac{2x}{5} = 63$

Multiply each term by 5 in order to remove the fraction. This gives:

$\qquad 5x + 5\frac{2x}{5} = 5 \times 63$

$\qquad 5x + 2x = 315$

$\qquad 7x = 315$

$\therefore \qquad x = \frac{315}{7} \qquad\qquad 7$

$\qquad x = 45$

From equation (1), y = 63 − x

$\qquad y = 63 - 45$

$\qquad y = 18$

∴ The two parts are 45 and 18.

4.  A bicycle moves for p hours at 10km/h and q hours at 15km/h. If his total journey is 80km in 7 hours, find the values of p and q.

Solution

Recall that:  Speed = $\frac{\text{Distance}}{\text{Time}}$

∴   At 10km/h, we have: $10 = \frac{\text{Distance}}{p}$     (Since time = p)

∴   Distance = 10p     (After cross multiplication)

Similarly, at 15km/hr, we have:

$\qquad$ Distance = 15q     (Since time is q in this case)

∴  Total distance is given by:

$\qquad$ 10p + 15q = 80 .......................Equation (1)

The total time is:   p + q = 7 ......................Equation (2)   (Note that the journey took 7 hours)

Multiply equation (2) by 10 in order to make the coefficient of p to be 10 in the two equations. This gives:

$\qquad$ 10(p + q = 7)

$\qquad$ 10p + 10q = 70 .......................Equation (3)

Bringing equations (1) and (3) together in order to solve them simultaneously gives:

$\qquad\qquad\qquad$ 10p + 15q = 80 .....................Equation (1)

$\qquad\qquad\qquad$ 10p + 10q = 70 ....................Equation (3)

Equation (1) − equation (3):$\qquad$ 5q = 10

$\qquad\qquad\qquad\qquad\qquad q = \frac{10}{5}$

$\qquad\qquad\qquad\qquad\qquad$ q = 2

From equation (2), p can be expressed as:

$\qquad$ p = 7 − q

$\qquad$ p = 7 − 2     (Since q = 2)

p = 5

∴ p = 5 hours and q = 2 hours.

5. The perimeter of an isosceles triangle is 56cm. If the two equal sides are (8y)cm and $(12y − 4x + 2)$cm, while the third side is $(2x + 4y)$cm, find the values of x and y.

Solution

Equating the two equal sides since they are equal gives:

$8y = 12y − 4x + 2$

∴ $4x + 8y − 12y = 2$ (By collecting the unknown terms on the left hand side)

$4x − 4y = 2$

Dividing each term by 2 in order to simplify the equation gives:

$2x − 2y = 1$ ........................Equation (1)

Adding the three sides of the triangle gives the perimeter as follows:

$8y + 12y − 4x + 2 + 2x + 4y = 56$

$24y − 2x = 56 − 2$

$−2x + 24y = 54$ ....................Equation (2)

Bringing equations (1) and (2) together gives:

$2x − 2y = 1$ ........................Equation (1)

$−2x + 24y = 54$ ....................Equation (2)

In order to eliminate $x$, we add equation (1) and (2): $22y = 55$

$$y = \frac{55}{22}$$

$$y = \frac{5}{2}$$ (After equal division by 11)

$$y = 2\frac{1}{2}$$

Substitute $\frac{5}{2}$ for y in equation (1)

$2x − 2y = 1$ ..................(Equation (1)

$2x − 2(\frac{5}{2}) = 1$

$2x − 5 = 1$

$2x = 1 + 5$

$2x = 6$

∴ $$x = \frac{6}{2}$$

$x = 3$

∴ $x = 3$ and $y = 2\frac{1}{2}$

6. The three sides of an equilateral triangle are $(3x)$cm, $(4y)$cm and $(x + y + 3)$cm. Find:

a. The values of $x$ and y.

b. The sides of the triangle.

<u>Solutions</u>

a.  Since the three sides of an equilateral triangle are equal, then it follows that: $3x = 4y = x + y + 3$

Equating any two of the terms above can be used to obtain two equations as follows:

$\qquad 3x = 4y$

$\therefore \quad 3x - 4y = 0$ ....................Equation (1)

and $\quad 3x = x + y + 3$

$\therefore \quad 3x - x - y = 3$

$\qquad 2x - y = 3$ ....................Equation (2)

From equation (2), $y = 2x - 3$ ....................Equation (3)

Substitute $2x - 3$ for y in equation (1)

$\therefore \quad 3x - 4y = 0$ ................Equation (1)

$\qquad 3x - 4(2x - 3) = 0$

$\qquad 3x - 8x + 12 = 0 \qquad$ (Note that $-4 \times -3 = +12$)

$\therefore \quad -5x = -12$

$\qquad x = \dfrac{-12}{-5}$

$\qquad x = \dfrac{12}{5} \qquad$ (The negative sign has cancelled out)

$\therefore \quad x = 2\dfrac{2}{5}$

From equation (3) substitute: $x = \dfrac{12}{5}$

$\therefore \quad y = 2x - 3$ ..................Equation (3)

$\qquad y = 2(\dfrac{12}{5}) - 3$

$\qquad = \dfrac{24}{5} - 3$

$\qquad = \dfrac{24 - 15}{5}$

$\qquad y = \dfrac{9}{5}$

$\qquad y = 1\dfrac{4}{5}$

Therefore, $x = 2\dfrac{2}{5}$ and $y = 1\dfrac{4}{5}$

b.  Each side of the triangle is given by:

$3x$ or $4y$ or $x + y + 3$

$\therefore \quad 3x$ will give:

$\qquad 3 \times \dfrac{12}{5} = \dfrac{36}{5}$

$\qquad = 7\dfrac{1}{5}$cm

$\therefore \quad$ Each side of the triangle is $7\dfrac{1}{5}$cm

7.  The sum of the digits of a two digits number is 13. If the digits are interchanged, the number is decreased by 45. Find the number.

<u>Solution</u>
Let the two digits number be $xy$. Therefore, the number itself is obtained as follows:

$10x + y$    (For example 24 = (2 x 10) + 4 = 20 + 4 = 24)

Note that $x$ is the tens digit while y is the unit digit. When the digits are interchanged to obtain $yx$, then the new number is: $10y + x$

Since the original number is decreased by 45 after the digits are interchanged, it means that the first number is greater than the second number by 45.

$\therefore$   $(10x + y) - (10y + x) = 45$

        $10x + y - 10y - x = 45$

             $9x - 9y = 45$

$\therefore$                    $x - y = 5$    (After dividing each term by 9 in order to simplify the equation)

$\therefore$   $x - y = 5$ .....................Equation (1)

Recall from the first sentence that:

               $x + y = 13$ .............Equation (2)

               <u>$x - y = 5$</u> ...............Equation (1)    (This is simply equation 1 brought down)

Equation (2) – (1):    $2y = 8$    ($x$ has been eliminated by the subtraction. Also +y − (−y) = 2y)

$\therefore$                        $y = \dfrac{8}{2}$

                       $y = 4$

From equation (1), substitute 4 for y. This gives:

    $x - y = 5$ ...................Equation (1)

    $x - 4 = 5$

      $x = 5 + 4$

      $x = 9$

$\therefore$   The number is 94 (i.e. $xy$)

**Check:** The number is 94. When the digits are interchanged it becomes 49. Their difference is 94 – 49 = 45. Hence it is correct as given in the question.


8. In a two digits number, the difference between the digits is 1. The number is 1 more than 5 times the sum of the digits. Find the number if the tens digit is less than the unit digit.


<u>Solution</u>
Let the number be $xy$ ($x$ is the tens digit)

    $y - x = 1$ ..............Equation (1)   (Note that the unit digit is greater than the tens digit). This equation can also be expressed as:

    $-x + y = 1$

The sum of the digits is $x + y$. And 5 times the sum of the digits is $5(x + y)$

The number itself is:

    $10x + y$

Since the number is 1 more than 5 times the sum of the digits, then it follows that:

    $10x + y - 5(x + y) = 1$

    $10x + y - 5x - 5y = 1$

    $5x - 4y = 1$ ..................Equation (2)

Bringing equation (1) and (2) together in order to solve them simultaneously gives:

$-x + y = 1$ .....................Equation (1)

$5x - 4y = 1$ ..................... equation (2)

Multiply equation (1) by 5. This is done in order to make the coefficient of x in the two equations to be 5. This gives:

$-5x + 5y = 5$ ..................Equation (3)

Bringing equation (2) and (3) together gives:

$5x - 4y = 1$ .......................Equation (2)

$-5x + 5y = 5$ .......................Equation (3)

Equation (2) + equation (3):    $y = 6$

Substitute 6 for y in equation (1)

$\therefore$    $-x + y = 1$ ................Equation (1)

$-x + 6 = 1$

$\therefore$    $x = 6 - 1$

$x = 5$

$\therefore$  The number is 56 (i.e. $xy$)


9.  The sum of a man's age and his son's age is 45 years. Five years ago, the man was 6 times as old as his son. How old was the man when his son was born?


Solution

Let the man's present age be $x$, and his son's present age be y.

$\therefore$    $x + y = 45$ ..................Equation (1)

Five years ago, the man's age was $x - 5$, while his son's age was y – 5. At that time, his age was 6 times that of his son. This means that:

$x - 5 = 6(y - 5)$

$x - 5 = 6y - 30$

$x - 6y = -30 + 5$

$x - 6y = -25$ ....................Equation (2)

Bringing equations (1) and (2) together gives:

$x + y = 45$ ....................Equation (1)

$x - 6y = -25$ ....................Equation (2)

Equation (1) – (2)     $7y = 70$     (Note that $+y -(-6y) = 7y$, and $45 -(-25) = 70$)

$y = \dfrac{70}{7}$

$\therefore$                $y = 10$

Substitute 10 for y in equation (1). This gives:

$x + y = 45$ ................Equation (1)

$x + 10 = 45$

$x = 45 - 10$

$x = 35$

$\therefore$  The man's present age is 35 years while his son's present age is 10 years. This means that the son was born 10 years ago. Therefore, 10 years ago, the man's age was $35 - 10 = 25$ years.

$\therefore$    The man was 25 years when his son was born.

10. If 1 is added to the numerator of a fraction, and 2 is added to the denominator, the fraction becomes $\frac{1}{3}$. If 3 is added to both the numerator and denominator of the fraction, the fraction becomes $\frac{1}{2}$. Find the fraction.

Solution

Let the numerator be $x$ and the denominator be y

∴   The fraction is $\frac{x}{y}$

From the first sentence:

$$\frac{x+1}{y+2} = \frac{1}{3}$$

∴   $3(x + 1) = 1(y + 2)$     (When we cross multiply)

   $3x + 3 = y + 2$

   $3x - y = 2 - 3$

∴   $3x - y = -1$ ....................Equation (1)

From the second sentence:

$$\frac{x+3}{y+3} = \frac{1}{2}$$

∴   $2(x + 3) = 1(y + 3)$

   $2x + 6 = y + 3$

   $2x - y = 3 - 6$

∴   $2x - y = -3$ .................Equation (2)

Bringing equations (1) and (2) together gives:

   $3x - y = -1$ .................Equation (1)

   $\underline{2x - y = -3}$ ......................Equation (2)

Equation (1) – (2)   $x$   $= 2$

Hence $x = 2$

Substitute 2 for $x$ in equation (1)

∴   $3x - y = -1$ ...............Equation (1)

   $3(2) - y = -1$

   $6 - y = -1$

   $6 + 1 = y$

∴      $y = 7$

∴ The fraction is $\frac{2}{7}$ (i.e. $\frac{x}{y}$)

Exercise 7

1. Five books and eight pens cost $90. Six books and four pens cost $80. Find the cost of a book and a pen.

2.  One-third of a girl's score in Mathematics plus two-seventh of her score in Biology makes 30. One-fifth of her score in Biology added to half of her score in Mathematics gives 29. What was her score in each subject?

3.  Divide 75 into two parts such that one part is one-third of the other.

4.  A man walks for m hours at 12km/h and n hours at 18km/h. If his total journey is 114km in 8 hours, find the values of m and n.

5.  The perimeter of an isosceles triangle is 50cm. If the two equal sides are (6y)cm and $(2y - 4x + 4)$cm, while the third side is $(2x + 6y)$cm, find the values of x and y. Hence, determine the length of each side of the triangle.

6.  The three sides of an equilateral triangle are $(5x)$cm, $(4y)$cm and $(x + 2y + 9)$cm. Find:
a. The values of $x$ and y.
b. Each side of the triangle.

7.  The sum of the digits of a two digits number is 11. If the digits are interchanged, the number is decreased by 27. Find the number.

8.  In a two digits number, the difference between the digits is 7. The number is 4 more than 8 times the sum of the digits. Find the number if the tens digit is greater than the unit digit.

9.  The sum of a woman's age and her daughter's age is 65 years. Ten years ago, the woman was 8 times as old as her daughter. How old was the woman when her daughter was born?

10.  If 5 is added to the numerator of a fraction, and 2 is added to the denominator, the fraction becomes $\frac{4}{5}$. If 2 is added to both the numerator and denominator of the fraction, the fraction becomes $\frac{1}{2}$. Find the fraction.

# CHAPTER 8
# LOGICAL REASONING

## Statements

In logical reasoning, a statement is a spoken or written sentence that gives information about something. The statement may be true or false.

Some statement may be either true or false. They are called closed statements. Statements that we do not know whether they are true or false, are called open statements.

**Examples**

Let us classify the following as closed or open statements or neither, and also say if they are true or false.

1. X: England is a country.  (Closed, and true)
2. Y: w is a prime number.  (Open. Impossible to say)
3. Z: How old are you? (Neither. Not applicable)

## Simple and Compound Statements

In logical reasoning a simple statement is a sentence which has only one main verb.

A compound statement is a sentence that contains two or more simple statements.

A statement can either be true or false. In logical context, questions, commands, sayings and exclamations are not considered as statement.

John is a clever and brave boy, is a compound statement. Jane is a geologist or a teacher is also a compound statement. When 'and' is used in a compound statement we call it a conjunction. When 'or' is used in a compound statement we call it a disjunction.

## Symbols for Conjunction and Disjunction

Let us examine two simple statements as given below:

Los Angeles is in USA

USA is in North America

Let the first statement be represented by p and the second statement be represented by q. A compound statement can be obtained from the two simple statement as follows:

a.  Los Angeles is in USA or USA is in North America. This can be expressed in symbol by:  p or q which also means p v q, where the connective v means 'or'

b.  Los Angeles is in USA and USA is in North America. This can be expressed in symbol by: p and q which also means p $\land$ q, where the connective $\land$ means 'and'.

## Truth Table

The truth table can be used to determine the truth value of a compound statement. A compound statement p v q can be read as; either p or q or both. The truth table for p v q is illustrated below.

| p | q | p ∨ q |
|---|---|---|
| T | T | T |
| T | F | T |
| F | T | T |
| F | F | F |

This shows that the statement p ∨ q is false only when p and q are false.
The truth table for p ∧ q is illustrated below.

| p | q | p ∧ q |
|---|---|---|
| T | T | T |
| T | F | F |
| F | T | F |
| F | F | F |

This shows that the statement p ∧ q is true only when both p and q are true.

## Negation

Negation simply means 'not'. For a statement, s, the negation of s can be written as ~s.
If s is true, then ~s is false, and if s is false, then, ~s is true.

Examples
The negation of the following statements are given in brackets below.
1.  Peter is brilliant (Peter is not brilliant)
2.  Jane is not a sailor (Jane is a sailor)
3.  $x = 12$ ($x \neq 12$)

## Implication (Conditional Statements)

Let us take a statement, s, as stated below:
    s: If a student plays football then the student takes part in sports.
Statement, s, is actually a combination of the following statements.
    p: A student in a school plays football
    q: A student in a school takes part in sports
s, states that if p is true, then q will also be true. We call a compound statement like this an implication.
Symbolically, we write the implication s as $p \Rightarrow q$, which reads as 'p implies q'. It can also be interpreted as any of the following:
1.  If p then q,
2.  p only if q

3. q follows from p
4. q if p

In the implication s: p ⇒ q, the substatement p is called the antecedent, while q is called the consequent. The statement p ⇒ q is sometimes called a conditional statement.
The truth table for p ⇒ q is illustrated below.

| p | q | p ⇒ q |
|---|---|-------|
| T | T | T |
| T | F | F |
| F | T | T |
| F | F | T |

This shows that the statement p ⇒ q is always true except when the first statement is true and the second statement is false.F

**Examples**
1. The statements, A, B, C are given below.
   A: Lawyers are liars
   B: Lawyers are brilliant
   C: Lawyers are neat
Write the following statement as implications
a. Brilliant lawyers are liars
b. No neat lawyers are brilliant

Solution
a. B ⇒ A   (This means B implies A)
b. ~C ⇒ B  (This means ~C implies B)

2. Write the following implications using ordinary language.
a. m is negative ⇒ $m^2$ is positive
b. $a^2 = 36 ⇒ a = +6$ or $-6$

Solution
a. If m is negative then $m^2$ is positive
b. If $a^2$ is equal to 36 then a is equal to +6 or −6

## Equivalent Statements
Two compound statements are said to be logically equivalent if they have the same truth value.
If in an implication, p ⇒ q and q ⇒ p, i.e. p implies q and q implies p, then we write p ⇔ q and say that p and q form equivalent statements.
From an implication p ⇒ q, we can deduce the equivalent implication as, ~p ⇒ ~q. The converse of p ⇒ q is q ⇒ p.

**Examples**
1. Write down the converse of the implication:
    A: John is smart $\Rightarrow$ John plays football

Solution
The converse of A is B where B is given by:
    B: John plays football $\Rightarrow$ John is smart

2. Write statements which are equivalent to the following:
a. If you play a game the computer will be hot
b. No student who offer physics is in the technical class

Solution
The equivalent statements are:
a. If you do not play a game the computer will not be hot
b. Those students who offer physics are not in the technical class

3. Use the truth table to prove that: $\sim (p \vee q) = \sim p \wedge \sim q$

Solution
The tables are as shown below.

| p | Q | p v q | ~ (p v q) |
|---|---|---|---|
| T | T | T | F |
| T | F | T | F |
| F | T | T | F |
| F | F | F | T |

| p | q | ~p | ~q | ~p ∧ ~q |
|---|---|---|---|---|
| T | T | F | F | F |
| T | F | F | T | F |
| F | T | T | F | F |
| F | F | T | T | T |

The last columns in each of the two tables above have the same truth values. Hence, $\sim (p \vee q) = \sim p \wedge \sim q$

## Valid Argument

A conclusion made from a set of statements (call the premises) can either be valid or invalid. The argument (the premises and conclusion) is valid if the fact that the premises are true, forces the conclusion to be true.

## Use of Venn Diagrams in Logical Operations

**Examples**

1. Determine if the following argument is valid.

   All Asian people are kind

   Chan is an Asian

   Chan is kind

<u>Solution</u>

First draw the Venn diagram from the first premise. Kind people should be the overall region. Since all Asian people are kind, the region for all Asian people should go into the region for kind people. The second premise, Chan is an Asian, (represented by C) will go inside the region representing all Asian. Hence, the two premises are true since they fall in the same circle. Therefore the conclusion is true, i.e. the argument is valid. Note that the last statement is the conclusion.

The Venn diagram is as shown below.

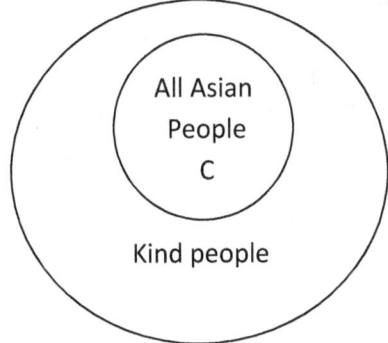

2. Is the following argument valid?

   Blue buildings are beautiful buildings

   This building is Blue

   This building is beautiful

<u>Solution</u>

Let B represent 'This building'. The Venn diagram below shows that B is in blue buildings and blue buildings is in beautiful buildings. Note that the overall circle represents beautiful buildings, which is the last part of the first statement.

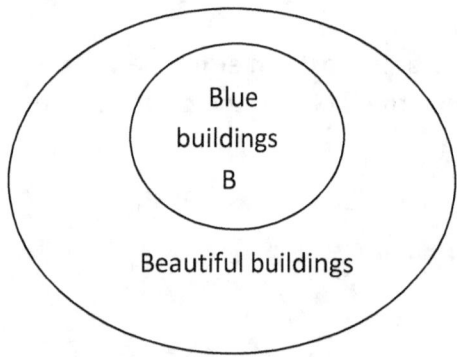

The Venn diagram shows that the two premises are true since they fall in the same circle. Therefore the argument is valid.

3. Determine if the argument below is valid.
    All workers in the factory eat bread
    Daniel eats bread
    Daniel is a worker

<u>Solution</u>
Let D represents Daniel. The Venn diagram below shows that D can be outside 'All workers in the factory' since we do not know if Daniel is a worker in the factory. Hence, D is in 'People who eat bread', but it can be outside 'All workers in the factory'. Therefore the conclusion that Daniel is a worker in the factory is not true. This shows that the argument is not valid.
Note that the overall region of the circle represents 'People who eat Bread' which is the last part of the first statement.

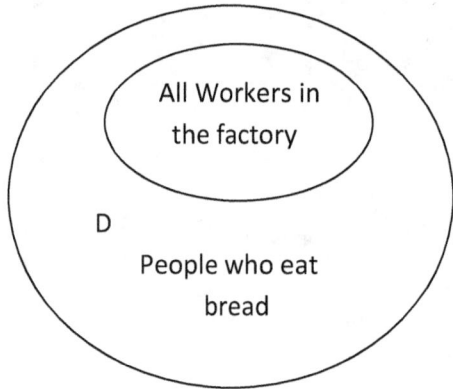

The Venn diagram shows that the two premises are not true since they do not fall in the same circle. Therefore the argument is invalid.

4. Determine if the following argument is valid.

Smith is a lawyer
Smith is a liar
All lawyers are liars

Solution

This is a situation of two groups of people (i.e. lawyers and liars. Smith (represented by S) fall into the two groups. However, the two groups cannot be put together since the statements have not indicated that they can be together. The Venn diagram is as shown below.

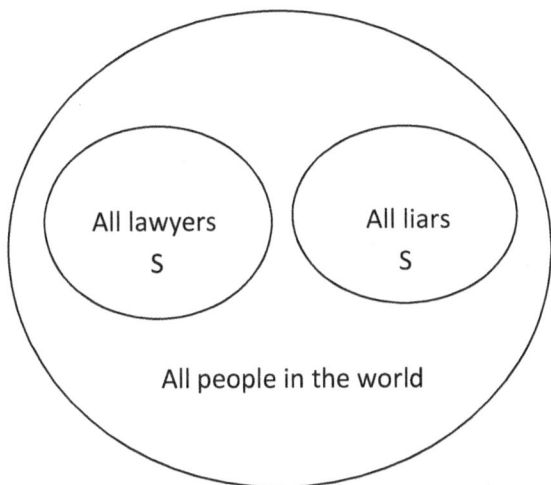

The Venn diagram above shows that these two groups of people in the world cannot be completely put together. Hence the conclusion drawn from them is not true. Therefore the argument is not valid.

## Exercise 8

1.  Classify the following into closed or open statements or neither, and also say if they are true or false.
a.  P: New York is a state.
b.  Q: $x$ is an odd number.
c.  R: What is your nationality?

2.  State the negation of the following statements:
a.  Stanley is neat
b.  Jane is not a doctor
c.  m = -5

3.  The statements, X, Y, Z are given below.

X: Politicians are liars
Y: Politicians are smart
Z: Politicians are rich

Write the following statement as implications
a.  Smart politicians are liars
b.  No rich politician is smart

4.  Write the following implications using ordinary language.
a.  r is negative $\Rightarrow r^3$ is negative
b.  $y^3 = -64 \Rightarrow y = -4$

5.  Write down the converse of the implication:
    X: Abel is clever $\Rightarrow$ Abel is intelligent

6.  Write statements which are equivalent to the following:
a.  If you travel by road it will take you 11 hours
b.  No student who offer Mathematics will score above 80

7.  Use the truth table to prove that: $\sim(\sim p \wedge \sim q) = p \vee q$

8.  Determine if the following argument is valid.
    All Asian people are intelligent
    Bill is an Asian
    Bill is intelligent

9.  Is the following argument valid?
    White buildings are attractive buildings
    This building is white
    This building is attractive

10.  Determine if the argument below is valid.
    All the people in the family are dark in complexion
    Franca is dark in complexion
    Franca is a member of the family

11.  Determine if the following argument is valid.
    Joy is a Doctor
    Joy works in a hospital
    All doctors work in hospitals

# CHAPTER 9
## QUADRATIC EQUATION

An equation such as $x^2 + 7x - 18 = 0$, in which 2 is the highest power of the unknown variable is called quadratic equation. There are four methods of solving quadratic equations. They are:
1. By factorization
2. By Completing the square
3. By use of quadratic equation formula
4. By graphical method

We are going to look at the first three methods.

## Factorization of Quadratic Expression

**Examples**
1. Factorize the following quadratic expressions
a. $3a^2 + 5a$
b. $y^2 - 81$
c. $4m^2 - 49$
d. $18c^2 - 72$

Solutions
a. $3a^2 + 5a$
In this case the common factor (H.C.F) of $3a^2$ and $5a$ is 'a'. Hence we factorize directly as follows:
$3a^2 + 5a = a(3a + 5)$
The terms in the bracket are obtained by dividing each of $3a^2$ and $5a$ by 'a' which is the common factor.

b. $y^2 - 81$
$y^2 - 81$ is a difference (minus) of two squares. So, recall that the difference of two squares such as $a^2 - b^2$ is factorized as follows: $a^2 - b^2 = (a + b)(a - b)$
Therefore, $y^2 - 81 = (y)^2 - (9)^2$
$$= (y + 9)(y - 9)$$

c. $4m^2 - 49$
Each term in this expression is a square. The square root of $4m^2$ is $2m$, while the square root of 49 is 7. Hence, $4m^2 - 49$ is a difference (minus) of two squares. So, recall that the difference of two squares such as $a^2 - b^2$ is factorized to obtain: $(a + b)(a - b)$
Therefore, $4m^2 - 49 = (2m)^2 - (7)^2$
$$= (2m + 7)(2m - 7) \quad \text{[In the same way as } a^2 - b^2 = (a + b)(a - b)]$$

d. $18c^2 - 72$
This is not exactly a difference of two squares, but it can be made a difference of two squares by first factorizing with 18 as a common factor. This gives:

$18c^2 - 72 = 18(c^2 - 4)$

The terms in the bracket are now difference of two squares. The square root of $c^2 = c$, while the square root of $4 = 2$. Hence, we now have:

$$18(c^2 - 4) = 18[(c)^2 - (2)^2]$$
$$= 18[(c + 2)(c - 2)]$$

Therefore, $18c^2 - 72 = 18[(c + 2)(c - 2)]$

In the four examples above, none of the quadratic expressions has a complete three terms of the usual quadratic expression such as $3x^2 + 7x - 4$. Hence it is easy to factorize such expression. However, when we have a complete quadratic expression such as $2x^2 - 11x - 30$, then we have to apply a different method of factorization. This method is explained in the examples below.

2. Factorize $2x^2 - 11x - 30$.

<u>Solution</u>

This is a complete quadratic expression with three terms. In order to factorize this expression, we first multiply the first and last terms. This gives:

$2x^2(-30) = -60x^2$

Then we look for two factors of $-60x^2$ (i.e. two numbers whose product is $-60x^2$) such that their sum will give $-11x$ (i.e. the middle term in the original quadratic expression).

Some simple rules that should be noted when looking for the two factors are as given below:

1. In a case such as $2x^2 + 7x + 5$, where the sign of the constant term (i.e. +5) and the middle term (i.e. +7) are both positive, then the two factors should be positive.

2. In a case such as $2x^2 - 7x + 5$, where the sign of the constant term is positive (i.e. +5) and the sign of the middle term is negative (i.e. −7), then the two factors should be negative.

3. In a case such as $2x^2 - 3x - 5$, where the sign of the constant term is negative (i.e. −5) and the sign of the middle term is negative (i.e. −3), then one factor will be positive and the other will be negative. However, the higher value of the two factors will take the negative sign (i.e. the sign of the middle term).

4. In a case such as $2x^2 + 3x - 5$, where the sign of the constant term is negative (i.e. −5) and the sign of the middle term is positive (i.e. +3), then one factor will be positive and the other will be negative. However, the higher value of the two factors will take the positive sign (i.e. from the middle term).

Let us summarize these four rules as shown below:

| S/N | Sign of product | Sign of middle term | Sign of factors |
|-----|-----------------|---------------------|-----------------|
| 1 | + | + | + and + |
| 2 | + | - | - and - |
| 3 | - | - | + and - with the larger value having - |
| 4 | - | + | + and - with the larger value having + |

With these four rules in mind, finding the factors in a quadratic equation becomes easy.

Let us continue with the solution to the question above.

The question at hand which is $2x^2 - 11x - 30$, is applicable to rule 3 above. This means that our two factors will have different signs with the higher value having the negative sign.

From the $-60x^2$ obtained above, a simple way of finding the factors of 60 (first ignore the minus sign and $x^2$) is to break 60 into its prime factors as follows:

| 2 | 60 |
|---|----|
| 2 | 30 |
| 3 | 15 |
| 5 | 5  |
|   | 1  |

The above division can help us obtain the twelve factors of 60 as follows: 1, 2, 3, 4 (from 2 x 2 in the division above), 5, 6 (from 2 x 3 in the division above), 10 (from 2 x 5 in the division above), 12 (from 2 x 2 x 3 in the division above), 15, 20 (from 2 x 2 x 5 in the division above), 30 and 60.

Note that only 1, 2, 3, 5, 15, 30 and 60 are visible from the division above. The other factors (4, 6, 10, 12 and 20) have to be obtained by multiplying any two or more of the visible factors. Let us rewrite the factors as follows: 1, 2, 3, 4, 5, 6, 10, 12, 15, 20, 30, and 60.

Now that we have our factors, we can pair them up to see which of them will give the required sum of -11. According to rule 3 above, the higher factor should have a negative sign. Let us now write out the pairs and assign a negative sign to the larger factor. Hence, the pair whose product will give us -60 are:

1 and -60  (This pair is out. Their sum cannot give -11)

2 and -30  (This pair is out)

3 and -20  (This pair is out)

5 and -12  (This pair is out)

6 and -10  (This pair is out)

4 and -15  (This is the required pair. Their sum gives -11)

Hence the only possible factors that can give a sum of -11 are 4 and -15. Hence we now include $x$ on them so that their product will give $-60x^2$ and their sum will give $-11x$. Hence the numbers are $-15x$ and $+4x$. We now substitute these two numbers for $-11x$ in the original expression. This gives:

$2x^2 - 11x - 30 = 2x^2 + 4x - 15x - 30$    (Note that any of the two numbers can be written first).

We now factorize each pair of terms in this new expression (i.e. factorization by grouping). This gives:
$2x^2 + 4x - 15x - 30 = 2x(x + 2) - 15(x + 2)$
Note that $-15x - 30 = -15(x + 2)$. The negative signs from the question have change to + in the bracket. This is what happens when a negative number is used for the factorization. A positive sign in the question will also change to a negative sign in the bracket. We now have:
$2x^2 + 4x - 15x - 30 = 2x(x + 2) - 15(x + 2)$    [Here the common factor is $(x + 2)$]
    $= (x + 2)(2x - 15)$    (Simply take one of the common brackets i.e. $(x + 2)$, and then the other two terms outside each of the bracket and enclose them in a bracket).
Therefore, $2x^2 - 11x - 30 = (x + 2)(2x - 15)$

3.  Factorize $x^2 + 13x - 30$
Here, you will observe that the coefficient of $x^2$ is 1 (i.e. $1x^2$). This is not like the previous example, i.e. $2x^2 - 11x - 30$, where the coefficient of $x^2$ is 2. So, a short method of factorizing this expression is to go ahead and look for two numbers whose product is $-30$ and whose sum is $+13$, without any need of multiplying the first and last terms. This is explained as follows:
$x^2 + 13x - 30 = (x \quad )(x \quad )$    (simply put the variable $x$ into the brackets)
In order to complete the bracket, we find two numbers whose product is $-30$ and whose sum is $+13$.  This means that we have to look for the factors of 30.
The factors of 30 are: 1, 2, 3, 5, 6, 10, 15 and 30.
Now that we have our factors, we can pair them up to see which of them will give a possible sum of $+13$. According to rule 4 above, the higher factor should have a positive sign. Let us now write out the pairs and assign a positive sign to the larger factor. Hence the pair whose product will give us $-30$ are:
    $-1$ and $+30$  (This is out. Their sum cannot give $+13$)
    $-3$ and $+10$  (This is out. Their sum cannot give $+13$)
    $-5$ and $+6$   (This is out.)
    $-2$ and $+15$  (These are the required factors. Their sum will give $+13$)
Therefore, we now put them into the brackets above as follows:
    $x^2 + 13x - 30 = (x + 15)(x - 2)$

4.  Factorize $3x^2 + 11x + 6$

Solution
    $3x^2 + 11x + 6$
The first thing to do is to multiply the first and last term. This gives:
    $3x^2(6) = 18x^2$
We now look for two factors of 18 whose product is 18 and sum is $+11$. This is a case of rule 1 above where the constant term is positive and the middle term is also positive. Hence the two factors will be positive. The factors of 18 are 1, 2, 3 , 6, 9, 18. If we pair them up based on any

two of the factors whose product is 18, we have:

(1 and 18), (2 and 9), (3 and 6).

Out of these pairs, only 2 and 9 can give us a sum of 11. Their signs must be positive based on rule 1 above. We now include $x$ on them so that their product will give $18x^2$ and their sum will give $11x$. Hence the numbers are $+9x$ and $+2x$.

We now substitute these two numbers for $11x$ in the original expression. This gives:

$3x^2 + 11x + 6 = 3x^2 + 9x + 2x + 6$     (Note that any of the two numbers can be written first).

We now factorize by grouping. This gives:

$$3x^2 + 9x + 2x + 6 = 3x(x + 3) + 2(x + 3)$$
$$= (x + 3)(3x + 2)$$

Therefore, $3x^2 + 11x + 6 = (x + 3)(3x + 2)$

5. Factorize $5x^2 - 14x + 8$

<u>Solution</u>

$5x^2 - 14x + 8$

The first thing to do is to multiply the first and last terms. This gives:

$5x^2(8) = 40x^2$

We now find two factors of 40 whose product is 40 and sum is $-14$ (i.e. the middle term). This is a case of rule 2 above where the constant term is positive and the middle term is negative. Hence the two factors will be negative. The factors of 40 are 1, 2, 4 , 5, 8, 10, 20 and 40. According to rule 2 above, the factors will both have a negative sign. Let us now write out the pairs and assign a negative sign to each factor. Hence the pairs whose product will give us 40 are:

$(-1$ and $-40)$, $(-2$ and $-20)$, $(-4$ and $-10)$, $(-5$ and $-8)$. Note that the product of two negative numbers gives a positive value.

Out of these pairs, only $-4$ and $-10$ can give us a sum of $-14$. Hence we now include $x$ on them so that their product will give $40x^2$ and their sum will give $-14x$. Hence the numbers are $-4x$ and $-10x$.

We now substitute these two numbers for $-14x$ in the original expression. This gives:

$5x^2 - 14x + 8 = 5x^2 - 4x - 10x + 8$     (Note that any of the two numbers can be written first).

We now factorize by grouping. This gives:

$$5x^2 - 4x - 10x + 8 = x(5x - 4) - 2(5x - 4)$$

Note that in the second pair (i.e. $-10x + 8$) the sign of the first number must be used. Hence the use of $-2$ to obtain $-2(5x - 4)$.

Therefore with $(5x - 4)$ as a common factor we proceed with the factorization as follows:

$$x(5x - 4) - 2(5x - 4) = (5x - 4)(x - 2)$$

Therefore, $5x^2 - 14x + 8 = (5x - 4)(x - 2)$

## Solving quadratic equations by by Factorization

In solving quadratic equation by factorization, we first factorize the equation and then equate each of the term or bracket to zero to solve the linear equation formed. The general form of a quadratic equation is given by: $ax^2 + bx + c = 0$. Hence if an equation is given by $3x^2 - 11x + 6$, we can compare it with $ax^2 + bx + c = 0$, and obtain:

a = 3, b = −11 and c = 6.

With these values of a, b and c, we can determine if a quadratic expression can be factorized or not. Hence, a quadratic equation can be factorized if:
$b^2 - 4ac$ gives a perfect square.
For example, let us determine which of the following can be factorized:

a.  $x^2 + 5x + 6 = 0$

b.  $4x^2 - 11x + 5 = 0$

c.  $2x^2 + x - 6 = 0$

d.  $5x^2 - 6x - 8 = 0$

e.  $3x^2 - 13x + 8 = 0$

Now, let us see which of them will give $b^2 - 4ac$ as a perfect square.

a.  $x^2 + 5x + 6 = 0$.
    a = 1, b = 5, c = 6
Hence, $b^2 - 4ac = 5^2 - (4 \times 1 \times 6)$
$$= 25 - 24$$
$$= 1$$
Therefore, 1 is a perfect square whose square root is 1. Hence, $x^2 + 5x + 6 = 0$ can be factorized.

b.  $4x^2 - 11x + 5 = 0$.
In this case, a = 4, b = −11, c = 5
Hence, $b^2 - 4ac = (-11)^2 - (4 \times 4 \times 5)$
$$= 121 - 80$$
$$= 41$$
Therefore, 41 is not a perfect square. Its square root will not give us a whole number. Hence, $4x^2 - 11x + 5 = 0$ cannot be factorized.

c.  $2x^2 + x - 6 = 0$.
    a = 2, b = 1, c = −6
Hence, $b^2 - 4ac = 1^2 - [4 \times 2 \times (-6)]$
$$= 1 - (-48)$$
$$= 1 + 48$$
$$= 49$$
Therefore, 49 is a perfect square whose square root is 7. Hence, $2x^2 + x - 6 = 0$ can be factorized.

d.  $5x^2 - 6x - 8 = 0$.
    a = 5, b = −6, c = −8

Hence, $b^2 - 4ac = (-6)^2 - [4 \times 5 \times (-8)]$

$\qquad = 36 - (-160)$

$\qquad = 36 + 160$

$\qquad = 196$

Therefore, 196 is a perfect square whose square root is 14. Hence, $5x^2 - 6x - 8 = 0$ can be factorized.

e.  $3x^2 - 13x + 8 = 0$.

  $a = 3, b = -13, c = 8$

Hence, $b^2 - 4ac = (-13)^2 - (4 \times 3 \times 8)$

$\qquad = 169 - 96$

$\qquad = 73$

Therefore, 73 is not a perfect square. Its square root will not give us a whole number. Hence, $3x^2 - 13x + 8 = 0$ cannot be factorized.

**Examples**

1.  Solve the following equations by factorization:

a.  $2a^2 + 8a = 0$

b.  $r^2 - 64 = 0$

c.  $9m^2 - 25 = 0$

d.  $(3x + 2)^2 = 16$

Solutions

a.  $2a^2 + 8a = 0$

Factorizing this quadratic equation by taking 2a as the common term gives:

  $2a(a + 4) = 0$

We now equate each term to zero and solve for 'a'. We equate each term to zero because the product of two or more terms is zero if at least one of the terms is equal to zero. Therefore equating each term above to zero and solving for 'a' gives:

Either,

  $2a = 0$

Dividing both sides by 2 gives:

  $a = \dfrac{0}{2}$

  $a = 0$

Or,  $(a + 4) = 0$

  $a + 4 = 0$

  $a = -4$

Therefore, a = 0 or a = −4

b.  $r^2 - 64 = 0$

This can be solved directly as follows:

  $r^2 - 64 = 0$

  $r^2 = 64$

  $r = \sqrt{64}$

$$r = \pm 8$$

Therefore, $r = +8$ or $r = -8$

Note that $\pm$ is used to obtain two values where one is given a positive value and the other is given a negative value.

Hence, $r = 8$ or $r = -8$

c.    $9m^2 - 25 = 0$

     $9m^2 = 25$

Divide both sides by 9. This gives:

$$m^2 = \frac{25}{9}$$

$$m = \sqrt{\frac{25}{9}}$$

$$m = \pm \frac{5}{3}$$

Therefore, $m = \frac{5}{3}$ or $m = -\frac{5}{3}$

d.    $(3x + 2)^2 = 16$

Let us take the square root of both sides. This gives:

$$\sqrt{(3x + 2)^2} = \sqrt{16}$$

$$(3x + 2) = \pm 4$$

$$3x = \pm 4 - 2$$

$$x = \frac{+4 - 2}{3} \quad \text{or} \quad x = \frac{-4 - 2}{3}$$

$$x = \frac{2}{3} \quad \text{or} \quad x = \frac{-6}{3}$$

$$x = \frac{2}{3} \quad \text{or} \quad x = -2$$

2.   Solve the equation $2x^2 - 5x - 3 = 0$

<u>Solution</u>

    $2x^2 - 5x - 3 = 0$

Multiply the first and last term on the left hand side. This gives:

    $2x^2(-3) = -6x^2$

Now look for two numbers in $x$ such that their product is $-6x^2$ and their sum is $-5x$ (i.e. the middle term). For us to obtain the two factors let us recall our rules as shown below:

| S/N | Sign of product | Sign of middle term | Sign of factors |
|---|---|---|---|
| 1 | + | + | + and + |
| 2 | + | - | - and - |
| 3 | - | - | + and - with the larger value having - |
| 4 | - | + | + and - with the larger value having + |

Based on the rules, since the product of the terms is negative (i.e. $-6x^2$) and their sum is also negative (i.e. $-5x$), as in rule 3 above, then the two factors of 6 that we need must have different signs, and with the higher of the factors having a negative sign. The pairs of factors of 6 whose product is 6 are (1 and 6) and (2 and 3). The larger factor should have a negative sign. If we assign a negative sign to the larger factors, then the pair can be written as: (+1 and –6), (2 and –3). Out of these two pairs, the one whose sum is –5 is (+1 and –6). Hence, we now include $x$ on them so that their product will give $-6x^2$ and their sum will give $-5x$. Hence the numbers are $+x$ and $-6x$. Note that $+1x$ is also $+x$.

We now substitute these two numbers for $-5x$ in the original expression. This gives:

$2x^2 - 6x + x - 3 = 0$

We now factorize by grouping. This gives:

$2x(x - 3) + 1(x - 3) = 0$

With $(x - 3)$ as a common factor we proceed as follows:

$(x - 3)(2x + 1) = 0$

Now that we have factorized the equation, we equate each of the bracket to zero and solve for $x$ in each of the linear equation formed. This gives:

Either, $x - 3 = 0$

$x = 3$

Or, $2x + 1 = 0$

$2x = -1$

$x = -\dfrac{1}{2}$  (After dividing both sides by 2)

Therefore, $x = 3$ or $x = -\dfrac{1}{2}$


3. Solve the quadratic equation $y^2 + y - 12 = 0$

Solution

$y^2 + y - 12 = 0$

Since the coefficient of $y^2$ in this equation is 1, then we can factorize the equation by putting directly the appropriate factors into bracket.

$y^2 + y - 12 = 0$

$(y\quad)(y\quad) = 0$

Here we look for the factors of 12 whose product is –12 and whose sum is +1 (i.e. +1 from +y which also means +1y). Since the product is negative and the sum is positive, then the two factors must have different sign, and with the positive sign going to the larger factor. The factors of 12 whose product will give -12 and whose sum is +1 are +4 and –3. Note that the larger value has the positive sign (i.e. from +1). We now fill these two values into the bracket above. This gives:

$$y^2 + y - 12 = 0$$
$$(y + 4)(y - 3) = 0$$

Equating each bracket to zero gives:

$$y + 4 = 0$$
$$y = -4$$
$$y - 3 = 0$$
$$y = 3$$

Hence, $y = 3$ or $y = -4$

4. Solve the quadratic equation $m^2 - 9m + 18 = 0$

<u>Solution</u>

$$m^2 - 9m + 18 = 0$$

Since the coefficient of $m^2$ in this equation is 1, then we can factorize the equation by putting directly the appropriate factors into bracket.

$$m^2 - 9m + 18 = 0$$
$$(m \quad )(m \quad ) = 0$$

Look for two factors of 18 whose product is +18 and whose sum is −9. Since the product is positive and the sum is negative, then the two factors must be negative. The factors of 18 whose product is +18 and whose sum is −9 are −6 and −3. We now fill these two values into the bracket above. This gives:

$$m^2 - 9m + 18 = 0$$
$$(m - 6)(m - 3) = 0$$

Equating each bracket to zero gives:

$$m - 6 = 0$$
$$m = 6$$
$$m - 3 = 0$$

$$m = 3$$

Hence, $m = 6$ or $m = 3$

5. Solve the equation $6n^2 = 5 - 13n$

<u>Solution</u>

$$6n^2 = 5 - 13n$$

Rearranging the terms in order to equate the equation to zero (i.e. make the right hand side zero) gives:

$$6n^2 + 13n - 5 = 0$$

Multiply the first and last term on the left hand side. This gives:

$$6n^2(-5) = -30n^2$$

Now look for two numbers in n such that their product is $-30n^2$ and their sum is $+13n$ (i.e. the middle term). Based on the rules given above for obtaining factors, and since the product of the terms is negative (i.e. $-30n^2$) and their sum is positive (i.e. $+13n$), then the two factors of 30 that we need must have different signs, and with the higher of the factors having a positive sign (i.e. from +13). The pairs of factors of 30 are: (1 and 30), (2 and 15), (3 and 10), (6 and 5). The larger factor should have a positive sign. If we assign a positive sign to the larger number of each pair of the factors, then the pairs can be written as: (−1 and 30), (−2 and 15), (−3 and 10), (−5 and 6)

Out of these pairs, the one whose sum is +13 is (−2 and 15).

Hence, we now include 'n' on them so that their product will give $-30n^2$ and their sum will give $+13n$. Hence the numbers are $-2n$ and $+15n$.

We now substitute these two numbers for $+13n$ in the original expression. This gives:

$6n^2 + 15n - 2n - 5 = 0$

We now factorize by grouping. This gives:

$(6n^2 + 15n) - (2n - 5) = 0$

$3n(2n + 5) - 1(2n + 5) = 0$

With $(2x + 5)$ as a common factor, we proceed as follows:

$(2n + 5)(3n - 1) = 0$

Now that we have factorized the equation, we equate each of the bracket to zero and solve for n in each of the linear equation formed. This gives:

Either, $2n + 5 = 0$

$\qquad 2n = -5$

$\qquad n = -\dfrac{5}{2}$

Or, $\quad 3n - 1 = 0$

$\qquad 3n = 1$

$\qquad n = \dfrac{1}{3}$    (After dividing both sides by 3)

Therefore, $n = -\dfrac{5}{2}$ or $n = \dfrac{1}{3}$

6. Solve the quadratic equation: $8 - 6x - 5x^2 = 0$

Solution

$\quad 8 - 6x - 5x^2 = 0$

Multiply the first and last term on the left hand side. This gives:

$\quad 8(-5x^2) = -40x^2$

Now look for two numbers in $x$ such that their product is $-40x^2$ and their sum is $-6x$ (i.e. the middle term). Based on the rules given above for obtaining factors, and since the product of the terms is negative (i.e. $-40x^2$) and their sum is negative (i.e. $-6x$), then the two factors of 40 that we need must have different signs. The higher of the factors will have a negative sign (i.e. from −6, the middle number). The pairs of factors whose product is 40 are (1 and 40), (2 and 20), (4 and 10), (5 and 8). The larger factor should have a negative sign. If we assign a negative sign to the larger number of each pair of the factors, then the pairs can be written as: (1 and −40), (2 and −20), (4 and −10), (5 and −8)

95

Out of these pairs, the one whose sum is −6 is (4 and −10), since −10 + 4 = −6.
Hence, we now include '$x$' on them so that their product will give −40$x^2$ and their sum will give −6$x$. Hence the numbers are +4$x$ and −10$x$.
We now substitute these two numbers for −6$x$ in the original equation. This gives:

$8 + 4x − 10x − 5x^2 = 0$

We now factorize by grouping. This gives:

$(8 + 4x) − (10x − 5x^2) = 0$

$4(2 + x) − 5x(2 + x) = 0$

With (2 + $x$) as a common factor, we proceed as follows:

$(2 + x)(4 − 5x) = 0$

Now that we have factorized the equation, we equate each of the brackets to zero and solve for $x$ in each of the linear equation formed. This gives:

Either, $2 + x = 0$

$x = −2$

Or, $4 − 5x = 0$

$4 = 5x$

$x = \dfrac{4}{5}$   (After dividing both sides by 5)

Therefore, $x = −2$ or $x = \dfrac{4}{5}$

## Construction of Quadratic Equation from Given Roots

The solutions of a quadratic equation are called roots. Since the roots are obtained from the factors, it follows that the factors can also be obtained from the roots. The multiplication of the factors gives the quadratic equation.

Consider the quadratic equation: $ax^2 + bx + c = 0$

If we divide each term by 'a', then we will obtain the quadratic equation in terms of its sum of roots and product of roots as follows:

$$x^2 + \dfrac{b}{a}x + \dfrac{c}{a} = 0$$

With this simplification, the sum of roots of the quadratic equation is:

$$−\dfrac{b}{a}$$

while the product of roots of the quadratic equation is:

$$\dfrac{c}{a}$$

Therefore, a quadratic equation can be generally expressed as follows:

$$x^2 − (\text{sum of roots})x + (\text{product of roots}) = 0$$

If the roots of a quadratic equation are given as $\alpha$ and $\beta$, then the quadratic equation can be represented as follows:

$$x^2 − (\alpha + \beta)x + (\alpha\beta) = 0$$

Let us now apply these principles with the following examples.

**Examples**
1. Find the quadratic equation whose roots are 4 and −7.

## Solution

### Method 1

Since the roots/solutions of the quadratic equation are 4 and −7, it follows that:

$x = 4$ or $x = -7$

Equating each solution to zero gives us the factors as follows:

$x - 4 = 0$ or $x + 7 = 0$ (Note that from $x = 4$, we have, $x - 4 = 0$, and from $x = -7$, $x + 7 = 0$)

The multiplication of these two factors gives us the equation as follows:

$(x - 4)(x + 7) = 0$

Expanding the brackets gives:

$x^2 + 7x - 4x - 28 = 0$

$x^2 + 3x - 28 = 0$ (Note that $7x - 4x = 3x$)

The quadratic equation is $x^2 + 3x - 28 = 0$

### Method 2

Recall that a quadratic equation is given in term of its roots as follows:

$x^2 - (\text{sum of roots})x + (\text{product of roots}) = 0$

Hence, we can now use 4 and −7 to obtain the sum and product of roots as follows:

$x^2 - (\text{sum of roots})x + (\text{product of roots}) = 0$

$x^2 - (-7 + 4)x + [-7(4)] = 0$

$x^2 - (-3)x + (-28) = 0$

$x^2 + 3x - 28 = 0$

Hence the quadratic equation is $x^2 + 3x - 28 = 0$.

2. Find the quadratic equation whose roots are $\dfrac{2}{3}$ and 5.

## Solution
### Method 1

Since the roots/solutions of the quadratic equation are $\dfrac{2}{3}$ and 5, it follows that:

$x = \dfrac{2}{3}$ or $x = 5$

Equating each solution to zero gives us the factors as follows:

$x - \dfrac{2}{3} = 0$ or $x - 5 = 0$

In order to clear out the fraction in $x - \dfrac{2}{3} = 0$, we simply multiply each term by 3. This gives:

$3(x) - 3(\dfrac{2}{3}) = 3(0)$

$3x - 2 = 0$

Hence the two factors are:

$3x - 2 = 0$ and $x - 5 = 0$

The multiplication of these two factors gives us the equation as follows:

$(3x - 2)(x - 5) = 0$

Expanding the brackets gives:

$3x^2 - 15x - 2x + 10 = 0$

$3x^2 - 17x + 10 = 0$

Hence, the quadratic equation is $3x^2 - 17x + 10 = 0$

## Method 2

Recall that a quadratic equation is given in term of its roots as follows:

$x^2 - $ (sum of roots)$x + $ (product of roots) $= 0$

Hence, we can now use $\frac{2}{3}$ and 5 to obtain the sum and product of roots as follows:

$x^2 - $ (sum of roots)$x + $ (product of roots) $= 0$

$x^2 - (\frac{2}{3} + 5)x + [\frac{2}{3}(5)] = 0$

$x^2 - (\frac{2 + 15}{3})x + (\frac{10}{3}) = 0$

$x^2 - (\frac{17}{3})x + (\frac{10}{3}) = 0$

In order to clear out the fractions, we multiply each term by 3 (i.e. the LCM of the denominators). This gives:

$3(x^2) - 3(\frac{17}{3})x + 3(\frac{10}{3}) = 0$

The 3 will cancel out each other in the fractions to give:

$3x^2 - 17x + 10 = 0$

Hence the quadratic equation is $3x^2 - 17x + 10 = 0$.

3.  Find the quadratic equation whose roots are $-2$ and $-\frac{3}{4}$

Solution

## Method 1

Since the roots of the quadratic equation are $-2$ and $-\frac{3}{4}$, it follows that:

$x = -2$ or $x = -\frac{3}{4}$

Equating each solution to zero gives us the factors as follows:

$x + 2 = 0$ or $x + \frac{3}{4} = 0$

In order to clear out the fraction in $x + \frac{3}{4} = 0$, we simply multiply each term by 4. This gives:

$4(x) + 4(\frac{3}{4}) = 4(0)$

$4x + 3 = 0$

Hence the two factors are:

$x + 2 = 0$ and $4x + 3 = 0$

The multiplication of these two factors gives us the equation as follows:

$(x + 2)(4x + 3) = 0$

Expanding the brackets gives:

$4x^2 + 3x + 8x + 6 = 0$

$4x^2 + 11x + 6 = 0$

Hence, the quadratic equation is $4x^2 + 11x + 6 = 0$

## Method 2

Recall that a quadratic equation is given in term of its roots as follows:

$x^2 - $ (sum of roots)$x + $ (product of roots) $= 0$

Hence, we can now use $-2$ and $-\dfrac{3}{4}$ to obtain the sum and product of roots as follows:

$x^2 - $ (sum of roots)$x + $ (product of roots) $= 0$

$x^2 - (-2 - \dfrac{3}{4})x + [-2(-\dfrac{3}{4})] = 0$

$x^2 - (\dfrac{-8 - 3}{4})x + [-2(-\dfrac{3}{4}) = 0$

$x^2 - (-\dfrac{11}{4})x + (\dfrac{6}{4}) = 0$

$x^2 - (-\dfrac{11}{4})x + (\dfrac{3}{2}) = 0$    (Note that $\dfrac{6}{4} = \dfrac{3}{2}$ in its lowest term)

In order to clear out the fractions, we multiply each term by 4 (i.e. the LCM of the denominators). This gives:

$4(x^2) - 4(-\dfrac{11}{4})x + 4(\dfrac{3}{2}) = 0$

This will now give:

$4x^2 + 11x + 6 = 0$    (Note that $-4(-\dfrac{11}{4}) = +11$)

Hence the quadratic equation is $4x^2 + 11x + 6 = 0$.


4. A quadratic equation is given by: $x^2 - 7x - 10 = 0$. Find:

a. the sum of the roots of the equation

b. the product of the roots of the equation

<u>Solution</u>

a. $x^2 - 7x - 10 = 0$

Comparing this equation with $ax^2 + bx + c = 0$, shows that:

   $a = 1$, $b = -7$, and $c = -10$    (Note that $x^2 = 1x^2$, hence $a = 1$)

Therefore, the sum of roots is given by:

$-(\dfrac{b}{a}) = -(-\dfrac{7}{1})$

$\qquad = 7$

Hence, the sum of the roots is 7


b. Also, the product of roots is given by:

$\dfrac{c}{a} = \dfrac{-10}{1} = -10$

Hence the product of the roots is −10

5.  A quadratic equation is given by: $2x^2 + 13x - 26 = 0$. Find:
a. the sum of the roots of the equation
b.  the product of the roots of the equation

<u>Solution</u>
a.    $2x^2 + 13x - 26 = 0$
Comparing this equation to $ax^2 + bx + c = 0$, shows that:
    a = 2, b = 13, c = −26

Recall that the sum of roots of a quadratic equation is given by $-\dfrac{b}{a}$

Hence, sum of roots of the equation above is:
$$-\frac{b}{a} = -\frac{13}{2}$$
Therefore sum of roots of the equation is $-\dfrac{13}{2}$

b.   Recall that the product of the roots of a quadratic equation is given by: $\dfrac{c}{a}$

Hence product of roots of the equation above is:
$$= -\frac{26}{2}$$
$$= -13$$
Therefore, product of roots of the equation is −13.

6.  If 3 is a root of the equation $2x^2 + Kx - 21 = 0$, find the value of K. Hence find the other root.

<u>Solution</u>
    $2x^2 + Kx - 21 = 0$
In order to find K, substitute 3 for $x$ in the equation. This gives:
    $2x^2 + Kx - 21 = 0$
    $2(3)^2 + K(3) - 21 = 0$
    $2(9) + 3K - 21 = 0$
    $18 + 3K - 21 = 0$
    $3K = 21 - 18$
    $3K = 3$
    $K = \dfrac{3}{3}$
    $K = 1$
Hence the value of K is 1.
The equation can now be written as $2x^2 + x - 21 = 0$
From this equation, a = 2, b = 1 and c = −21
Hence, product of roots $= \dfrac{c}{a}$

$$= -\frac{21}{2}$$

Let the second root be m.

Since one of the roots is 3, then product of roots which is $-\frac{21}{2}$ is given by:

$3(m) = -\frac{21}{2}$  (Note that the two roots are 3 and m)

$3m = -\frac{21}{2}$

$m = \dfrac{-\frac{21}{2}}{3}$

$= -\frac{21}{2} \times \frac{1}{3}$

$m = -\frac{21}{6}$

$m = -\frac{7}{2}$  (In its lowest term)

Therefore the other root is $-\frac{7}{2}$

Note that we can also factorize the equation $2x^2 + x - 21 = 0$, to obtain the two roots, and hence the other root.


7. If 1 is a root of the equation $5x^2 - kx - 3 = 0$, find the value of K. Hence find the other root.

<u>Solution</u>
$5x^2 - kx - 3 = 0$
In order to find K, substitute 1 for $x$ in the equation. This gives:
$5x^2 - kx - 3 = 0$
$5(1)^2 - k(1) - 3 = 0$
$5(1) - k - 3 = 0$
$5 - k - 3 = 0$
$5 - 3 = k$
$k = 2$
Hence the value of k is 2.
The equation can now be written as $5x^2 - 2x - 3 = 0$
From this equation, a = 5, b = −2 and c = −3
Hence, product of roots $= \dfrac{c}{a}$

$$= -\frac{3}{5}$$

Let the second root be y.

Since one of the roots is 1, then product of roots which is $-\dfrac{3}{5}$ is given by:

$\qquad 1(y) = -\dfrac{3}{5}$    (Note that the two roots are 1 and y, hence their product is 1y)

$\qquad y = -\dfrac{3}{5}$

Therefore the other root is $-\dfrac{3}{5}$

8. 5 and −7 are the roots of the quadratic equation $y^2 - ky + M = 0$. Find the values of k and M.

<u>Solution</u>

$\qquad y^2 - ky + M = 0$

From this equation, a = 1, b = −k, c = M

Hence sum of roots $= -\dfrac{b}{a}$

$\qquad\qquad\qquad = -\left(-\dfrac{k}{1}\right)$    (Since b = −k)

$\qquad\qquad\qquad = k$

Since the two roots are 5 and −7, and their sum is k, then it follows that:

$\qquad -7 + 5 = k$

$\qquad\quad -2 = k$

$\qquad\quad\; k = -2$

Also, product of roots $= \dfrac{c}{a}$

$\qquad\qquad\qquad\qquad = \dfrac{M}{1}$    (Since c = M)

$\qquad\qquad\qquad\qquad = M$

Since the two roots are 5 and −7, and their product is M, then it follows that:

$\qquad 5(-7) = M$

$\qquad\quad -35 = M$

$\qquad\qquad M = -35$

Therefore, k = −2 and M = −35

9. If −3 and 8 are the roots of the quadratic equation $2x^2 + Dx - E = 0$. Find the values of the constants D and E.

<u>Solution</u>

$\qquad 2x^2 + Dx - E = 0$

From this equation, a = 2, b = D, c = −E

Hence sum of roots $= -\dfrac{b}{a}$

$$= -\frac{D}{2} \quad \text{(Since b = D)}$$

Since the two roots are –3 and 8, and their sum is $-\frac{D}{2}$, then it follows that:

$$-3 + 8 = -\frac{D}{2}$$

$$5 = -\frac{D}{2}$$

$$-D = 2 \times 5$$
$$-D = 10$$
$$D = -10 \quad \text{(After dividing both sides by –1)}$$

Also, product of roots = $\frac{c}{a}$

$$= -\frac{E}{2} \quad \text{(Since c = –E and a = 2)}$$

Since the two roots are –3 and 8, and their product is $-\frac{E}{2}$, then it follows that:

$$-3(8) = -\frac{E}{2}$$

$$-24 = -\frac{E}{2}$$

$$-E = -24 \times 2$$
$$-E = -48$$
$$E = 48 \quad \text{(After dividing both sides by –1)}$$

Therefore, D = –10 and E = 48

10. If one root of the equation $y^2 - ky + 18 = 0$ is twice the other, find the two possible values of k. Hence, what are the two possible equations?

<u>Solution</u>
Let one root of the equation be m. Therefore, the other root will be 2m since one root is twice the other.
Hence product of roots = m(2m)
$$= 2m^2$$
The equation is $y^2 - ky + 18 = 0$
From this equation, a = 1, b = –k and c = 18
Product of roots = $\frac{c}{a}$

$$= \frac{18}{1} = 18$$

Hence, $2m^2 = 18 \quad$ (Note that product of roots is also $2m^2$)
$$m^2 = \frac{18}{2}$$
$$m^2 = 9$$

103

$$m = \sqrt{9}$$
$$m = \pm 3$$
$$m = 3 \text{ or } -3$$

When m = 3, the other root, 2m, is 2(3) = 6    (Since one root is twice the other)

When m = −3, then the other root, 2m, is 2(−3) = −6

Hence, the roots are 3 and 6, or −3 and −6.

Recall that the equation is $y^2 - ky + 18 = 0$, and a = 1, b = −k, and c = 18.

Hence, sum of roots $= -\dfrac{b}{a}$

$$= -(-\dfrac{k}{1}) \qquad \text{(Since b = −k and a = 1)}$$

$$= k$$

Therefore, when the roots are 3 and 6, then sum of roots = 3 + 6 = 9. But sum of roots is also equal to k.

Hence, k = 9.

When the roots are −3 and −6, then sum of roots = −3 − 6 = −9

Hence, k = −9

Therefore, k = 9 or −9

When k = 9, the equation is: $y^2 - ky + 18 = 0$, which gives:

$$y^2 - 9y + 18 = 0$$

When k = −9, the equation is:

$$y^2 - (-9)y + 18 = 0, \text{ which gives:}$$
$$y^2 + 9y + 18 = 0$$

**Completing the Square of Quadratic expressions**

Another method of solving quadratic equation is by completing the square. In order to make a quadratic expression a perfect square, simply add the square of half the coefficient of $x$ of the quadratic expression.

Hence, completing the square of a quadratic expression means to make the quadratic expression a perfect square.

**Examples**

1.  What must be added to $x^2 + 10x$ to make it a perfect square?

Solution

$$x^2 + 10x$$

Half the coefficient of $x$ is given by: $\dfrac{10}{2} = 5$.

Square this value obtained. This gives:

$$5^2 = 25.$$

Hence, 25 must be added to the quadratic expression to make it a perfect square.

If we add 25 to the quadratic expression it gives:

$$x^2 + 10x + 25$$

This means that $x^2 + 10x + 25$ is a perfect square. Its square root is obtained by taking the square root of the first and last terms and adding them together. The square root of $x^2$ is $x$

while the square root of 25 is $\pm 5$, but we take $+5$ (take the sign of the coefficient of $x$ in the expression). Therefore the square root of $x^2 + 10x + 25$ is $(x + 5)$. If we square the square root, it gives the perfect square.

Hence, $x^2 + 10x + 25 = (x + 5)^2$

2. What must be added to $x^2 - 18x$ to make it a perfect square?

Solution

$x^2 - 18x$

The coefficient of $x$ is $-18$. Divide this coefficient by 2. This gives:

$-\dfrac{18}{2} = -9$

Square this value obtained. This gives:

$-9^2 = 81$.

Hence, 81 must be added to the quadratic expression to make it a perfect square.

If we add 81 to the quadratic expression it gives:

$x^2 - 18x + 81$

This means that $x^2 - 18x + 81$ is a perfect square. Its square root is obtained by taking the square roots of the first and last terms and adding them together. The square root of $x^2$ is $x$ while the square root of 81 is $\pm 9$, but we take $-9$ (take the sign of the coefficient of $x$). Therefore the square root of $x^2 - 18x + 81$ is $(x - 9)$. If we square the square root, it gives the perfect square.

Hence, $x^2 - 18x + 81 = (x - 9)^2$

3. Make $y^2 - 5y$ a perfect square.

Solution

$y^2 - 5y$

The coefficient of $y$ is $-5$. Divide this coefficient by 2. This gives:

$-\dfrac{5}{2}$

Square this value obtained. This gives:

$\left(-\dfrac{5}{2}\right)^2 = \dfrac{25}{4}$

Hence, $\dfrac{25}{4}$ must be added to the quadratic expression to make it a perfect square.

If we add $\dfrac{25}{4}$ to the quadratic expression it gives:

$y^2 - 5y + \dfrac{25}{4}$

This means that $y^2 - 5y + \dfrac{25}{4}$ is a perfect square.

Its square root is obtained by taking the square roots of the first and last terms and adding them together. The square root of $y^2$ is $y$ while the square root of $\dfrac{25}{4}$ is $\pm\dfrac{5}{2}$, but we take $-\dfrac{5}{2}$

since the coefficient of y in the expression is negative.

Therefore the square root of $y^2 - 5y + \dfrac{25}{4}$ is $(y - \dfrac{5}{2})$. If we square the root, it gives the perfect square.

Hence, $y^2 - 5y + \dfrac{25}{4} = (x - \dfrac{5}{2})^2$

4. Make $y^2 + \dfrac{9}{2}y$ a perfect square.

Solution

$$y^2 + \dfrac{9}{2}y$$

The coefficient of y is $\dfrac{9}{2}$. Divide this coefficient by 2. This gives:

$$\dfrac{9}{2} \div 2$$

$$= \dfrac{9}{2} \times \dfrac{1}{2}$$

$$= \dfrac{9}{4}$$

Square this value obtained. This gives:

$$(\dfrac{9}{4})^2 = \dfrac{81}{16}$$

Hence, $\dfrac{81}{16}$ must be added to the quadratic expression to make it a perfect square.

If we add $\dfrac{81}{16}$ to the quadratic expression it gives:

$$y^2 + \dfrac{9}{2}y + \dfrac{81}{16}$$

This means that $y^2 + \dfrac{9}{2}y + \dfrac{81}{16}$ is a perfect square.

Therefore the square root of $y^2 + \dfrac{9}{2}y + \dfrac{81}{16}$ is $(y + \dfrac{9}{4})$.

Hence, $y^2 + \dfrac{9}{2}y + \dfrac{81}{16} = (y + \dfrac{9}{4})^2$

5. If $x^2 - 6x + (-3)^2$ is a perfect square, find its square root.

Solution

$$x^2 - 6x + (-3)^2$$

Since this is a perfect square, we find its square root by taking each of the squared term and enclosing them in a bracket. The square root of $x^2$ is $x$, while the square root of $(-3)^2$ is $-3$. Hence, we put these two square root terms in a bracket. This gives: $(x - 3)$

Therefore the square root of $x^2 - 6x + (-3)^2$ is $(x - 3)$

This means that: $x^2 - 6x + (-3)^2 = (x - 3)^2$

6.  If $x^2 - \frac{7}{2}x + (-\frac{7}{4})^2$ is a perfect square, find its square root.

Solution

$$x^2 - \frac{7}{2}x + (-\frac{7}{4})^2$$

Since this is a perfect square, we find its square root by taking each of the squared term and enclosing them in a bracket. The square root of $x^2$ is $x$, while the square root of $(-\frac{7}{4})^2$ is $-\frac{7}{4}$. Hence, we put these

two square root terms in a bracket. This gives: $(x - \frac{7}{4})$

Therefore the square root of $x^2 - \frac{7}{2}x + (-\frac{7}{4})^2$ is $(x - \frac{7}{4})$

This means that: $x^2 - \frac{7}{2}x + (-\frac{7}{4})^2 = (x - \frac{7}{4})^2$


7.  If $x^2 + 1.5x + (0.75)^2$ is a perfect square, find its square root.

Solution

$$x^2 + 1.5x + (0.75)^2$$

The square root of $x^2$ is $x$, while the square root of $(0.75)^2$ is 0.75. Hence, we put these two square root terms in a bracket. This gives: $(x + 0.75)$

Therefore the square root of $x^2 + 1.5x + (0.75)^2$ is $(x + 0.75)$

## Solving Quadratic Equation by Completing the Square

**Examples**

1.  Use the method of completing the square to solve the quadratic equation $y^2 + 7y - 30 = 0$

Solution

$$y^2 + 7y - 30 = 0$$

Take the constant term (i.e. −30) to the right hand side of the equation. This gives:

$$y^2 + 7y = 30$$

The next step is to make the left hand side expression a perfect square. In doing this, whatever is added to the left hand side should also be added to the right hand side. Let us do this as follows.

The coefficient of y is 7. Divide this coefficient by 2. This gives:

$$\frac{7}{2}$$

Square the value obtained. This gives:

$$(\frac{7}{2})^2 = \frac{49}{4}$$

Hence, $\frac{49}{4}$ must be added to the left hand expression to make it a perfect square. However, we are not going to write it as $\frac{49}{4}$. We will leave it as $(\frac{7}{2})^2$. We will now add $(\frac{7}{2})^2$ to the left hand side, and also add it to the right hand side. This gives:

$$y^2 + 7y + \left(\frac{7}{2}\right)^2 = 30 + \left(\frac{7}{2}\right)^2$$

While keeping the left hand side as it is, so that we can easily express it as a square of its square root, we go ahead and simplify the right hand side. This gives:

$$y^2 + 7y + \left(\frac{7}{2}\right)^2 = \frac{30}{1} + \frac{49}{4}$$

$$y^2 + 7y + \left(\frac{7}{2}\right)^2 = \frac{120 + 49}{4} \qquad \text{(Note that 4 x 30 = 120)}$$

$$y^2 + 7y + \left(\frac{7}{2}\right)^2 = \frac{169}{4}$$

Take the square root of the left hand side and square it. This is done by taking the terms in square (i.e. y and $\frac{7}{2}$), enclosing them in a bracket and squaring the bracket. This gives:

$$\left(y + \frac{7}{2}\right)^2 = \frac{169}{4}$$

Take the square root of both sides in order to remove the square on the left hand side. This gives:

$$\sqrt{\left(y + \frac{7}{2}\right)^2} = \sqrt{\frac{169}{4}}$$

$$\left(y + \frac{7}{2}\right) = \pm \frac{13}{2}$$

(Note that $\sqrt{x^2} = x$, hence the removal of square from $\left(y + \frac{7}{2}\right)^2$ to get $\left(y + \frac{7}{2}\right)$. Also, remember to add $\pm$ to the square root of a number).

Let us continue with our solution as follows:

$$\left(y + \frac{7}{2}\right) = \pm \frac{13}{2}$$

Or, $\quad y + \dfrac{7}{2} = \pm \dfrac{13}{2}$

Take the constant term on the left hand side to the right. This gives:

$$y = -\frac{7}{2} \pm \frac{13}{2}$$

We now use $\pm$ to split the values above into two solutions. When we take only the positive sign of $\pm$, our value above gives:

$$y = -\frac{7}{2} + \frac{13}{2}$$

$$y = \frac{-7 + 13}{2}$$

$$y = \frac{6}{2}$$

$$y = 3$$

When we take only the negative sign of $\pm$, our value above gives:

$$y = -\frac{7}{2} - \frac{13}{2}$$

$$y = \frac{-7 - 13}{2}$$

$$y = -\frac{20}{2}$$

$$y = -10$$

$$\therefore \quad y = 3 \text{ or } y = -10$$


2. Use the method of completing the square to solve the quadratic equation $x^2 - 11x + 28 = 0$

Solution

$$x^2 - 11x + 28 = 0$$

Take the constant term (i.e. 28) to the right hand side of the equation. This gives:

$$x^2 - 11x = -28$$

The coefficient of $x$ is $-11$. Divide this coefficient by 2 to obtain:

$$-\frac{11}{2}$$

Square $-\frac{11}{2}$. This gives:

$$(-\frac{11}{2})^2$$

Add $(-\frac{11}{2})^2$ to both sides of the equation. This gives:

$$x^2 - 11x + (-\frac{11}{2})^2 = -28 + (-\frac{11}{2})^2$$

While keeping the left hand side, simplify the right hand side. This gives:

$$x^2 - 11x + (-\frac{11}{2})^2 = -28 + \frac{121}{4}$$

$$x^2 - 11x + (-\frac{11}{2})^2 = \frac{-112 + 121}{4} \qquad \text{(Note that 4 x } -28 = -112)$$

$$x^2 - 11x + (-\frac{11}{2})^2 = \frac{9}{4}$$

Take the square root of the left hand side and square it. This is done by taking the terms in square (i.e. $x$ and $-\frac{11}{2}$), enclosing them in a bracket and squaring the bracket. This gives:

$$(x - \frac{11}{2})^2 = \frac{9}{4}$$

Take the square root of both sides. This gives:

$$\sqrt{(x - \frac{11}{2})^2} = \sqrt{\frac{9}{4}}$$

$$(x - \frac{11}{2}) = \pm\frac{3}{2}$$

Or, $x - \frac{11}{2} = \pm\frac{3}{2}$

$$x = \frac{11}{2} \pm \frac{3}{2}$$

When we take only the positive sign of $\pm$, our value above gives:

$$x = \frac{11}{2} + \frac{3}{2}$$

$$x = \frac{11 + 3}{2}$$

$$x = \frac{14}{2}$$

$$x = 7$$

When we take only the negative sign of $\pm$, our value above gives:

$$x = \frac{11}{2} - \frac{3}{2}$$

$$x = \frac{11 - 3}{2}$$

$$x = \frac{8}{2}$$

$$x = 4$$

$\therefore \quad x = 7$ or $x = 4$

3.  Solve the quadratic equation $x^2 - 6x - 16 = 0$, by the method of completing the square.

<u>Solution</u>

$$x^2 - 6x - 16 = 0$$

Take the constant term (i.e. –16) to the right hand side of the equation. This gives:

$$x^2 - 6x = 16$$

The coefficient of $x$ is –6. Divide this coefficient by 2 to obtain:

$$-\frac{6}{2} = -3$$

Square the value obtained. This gives:

$$(-3)^2$$

Add $(-3)^2$ to both sides of the equation. This gives:

$$x^2 - 6x + (-3)^2 = 16 + (-3)^2$$
$$x^2 - 6x + (-3)^2 = 16 + 9$$
$$x^2 - 6x + (-3)^2 = 25$$

Take the square root of the left hand side and square it. This gives:

$$(x - 3)^2 = 25$$

Take the square root of both sides. This gives:

$$\sqrt{(x - 3)^2} = \sqrt{25}$$
$$x - 3 = \pm 5$$
$$x = 3 \pm 5$$
$$x = 3 + 5 \quad \text{or} \quad x = 3 - 5$$

$\therefore \quad x = 8$ or $x = -2$

4.  Solve the quadratic equation $2x^2 - 5x + 3 = 0$, by the method of completing the square.

## Solution

$$2x^2 - 5x + 3 = 0$$

In this case, the coefficient of $x^2$ is 2. It is not 1 like our first three examples above. Therefore our first step will be to make the coefficient of $x^2$ to be 1. This is done by dividing each term in the equation by the coefficient of $x^2$ which is 2. This gives:

$$\frac{2x^2}{2} - \frac{5x}{2} + \frac{3}{2} = \frac{0}{2}$$

$$x^2 - \frac{5}{2}x + \frac{3}{2} = 0$$

Now that we have made the coefficient of $x^2$ to be 1, we can proceed as usual.

Take the constant term (i.e. $\frac{3}{2}$) to the right hand side of the equation. This gives:

$$x^2 - \frac{5}{2}x = -\frac{3}{2}$$

The coefficient of $x$ is $-\frac{5}{2}$. Divide this coefficient by 2 to obtain:

$$-\frac{5}{2} \div 2 = -\frac{5}{2} \times \frac{1}{2}$$

$$= -\frac{5}{4}$$

Square the value obtained. This gives:

$$(-\frac{5}{4})^2$$

Add $(-\frac{5}{4})^2$ to both sides of the equation. This gives:

$$x^2 - \frac{5}{2}x + (-\frac{5}{4})^2 = -\frac{3}{2} + (-\frac{5}{4})^2$$

$$x^2 - \frac{5}{2}x + (-\frac{5}{4})^2 = -\frac{3}{2} + \frac{25}{16}$$

$$x^2 - \frac{5}{2}x + (-\frac{5}{4})^2 = \frac{-24 + 25}{16}$$

$$x^2 - \frac{5}{4}x + (-\frac{5}{4})^2 = \frac{1}{16}$$

Take the square root of the left hand side and square it. This gives:

$$(x - \frac{5}{4})^2 = \frac{1}{16}$$

Take the square root of both sides. This gives:

$$\sqrt{(x - \frac{5}{4})^2} = \sqrt{\frac{1}{16}}$$

$$x - \frac{5}{4} = \pm\frac{1}{4}$$

$$x = \frac{5}{4} \pm \frac{1}{4}$$

$$x = \frac{5}{4} + \frac{1}{4} \quad \text{or} \quad x = \frac{5}{4} - \frac{1}{4}$$

$$x = \frac{5+1}{4} \quad \text{or} \quad x = \frac{5-1}{4}$$

$$x = \frac{6}{4} \quad \text{or} \quad x = \frac{4}{4}$$

$$x = \frac{3}{2} \quad \text{or} \quad x = 1$$

5   Solve the quadratic equation $3y^2 + 9y + 5 = 0$, by the method of completing the square.

Solution
$$3y^2 + 9y + 5 = 0$$
In this case, the coefficient of $y^2$ is 3. Therefore our first step will be to make the coefficient of $y^2$ to be 1. This is done by dividing each term in the equation by 3. This gives:

$$\frac{3y^2}{3} + \frac{9y}{3} + \frac{5}{3} = \frac{0}{3}$$

$$y^2 + 3y + \frac{5}{3} = 0$$

Take the constant term to the right hand side of the equation. This gives:

$$y^2 + 3y = -\frac{5}{3}$$

The coefficient of y is 3. Divide this coefficient by 2 to obtain:

$$\frac{3}{2}$$

Square the value obtained. This gives:

$$\left(\frac{3}{2}\right)^2$$

Add $\left(\frac{3}{2}\right)^2$ to both sides of the equation. This gives:

$$y^2 + 3y + \left(\frac{3}{2}\right)^2 = -\frac{5}{3} + \left(\frac{3}{2}\right)^2$$

$$y^2 + 3y + \left(\frac{3}{2}\right)^2 = -\frac{5}{3} + \frac{9}{4}$$

$$y^2 + 3y + \left(\frac{3}{2}\right)^2 = \frac{-20 + 27}{12}$$

$$y^2 + 3y + \left(\frac{3}{2}\right)^2 = \frac{7}{12}$$

Take the square root of the left hand side and square it. This gives:

$$\left(y + \frac{3}{2}\right)^2 = \frac{7}{12}$$

Take the square root of both sides. This gives:

$$\sqrt{\left(y + \frac{3}{2}\right)^2} = \sqrt{0.5833}$$

$$y + \frac{3}{2} = \pm 0.764$$

$$y + 1.5 = \pm 0.764$$

$$y = -1.5 \pm 0.764$$

$$y = -1.5 + 0.764 \quad \text{or} \quad y = -1.5 - 0.764$$
$$y = -0.736 \quad \text{or} \quad y = -2.264$$

6. Solve the equation $5m^2 + 7m - 3 = 0$

Solution
$$5m^2 + 7m - 3 = 0$$
In this case, the coefficient of $m^2$ is 5. Therefore our first step will be to make the coefficient of $m^2$ to be 1. This is done by dividing each term in the equation by 5. This gives:
$$\frac{5m^2}{5} + \frac{7m}{5} - \frac{3}{5} = \frac{0}{5}$$
$$m^2 + \frac{7}{5}m - \frac{3}{5} = 0$$
Take the constant term to the right hand side of the equation. This gives:
$$m^2 + \frac{7}{5}m = \frac{3}{5}$$
The coefficient of m is $\frac{7}{5}$. Divide this coefficient by 2 to obtain:
$$\frac{7}{5} \div 2$$
$$= \frac{7}{5} \times \frac{1}{2}$$
$$= \frac{7}{10}$$
Square the value obtained. This gives:
$$(\frac{7}{10})^2$$
Add $(\frac{7}{10})^2$ to both sides of the equation. This gives:
$$m^2 + \frac{7}{5}m + (\frac{7}{10})^2 = \frac{3}{5} + (\frac{7}{10})^2$$
$$m^2 + \frac{7}{5}m + (\frac{7}{10})^2 = \frac{3}{5} + \frac{49}{100}$$
$$m^2 + \frac{7}{5}m + (\frac{7}{10})^2 = \frac{60+49}{100}$$
$$m^2 + \frac{7}{5}m + (\frac{7}{10})^2 = \frac{109}{100}$$
Take the square root of the left hand side and square it. This gives:
$$(m + \frac{7}{10})^2 = \frac{109}{100}$$
Take the square root of both sides. This gives:
$$\sqrt{(m + \frac{7}{10})^2} = \sqrt{1.09}$$

$$m + \frac{7}{10} = \pm 1.044$$
$$m + 0.7 = \pm 1.044$$
$$m = -0.7 \pm 1.044$$

Separating the sign $\pm$ into $+$ and $-$ gives the two solutions as follows:
$$m = -0.7 + 1.044 \quad \text{or} \quad m = -0.7 - 1.044$$
$$m = 0.344 \quad \text{or} \quad m = -1.744$$

7. Solve the quadratic equation $2x^2 - 8x + 5 = 0$

<u>Solution</u>
$$2x^2 - 8x + 5 = 0$$
The coefficient of $x^2$ is 2. Therefore our first step will be to make the coefficient of $x^2$ to be 1. This is done by dividing each term in the equation by 2. This gives:
$$\frac{2x^2}{2} - \frac{8x}{2} + \frac{5}{2} = \frac{0}{2}$$
$$x^2 - 4x + \frac{5}{2} = 0$$
Take the constant term to the right hand side of the equation. This gives:
$$x^2 - 4x = -\frac{5}{2}$$
The coefficient of $x$ is $-4$. Divide this coefficient by 2 to obtain:
$$-\frac{4}{2} = -2$$
Square the value obtained. This gives:
$$(-2)^2$$
Add $(-2)^2$ to both sides of the equation. This gives:
$$x^2 - 4x + (-2)^2 = -\frac{5}{2} + (-2)^2$$

$$x^2 - 4x + (-2)^2 = -\frac{5}{2} + 4$$

$$x^2 - 4x + (-2)^2 = \frac{-5 + 8}{2}$$

$$x^2 - 4x + (-2)^2 = \frac{3}{2}$$
Take the square root of the left hand side and square it. This gives:
$$(x - 2)^2 = \frac{3}{2}$$
Take the square root of both sides. This gives:
$$\sqrt{(x - 2)^2} = \sqrt{1.5}$$
$$x - 2 = \pm 1.225$$
$$x = 2 \pm 1.225$$
$$x = 2 + 1.225 \quad \text{or} \quad x = 2 - 1.225$$
$$x = 3.225 \quad \text{or} \quad x = 0.775$$

## Quadratic Formula

The general expression of a quadratic equation is given by: $ax^2 + bx + c = 0$, where a, b and c are constants. This equation can be solved by completing the square are follows:

$$ax^2 + bx + c = 0$$

The coefficient of $x^2$ is a. Therefore we make the coefficient of $x^2$ to be 1. This is done by dividing each term in the equation by a. This gives:

$$\frac{ax^2}{a} + \frac{bx}{a} + \frac{c}{a} = \frac{0}{a}$$

$$x^2 + \frac{b}{a}x + \frac{c}{a} = 0$$

Take the constant term to the right hand side of the equation. This gives:

$$x^2 + \frac{b}{a}x = -\frac{c}{a}$$

The coefficient of $x$ is $\frac{b}{a}$. Divide this coefficient by 2 to obtain:

$$\frac{b}{a} \div 2$$

$$= \frac{b}{a} \times \frac{1}{2}$$

$$= \frac{b}{2a}$$

Square the value obtained. This gives:

$$(\frac{b}{2a})^2$$

Add $(\frac{b}{2a})^2$ to both sides of the equation. This gives:

$$x^2 + \frac{b}{a}x + (\frac{b}{2a})^2 = -\frac{c}{a} + (\frac{b}{2a})^2$$

$$x^2 + \frac{b}{a}x + (\frac{b}{2a})^2 = -\frac{c}{a} + \frac{b^2}{4a^2} \qquad \text{(In } (\frac{b}{2a})^2, \text{ each term in the bracket is squared)}$$

$$x^2 + \frac{b}{a}x + (\frac{b}{2a})^2 = \frac{-4ac + b^2}{4a^2} \qquad \text{(Note that the LCM of } 4a^2 \text{ and a is } 4a^2)$$

$$x^2 + \frac{b}{a}x + (\frac{b}{2a})^2 = \frac{b^2 - 4ac}{4a^2}$$

Take the square root of the left hand side and square it. This gives:

$$(x + \frac{b}{2a})^2 = \frac{b^2 - 4ac}{4a^2}$$

Take the square root of both sides. This gives:

$$\sqrt{(x + \frac{b}{2a})^2} = \frac{b^2 - 4ac}{4a^2}$$

$$x + \frac{b}{2a} = \sqrt{\frac{b^2 - 4ac}{4a^2}} \qquad \text{(Note that } \sqrt{\frac{b}{a}} \text{ can also be expressed as } \frac{\sqrt{b}}{\sqrt{a}})$$

$$x + \frac{b}{2a} = \pm \frac{\sqrt{b^2 - 4ac}}{2a}$$ (Note that the square root of $4a^2$ is 2a)

$$x = -\frac{b}{2a} \pm \frac{\sqrt{b^2 - 4ac}}{2a}$$

Since their LCM is 2a, we simplify further as follows:

$$x = \frac{-b \pm \sqrt{b^2 - 4ac}}{2a}$$

The above formula is known as quadratic formula. It can be used to solve any quadratic equation.

**Examples**
1. Solve the following equations by using the quadratic formula:
a. $2x^2 + 3x - 9 = 0$
b. $4x^2 - 8x + 3 = 0$

Solutions
a. $2x^2 + 3x - 9 = 0$
When this equation is compared with $ax^2 + bx + c = 0$, it shows that:
   a = 2, b = 3 and c = –9
We now substitute these values into the quadratic formula as follows:

$$x = \frac{-b \pm \sqrt{b^2 - 4ac}}{2a}$$

$$= \frac{-3 \pm \sqrt{3^2 - [4(2)(-9)]}}{2(2)}$$

$$= \frac{-3 \pm \sqrt{9 - (-72)}}{2 \times 2}$$

$$= \frac{-3 \pm \sqrt{9 + 72}}{4}$$

$$= \frac{-3 \pm \sqrt{81}}{4}$$

$$x = \frac{-3 \pm 9}{4}$$

We now split the sign $\pm$ into + and – to obtain the two solutions as follows:

$$x = \frac{-3 + 9}{4} \quad \text{or} \quad x = \frac{-3 - 9}{4}$$

$$x = \frac{6}{4} \quad \text{or} \quad x = \frac{-12}{4}$$

$$x = \frac{3}{2} \qquad \text{or} \qquad x = -3$$

b.   $4x^2 - 8x + 3 = 0$

When this equation is compared with $ax^2 + bx + c = 0$, it shows that:

   a = 4, b = −8 and c = 3

We now substitute these values into the quadratic formula as follows:

$$x = \frac{-b \pm \sqrt{b^2 - 4ac}}{2a}$$

$$= \frac{-(-8) \pm \sqrt{(-8)^2 - [4(4)(3)]}}{2(4)} \qquad \text{(Note that } -b \text{ is } -(-8) \text{ since b is } -8)$$

$$= \frac{+8 \pm \sqrt{64 - 48}}{8}$$

$$= \frac{8 \pm \sqrt{16}}{8}$$

$$x = \frac{8 \pm 4}{8}$$

We now split the sign $\pm$ into + and − to obtain the two solutions as follows:

$$x = \frac{8 + 4}{8} \qquad \text{or} \qquad x = \frac{8 - 4}{8}$$

$$x = \frac{12}{8} \qquad \text{or} \qquad x = \frac{4}{8}$$

$$x = \frac{3}{2} \qquad \text{or} \qquad x = \frac{1}{2}$$

2.   Solve the following equations by using the quadratic formula:

a.   $x^2 - 4x - 2 = 0$

b.   $4 + x - 5x^2 = 0$

c.   $7 - 2x - 3x^2 = 0$

d.   $2n^2 - 15n + 18 = 0$

Solutions

a.   $x^2 - 4x - 2 = 0$

In this equation, a = 1, b = −4 and c = −2

We now substitute these values into the quadratic formula as follows:

$$x = \frac{-b \pm \sqrt{b^2 - 4ac}}{2a}$$

$$= \frac{-(-4) \pm \sqrt{(-4)^2 - [4(1)(-2)]}}{2(1)}$$

$$= \frac{4 \pm \sqrt{16 - (-8)}}{2}$$

$$= \frac{4 \pm \sqrt{16 + 8}}{2}$$

$$= \frac{4 \pm \sqrt{24}}{2}$$

$$x = \frac{4 \pm 4.90}{2}$$

We now split the sign $\pm$ into + and − to obtain the two solutions as follows:

$$x = \frac{4 + 4.90}{2} \quad \text{or} \quad x = \frac{4 - 4.90}{2}$$

$$x = \frac{8.90}{2} \quad \text{or} \quad x = \frac{-0.90}{2}$$

$$x = 4.45 \quad \text{or} \quad x = -0.45$$

b. $4 + x - 5x^2 = 0$

In this equation, a = −5, b = 1 and c = 4

Note that 'a' is the coefficient of $x^2$, b is the coefficient of $x$, while c is the constant term. So, do not be confused by the reversed arrangement of the quadratic equation.

We now substitute the values of a, b and c into the quadratic formula as follows:

$$x = \frac{-b \pm \sqrt{b^2 - 4ac}}{2a}$$

$$= \frac{-1 \pm \sqrt{1^2 - [4(-5)(4)]}}{2(-5)}$$

$$= \frac{-1 \pm \sqrt{1 - (-80)}}{-10}$$

$$= \frac{-1 \pm \sqrt{1 + 80}}{-10}$$

$$= \frac{-1 \pm \sqrt{81}}{-10}$$

$$x = \frac{-1 \pm 9}{-10}$$

We now split the sign $\pm$ into + and − to obtain the two solutions as follows:

$$x = \frac{-1 + 9}{-10} \quad \text{or} \quad x = \frac{-1 - 9}{-10}$$

$$x = \frac{8}{-10} \quad \text{or} \quad x = \frac{-10}{-10}$$

$$x = -\frac{4}{5} \quad \text{or} \quad x = 1$$

c.  $7 - 2x - 3x^2 = 0$

This arrangement is similar to that of question (b) above. However, let us solve this question by changing this arrangement to the usual arrangement of a quadratic equation. In order to do this, we divide each term in the equation by −1 so that the coefficient of $x^2$ will become a positive value. This gives:

$$\frac{7}{-1} - \frac{2x}{-1} - \frac{3x^2}{-1} = \frac{0}{-1}$$

$$-7 + 2x + 3x^2 = 0$$

Or,  $3x^2 + 2x - 7 = 0$

In this new arrangement, a = 3, b = 2 and c = −7

We now substitute the values of a, b and c into the quadratic formula as follows:

$$x = \frac{-b \pm \sqrt{b^2 - 4ac}}{2a}$$

$$= \frac{-2 \pm \sqrt{2^2 - [4(3)(-7)]}}{2(3)}$$

$$= \frac{-2 \pm \sqrt{4 - (-84)}}{6}$$

$$= \frac{-2 \pm \sqrt{4 + 84}}{6}$$

$$= \frac{-2 \pm \sqrt{88}}{6}$$

$$x = \frac{-2 \pm 9.38}{6}$$

$$x = \frac{-2 + 9.38}{6} \quad \text{or} \quad x = \frac{-2 - 9.38}{6}$$

$$x = \frac{7.38}{6} \quad \text{or} \quad x = \frac{-11.38}{6}$$

$$x = 1.23 \quad \text{or} \quad x = -1.90$$

d.  $2n^2 - 15n + 18 = 0$

$\qquad$ a = 2, b = −15 and c = 18

$$n = \frac{-b \pm \sqrt{b^2 - 4ac}}{2a}$$

$$= \frac{-(-15) \pm \sqrt{(-15)^2 - [4(2)(18)]}}{2(2)}$$

$$= \frac{15 \pm \sqrt{225 - 144}}{4}$$

$$= \frac{15 \pm \sqrt{81}}{4}$$

$$n = \frac{15 \pm 9}{4}$$

$$n = \frac{15 + 9}{4} \quad \text{or} \quad n = \frac{15 - 9}{4}$$

$$n = \frac{24}{4} \quad \text{or} \quad n = \frac{6}{4}$$

$$n = 6 \quad \text{or} \quad n = \frac{3}{2}$$

## Exercise 9

1. Factorize the following quadratic expressions
a. $7a^2 + 9a$
b. $m^2 - 16$
c. $9p^2 - 121$
d. $32b^2 - 200$

2. Factorize the following equations:
a. $3x^2 - 4x - 15$.
b. $x^2 + x - 72$
c. $5x^2 + 22x + 8$
d. $10x^2 - 4x - 6$

3. Determine if the following equations can be factorized or not:
a. $x^2 + 14x + 13 = 0$
b. $3x^2 - 11x + 6 = 0$
c. $5x^2 + 2x - 9 = 0$
d. $2x^2 - 7x - 11 = 0$
e. $8x^2 - 21x + 12 = 0$

4. Solve the following equations by factorization:
a. $3a^2 + 12a = 0$
b. $b^2 - 36 = 0$

c. $4y^2 - 49 = 0$
d. $(2x - 5)^2 = 9$

5. Solve the following equations by factorization:
 a. $3x^2 - x - 4 = 0$
b. $2y^2 + 2y - 12 = 0$
c. $p^2 - 20p + 36 = 0$
d. $10m^2 = 4 - 3m$
e. $6 - x - 7x^2 = 0$

6. Find the quadratic equation whose roots are:
a. 8 and –5.
b. $\frac{1}{4}$ and -7.
c. –6 and $-\frac{2}{5}$

7. A quadratic equation is given by: $2x^2 - 15x - 8 = 0$. Find:
a. the sum of the roots of the equation
b. the product of the roots of the equation

8. A quadratic equation is given by: $3x^2 + 4x - 11 = 0$. Find:
a. the sum of the roots of the equation
b. the product of the roots of the equation

9. If 2 is a root of the equation $2x^2 + Mx - 18 = 0$, find the value of M. Hence find the other root.
10. If 1 is a root of the equation $3x^2 - kx - 11 = 0$, find the value of K. Hence find the other root.
11. 3 and –9 are the roots of the quadratic equation $b^2 - kb + P = 0$. Find the values of k and P.
12. If –5 and 2 are the roots of the quadratic equation $3x^2 + Ax - B = 0$. Find the values of the constants A and B.
13. If one root of the equation $m^2 + km + 27 = 0$ is thrice the other, find the two possible values of k. Hence, what are the two possible equations?
14. What must be added to $x^2 + 16x$ to make it a perfect square?
15. What must be added to $x^2 - 13x$ to make it a perfect square?
16. Make $a^2 - 6a$ a perfect square.
17. Make $m^2 + \frac{5}{3}m$ a perfect square.
18. If $x^2 - 8x + 16$ is a perfect square, find its square root.
19. If $x^2 + 1.2x + (0.6)^2$ is a perfect square, find its square root.

20. Use the method of completing the square to solve the following quadratic equations:
a. $y^2 + 9y - 22 = 0$
b. $x^2 - 11x + 24 = 0$
c. $2x^2 - 5x - 12 = 0$
d. $3x^2 - 10x + 7 = 0$

121

e  $5y^2 + 11y + 6 = 0$

f.  $2m^2 + 3m - 14 = 0$

g.  $x^2 - 7x + 8 = 0$

21.  Solve the following equations by using the quadratic formula:

a.  $x^2 + x - 12 = 0$

b.  $7x^2 - 17x + 10 = 0$

c.  $2x^2 - 7x - 4 = 0$

d.  $9 + 5x - 3x^2 = 0$

e.  $5 - 7x - 5x^2 = 0$

f.  $4n^2 - 12n + 9 = 0$

# CHAPTER 10
## WORD PROBLEMS LEADING TO QUADRATIC EQUATIONS

**Examples**

1. Find two numbers whose difference is 3 and whose product is 54.

Solution

Let the smaller number be $x$.

Then the larger number will be $x + 3$     (The larger one is greater than the smaller one by 3)

∴ Their product is $x(x + 3) = 54$

$\quad x^2 + 3x = 54$

$\quad x^2 + 3x - 54 = 0$

Solving this quadratic equation by factorization gives:

$\quad (x + 9)(x - 6) = 0$     (Two numbers whose product is –54 and sum is 3 are 9 and –6)

∴  $x = -9$ or $x = 6$   (When each of the bracket above is equated to zero and solved)

When $x = -9$, then the larger number is:

$\quad x + 3$     (As stated at the beginning)

$\quad = -9 + 3$    (Since $x = -9$)

$\quad = -6$

When $x = 6$, then the larger number is:

$\quad x + 3$

$\quad = 6 + 3 = 9$

∴   The two numbers are –9 and –6 or 9 and 6.


2. The difference between two numbers is 12. The sum of their squares is 74. Find the numbers.

Solution

Let the smaller number be 'a'

∴  The larger number is $a + 12$

Their squares are $a^2$ and $(a + 12)^2$

∴  The sum of their squares is:

$\quad a^2 + (a + 12)^2 = 74$

$\quad a^2 + (a + 12)(a + 12) = 74$

$\quad a^2 + a^2 + 12a + 12a + 144 = 74$

$\quad 2a^2 + 24a + 144 - 74 = 0$

$\quad 2a^2 + 24a + 70 = 0$

Dividing throughout by 2 to simplify it further gives:

$\quad a^2 + 12a + 35 = 0$

Solving this quadratic equation by factorization gives:

$\quad (a + 7)(a + 5) = 0$

∴  $a = -7$ or $a = -5$

When $a = -7$, the larger number is:

a + 12

   = −7 + 12

   = 5

When a = −5, the larger number is:

   a + 12

   = −5 + 12

   = 7

∴  The two numbers are −7 and 5 or −5 and 7.

3.  A boy is 5 years older than his brother. The product of their ages is 204. Find their ages.

<u>Solution</u>

Let his brother's age be y years.   (Note that his brother's age is smaller since the boy is older)

∴  The boy's age = (y + 5)years

The product of their ages is given by:

   y(y + 5) = 204

∴  $y^2 + 5y = 204$

   $y^2 + 5y − 204 = 0$

Solving this quadratic equation by factorization gives:

   (y − 12)(y + 17) = 0   (Note: two numbers whose product is −204 and sum is 5 are −12 and 17)

∴  y = 12 or y = −17

Age cannot be negative, therefore, y = 12

∴  The brother's age is 12 years.

The boy's age is y + 5

              = 12 + 5

              = 17

The boy's age is 17 years.

4.  A paper measures 15cm by 10cm. A strip of equal width and having an 'L' shape is cut off from the ends of the paper. If the area of the remaining paper is 84cm$^2$, find the width of the strip removed.

<u>Solution</u>

Let the width of the strip removed be $x$ cm. Since the L shape affect the length and width, then the new length of the paper is $15 − x$, while the new width is $10 − x$. This is because a size of $x$ cm has been cut off along a length and along a width of the paper.

∴  The new area (length x width) is given by:

   $(15 − x)(10 − x) = 84$

   $150 − 15x − 10x + x^2 = 84$   (Note that $−x \times −x = +x^2$

   $x^2 − 15x − 10x + 150 − 84 = 0$

   $x^2 − 25x + 66 = 0$

Factorizing this quadratic equation gives:

$(x - 22)(x - 3) = 0$

∴ $x = 22$ or $3$

The value 22 is not possible since it is greater than the original dimension of the paper.

∴ $x = 3$.

∴ The width of the strip removed is 3cm.


5.   A rectangular field is 16m long and 8m wide. A path of equal width runs along one side and one end thereby forming an 'L' shape path. If the total area of the field and path is 180m$^2$, find the width of the path.

<u>Solution</u>

Let the width of the path be y m.

∴ The new length of the field including the path is $(16 + y)$m while the new width is $(8 + y)$m.

∴ The new area (i.e. total area) is:

$$(16 + y)(8 + y) = 180$$
$$128 + 16y + 8y + y^2 = 180$$
$$y^2 + 16y + 8y + 128 - 180 = 0$$
$$y^2 + 24y - 52 = 0$$

Factorizing this equation gives:

$$(y + 26)(y - 2) = 0$$

∴   $y = -26$ or $y = 2$

   $y = -26$ is not possible since length/width cannot be negative

∴   $y = 2$

∴ The width of the path is 2m.


6.   The ages of two brothers are 7 and 2 years. In how many years time will the product of their ages be 234?

<u>Solution</u>

Let the product of their ages be 234 in $x$ years time.

∴   In $x$ years time, their ages will be $(7 + x)$ years and $(2 + x)$ years

∴   Product of their ages then will be

$$(7 + x)(2 + x) = 234$$
$$14 + 7x + 2x + x^2 = 234$$
$$x^2 + 7x + 2x + 14 - 234 = 0$$
$$x^2 + 9x - 220 = 0$$

Factorizing this equation and solving it gives:

$$(x + 20)(x - 11) = 0$$

∴   $x = -20$ or $x = 11$

$x = -20$ is not correct since years cannot be negative.

∴   $x = 11$years.

∴ In 11 years time, the product of their ages will be 234.

7. Two sisters are 25 years and 18 years old. How many years ago was the product of their ages 144?

Solution
Let the product of their ages be 144, $x$ years ago.
∴   $x$ years ago, their ages were $(25 - x)$ years and $(18 - x)$ years.
∴   The product of their ages then was:
$$(25 - x)(18 - x) = 144$$
$$450 - 25x - 18x + x^2 = 144$$
$$x^2 - 25x - 18x + 450 - 144 = 0$$
$$x^2 - 43x + 306 = 0$$
Factorizing this quadratic equation gives:
$$(x - 34)(x - 9) = 0$$
∴   $x = 34$ or $x = 9$
34 is not possible since it is greater than the sisters' ages
∴   $x = 9$
∴ 9 years ago, the product of their ages was 144.

8.   Find two consecutive odd numbers whose product is 255.

Solution
Let the smaller number be $x$. Therefore the larger number is $(x + 2)$
Note that any two consecutive odd numbers (or even numbers) differs by 2. For example, 5 and 7.
∴   Their product is given by:
$$x(x + 2) = 255$$
$$x^2 + 2x - 255 = 0$$
Solving the equation by factorizing method gives:
$$(x + 17)(x - 15) = 0$$
∴   $x = -17$ or $x = 15$
When $x = -17$, the larger number is $(x + 2)$. This gives:
$$-17 + 2 = -15$$
When $x = 15$, the larger number is $15 + 2 = 17$.
∴   The two consecutive odd numbers are −15 and −17 or 15 and 17.

9.   Twice a particular whole number is subtracted from 3 times the square of the number. The answer is 133. What is the number?

Solution
Let the number be $x$.
∴   Twice the number is $2x$.
The square of the number is $x^2$.
3 times the square of the number is $3x^2$.

∴ When twice the number is subtracted from 3 times the square of the number to give 133, then this forms an equation as follows:

$3x^2 - 2x = 133$

∴ $3x^2 - 2x - 133 = 0$

Apart from factorization method, this equation can also be solved by using quadratic equation formula as follows:

$3x^2 - 2x - 133 = 0$

Hence , a = 3, b = −2, and c = −133

∴ $x = \dfrac{-b \pm \sqrt{b^2 - 4ac}}{2a}$

$= \dfrac{-(-2) \pm \sqrt{(-2)^2 - [4 \times 3 \times (-133)]}}{2 \times 3}$

$= \dfrac{2 \pm \sqrt{4 - (-1596)}}{6}$

$= \dfrac{2 \pm \sqrt{4 + 1596)}}{6}$

$= \dfrac{2 \pm \sqrt{1600}}{6}$

$= \dfrac{2 \pm 40}{6}$

$x = \dfrac{2 + 40}{6}$ or $x = \dfrac{2 - 40}{6}$

$x = \dfrac{42}{6}$ or $x = \dfrac{-38}{6}$

$x = 7$ or $-6.3$

∴ $x = 7$ since the number is a whole number

∴ The number is 7.

10  A man is 4 times as old as his daughter. 5 years ago, the product of their ages was 175. Find their present ages.

Solution
Let the daughter's age be y years.

∴ The man's age is 4y years

5 years ago, the daughter's age was (y − 5) years, while the man's age was (4y − 5) years.

∴ The product of their ages then (i.e. 5 years ago) is given by:

$(4y - 5)(y - 5) = 175$

$4y^2 - 20y - 5y + 25 = 175$

$4y^2 - 25y + 25 - 175 = 0$

$4y^2 - 25y - 150 = 0$

In order to solve this equation by factorization, multiply the first and last terms on the left hand side. This gives:

$4y^2 \times (-150) = -600y^2$

Now look for two numbers in y whose product is $-600y^2$ and sum is −25y (i.e. the middle term in the equation). The two numbers are −40y and +15y. Replace −25y in the original equation by these two numbers. This gives:

$4y^2 - 40y + 15y - 150 = 0$

Factorizing by grouping gives:

$4y(y - 10) + 15(y - 10) = 0$

Take the terms outside each bracket, and then one of the two equal brackets. Note that $(y - 10)$ is common (i.e. the HCF) in the equation above. This gives:

$(y - 10)(4y + 15) = 0$

$\therefore$ y = 10 or $y = -\dfrac{15}{4}$

y = 10   (Ignore $y = -\dfrac{15}{4}$ since age cannot be negative or fraction)

$\therefore$ The daughter's age is 10 years, while the man's age is 40 years (i.e. 4y = 4 x 10 = 40)

## Exercise 10

1. Find two numbers whose difference is 14 and whose product is 72.
2. The difference between two numbers is 10. The sum of their squares is 58. Find the numbers.
3. A boy is 7 years older than his brother. The product of their ages is 198. Find their ages.
4. A field measures 21cm by 12cm. A path of equal width and having an 'L' shape is at the ends of the field. If the area of the remaining field is 165cm$^2$, find the width of the path.
5. A rectangular field is 16m long and 8m wide. A path of equal width runs along one side and one end thereby forming an 'L' shape path. If the total area of the field and path is 180m$^2$, find the width of the path.
6. The ages of two friends are 13 and 9 years. In how many years time will the product of their ages be 480?
7. Two brothers are 28 years and 10 years old. How many years ago was the product of their ages 144?
8. Find two consecutive even numbers whose product is 288.
9. Thrice a particular whole number is added to 2 times the square of the number. The answer is 230. What is the number?
10  A woman is 3 times as old as his daughter. 5 years ago, the product of their ages was 217. Find their present ages.

# CHAPTER 11
# VARIATION

## Direct Variation

Direct variation involves the relationship between two quantities whereby an increase or decrease in one of them leads to an increase or decrease respectively in the other.

The symbol $\alpha$ means 'varies with' or is 'proportional to'.

**Examples**

1. If $x$ varies directly as y and $x = 30$ when y=12, find:
a. the formula connecting $x$ and y
b. $x$ when y=10
c. y when $x = 20$

Solution

a. $x \, \alpha \, y$  (This means $x$ varies directly as y)

  $x = Ky$  (Replacing the proportionality sign with the equals sign introduces a constant K)

So, when $x = 30$ and y = 12, the equation above becomes:

  30 = K x 12

  30 = 12K

  $K = \dfrac{30}{12} = \dfrac{5}{2}$

The formula connecting $x$ and y is:

  $x = \dfrac{5}{2} y$  (This is obtained by substituting $\dfrac{5}{2}$ for K in the equation above, i.e. $x = Ky$)

b. When y = 10, $x$ is given by:

  $x = \dfrac{5}{2} y$

  $x = \dfrac{5}{2} \times 10 = \dfrac{50}{2} = 25$

  $\therefore x = 25$

c. When $x = 20$, y is given by:

  $x = \dfrac{5}{2} y$

  $20 = \dfrac{5}{2} y$

  $5y = 40$

  $y = \dfrac{40}{5}$

y = 8

2. If m varies as the square of n and m=27 when n=3, find:
a. the relationship between m and n
b. n when m = 48
c. m when n = $2\frac{1}{3}$

Solution
a. $m \, \alpha \, n^2$ (This means m varies as $n^2$)
  $m = Kn^2$ (Replacing the proportionality sign with the equals sign introduces a constant K)
So, when m = 27 and n = 3, the equation above becomes:
$27 = K \times 3^2$
$27 = 9K$
$K = \dfrac{27}{9} = 3$
The relationship between m and n is:
$m = 3n^2$ (This is obtained by substituting 3 for K in the equation $m = Kn^2$ above)

b. When m = 48, n is given by:
$m = 3n^2$
$48 = 3 \times n^2$
$48 = 3n^2$
$n^2 = \dfrac{48}{3} = 16$
$n = \sqrt{16}$
$n = 4$

c. When n = $2\frac{1}{3}$ , m is given by:
$m = 3n^2$
$m = 3 \times (\frac{7}{3})^2$ ( Note that $2\frac{1}{3}$ has been converted to $\frac{7}{3}$ )
$m = 3 \times \dfrac{49}{9}$
Canceling out gives:
$m = \dfrac{49}{3}$

Inverse Variation
In inverse or indirect variation, as one quantity increases the other decreases.

**Examples**
1. If c varies inversely as d and c=18 when y=4, find:
a. the formula connecting c and d
b. c when d=10
c. d when c=12

<u>Solution</u>

a. $c \alpha \dfrac{1}{d}$   (This means c varies inversely as d)

$c = \dfrac{K}{d}$   (Replacing the proportionality sign with the equals sign introduces a constant K)

So, when c = 18 and d = 4, the equation above becomes:

$18 = \dfrac{K}{4}$

$K = 18 \times 4$

$K = 72$

The formula connecting c and d is:

$c = \dfrac{72}{d}$   (This is obtained by substituting 72 for K in the equation above, i.e. $c = \dfrac{K}{d}$)

b. When d = 10, c is given by:

$c = \dfrac{72}{d}$

$c = \dfrac{72}{10}$

$c = 7.2$

c. When c = 12, d is given by:

$c = \dfrac{72}{d}$

$12 = \dfrac{72}{d}$

$12d = 72$

$d = \dfrac{72}{12}$

$d = 6$

2. If r varies inversely as the cube root of t and r=6 when t=64, find:
a. the relationship between r and t

b.  t when r = 16

c.  r when $t = \dfrac{8}{27}$

Solution

a.  $r \propto \dfrac{1}{\sqrt[3]{t}}$  (This means r varies inversely as the cube root of t)

$r = \dfrac{K}{\sqrt[3]{t}}$  (Replacing the proportionality sign with the equals sign introduces a constant K)

So, when r = 6 and t = 64, the equation above becomes:

$6 = \dfrac{K}{\sqrt[3]{64}}$

$6 = \dfrac{K}{4}$

K = 6 x 4 = 24

The relationship between r and t is:

$r = \dfrac{24}{\sqrt[3]{t}}$  (This is obtained by substituting 24 for K in the equation $r = \dfrac{K}{\sqrt[3]{t}}$ )

b.  When r = 16, t is given by:

$r = \dfrac{24}{\sqrt[3]{t}}$

$16 = \dfrac{24}{\sqrt[3]{t}}$

$\sqrt[3]{t} = \dfrac{24}{16}$

$\sqrt[3]{t} = \dfrac{3}{2}$

In order to remove the cube root, take the cube of both sides. This gives:

$(\sqrt[3]{t})^3 = (\dfrac{3}{2})^3$

$\therefore \ t = \dfrac{27}{8}$

c. When $t = \dfrac{8}{27}$, r is given by:

$$r = \frac{24}{\sqrt[3]{t}}$$

$$r = \frac{24}{\sqrt[3]{8/27}}$$

$$r = \frac{24}{2/3}$$

$$r = 24 \times \frac{3}{2}$$

After equal division by 2, it gives:

$$r = 12 \times 3$$
$$r = 36$$

## Joint Variation

In joint variation, three or more quantities are related directly or inversely or both.

### Examples

1. If m varies directly as the square of n and inversely as p, and m=3 when n=2 and p=8, find:

a. the relationship between m, n and p

b. m when n = 3 and p =27

c. p when $m = \dfrac{1}{2}$ and $n = \dfrac{3}{2}$

<u>Solutions</u>

$m \, \alpha \, \dfrac{n^2}{p}$     (This means m varies directly as the square of n and inversely as p)

$m = \dfrac{Kn^2}{p}$

So, when m = 3, n = 2, and p = 8, the equation above becomes:

$$3 = \frac{K \times 2^2}{8}$$

$$3 = \frac{4K}{8}$$

$$4K = 3 \times 8 = 24$$

$$K = \frac{24}{4} = 6$$

The relationship between m, n and p is:

$m = \dfrac{6n^2}{p}$     (This is obtained by substituting 6 for K in the equation $m = \dfrac{Kn^2}{p}$ )

b. When n = 3 and p = 27, then m is given by:

$$m = \frac{6n^2}{p}$$

$$m = \frac{6 \times 3^2}{27}$$

$$m = \frac{6 \times 9}{27}$$

$$m = \frac{54}{27}$$

$$m = 2$$

c. When $m = \frac{1}{2}$ and $n = \frac{3}{2}$, then p is given by:

$$m = \frac{6n^2}{p}$$

$$\frac{1}{2} = \frac{6 \times \frac{3}{2}}{p}$$

$$\frac{1}{2} = \frac{9}{p}$$

$$p = 9 \times 2$$
$$P = 18$$

2. The weight w of a rod varies jointly as its length L and the square root of its density d. If w =12 when L = 5 and d = 9, find:

a. L in terms of w and d
b. w when L = 8 and d = 25
c. d when L = 20 and w = 4

Solutions

a. $w \propto L\sqrt{d}$     (This means w varies jointly as L and the square root of d)

  $w = KL\sqrt{d}$

So, when w = 12, L = 5, and d = 9, the equation above becomes:

  $12 = K \times 5 \times \sqrt{9}$

  $12 = 15K$

$K = \frac{12}{15} = \frac{4}{5}$

The formula connecting w, L and d is:

  $w = \frac{4}{5}L\sqrt{d}$    (This is obtained by substituting $\frac{4}{5}$ for K in the equation $w = KL\sqrt{d}$ )

134

L can now be expressed in terms of w and d as follows:

$$w = \frac{4}{5}L\sqrt{d}$$

$$5w = 4L\sqrt{d}$$

Dividing both sides of the equation by $4\sqrt{d}$ gives:

$$L = \frac{5w}{4\sqrt{d}}$$

b. When L = 8 and d = 25, then w is given by:

$$w = \frac{4}{5}L\sqrt{d}$$

$$w = \frac{4}{5} \times 8 \times \sqrt{25}$$

$$w = \frac{4}{5} \times 8 \times 5$$

Cancelling out the 5 gives:

$$w = 4 \times 8$$

$$w = 32$$

c. When L = 20 and w = 4, then d is given by:

$$w = \frac{4}{5}L\sqrt{d}$$

$$4 = \frac{4}{5} \times 20 \times \sqrt{d}$$

$$4 \times 5 = 4 \times 20 \times \sqrt{d}$$

$$20 = 80\sqrt{d}$$

$$\sqrt{d} = \frac{20}{80} = \frac{1}{4}$$

Taking the square of both sides gives:

$$(\sqrt{d})^2 = (\frac{1}{4})^2$$

$$d = \frac{1}{16}$$

## Partial Variation

The fourth type of variation is called partial variation. In partial variation, one quantity is partly constant and partly varies with the other. Two constants are involved in partial variation.

**Examples**
1. $x$ is partly constant and partly varies as y. When y=2, $x$=30, and when y=6, $x$=50.
a. Find the formula which connects $x$ and y.
b. Find $x$ when y=3

<u>Solutions</u>
a. From the first sentence, we have:
   $x = C + Ky$    (Let this be equation 1) where C and K are constants.

Substituting y=2 and $x$=30 in this equation gives:

30 = C + 2K    (Let this be equation 2)

Similarly, when y = 6 and $x$ = 50, we have:

50 = C + 6K    (Let this be equation 3)

Bringing equation 2 and 3 together gives:

$$30 = C + 2K \quad \text{(Equation 2)}$$
$$\underline{50 = C + 6K} \quad \text{(Equation 3)}$$

Equation 3 – Equation 2:    20 =    4K

Divide both sides by 4. This gives:

$$K = \frac{20}{4} = 5$$

K = 5

Substitute 5 for K in equation 2.

30 = C + 2K

30 = C + (2 x 5)

30 = C + 10

30 – 10 = C

C = 20

We now substitute the values of C and K into equation 1 in order to obtain the formula connecting $x$ and y.

The formula connecting $x$ and y is now given by:

$x$ = 20 + 5y

b.  When y = 3, $x$ is obtained by substituting 3 for y in the formula connecting $x$ and y.

$x$ = 20 + 5y

= 20 + (5x3)

= 20 + 15

$x$ = 35

2.  m is partly constant and partly varies as n. When n = 4, m = 5, and when n = 12, m = 14.

a.  Find the formula which connects m and n.

b.  Find m when n = 16

c.  Find n when m = 9

Solutions

a.  From the first sentence, we have:

m = C + Kn    (Let this be equation 1) where C and K are constants.

Substituting n=4 and m=5 in equation 1 gives:

5 = C + 4K    (Let this be equation 2)

Similarly, when n=12 and m=14, we have:

14 = C + 12K    (Let this be equation 3)

Bringing equation 2 and 3 together gives:

$$5 = C + 4K \quad \text{(Equation 2)}$$

$$\underline{14 = C + 12K} \quad \text{(Equation 3)}$$

Equation 3 – Equation 2: $\qquad 9 = \qquad 8K$

Divide both sides by 8. This gives:

$$K = \frac{9}{8}$$

Substitute $\frac{9}{8}$ for K in equation 2.

$$5 = C + 4K$$

$$5 = C + (4 \times \frac{9}{8})$$

$$5 = C + \frac{9}{2}$$

$$5 - \frac{9}{2} = C$$

$$C = \frac{1}{2}$$

We now substitute the values of C and K into equation 1 in order to obtain the formula connecting m and n.

The formula connecting m and n is given by:

$$m = \frac{1}{2} + \frac{9}{8}n$$

b.  When n = 16, m is obtained by substituting 16 for n in the formula connecting m and n.

$$m = \frac{1}{2} + \frac{9}{8}n$$

$$m = \frac{1}{2} + (\frac{9}{8} \times 16)$$

$$= \frac{1}{2} + 18 = \frac{37}{2}$$

$$m = 18\frac{1}{2}$$

c.  When m = 9, n is obtained by substituting 9 for m in the formula connecting m and n.

$$m = \frac{1}{2} + \frac{9}{8}n$$

$$9 = \frac{1}{2} + (\frac{9n}{8})$$

$$9 - \frac{1}{2} = \frac{9n}{8}$$

$$\frac{17}{2} = \frac{9n}{8}$$

$$17 \times 8 = 9n \times 2$$

$$136 = 18n$$

$$n = \frac{136}{18} = \frac{68}{9}$$

$$n = 7\frac{5}{9}$$

## Exercise 11

1. If $x$ varies directly as y and $x$ =10 when y=8, find:
a. the formula connecting $x$ and y
b. $x$ when y=10
c. y when $x$ =16

2. If h varies as the square root of p and h=5 when p=9, find:
a. the relationship between h and p
b. p when h = 20
c. h when p = $6\frac{1}{4}$

3. If p varies inversely as q and p = 12 when q = 3, find:
a. the formula connecting p and q
b. q when p = 20
c. p when q = 5

4. If m varies inversely as the cube root of n and m=5 when n = 27, find:
a. the relationship between m and n
b. m when n = 8
c. n when m = $\frac{64}{125}$

5. If a varies directly as the square of b and inversely as c, and when a=4 when b=3 and c=6, find:
a. the formula connecting a, b and c
b. a when b = 5 and c = 10
c. b when a = $\frac{1}{2}$ and c = 8

6. The height h of a box varies jointly as its length L and the square of its width w. If h = 20 when L = 4 and w = 3, find:
a. w in terms of h and L
b. w when h = 12 and L = 4
c. L when h = 8 and w = 5

7. $x$ is partly constant and partly varies as y. When y = 4, $x$ = 14, and when y = 5, $x$ = 17.
a. Find the relationship between $x$ and y.
b. Find $x$ when y = 8

8.  E is partly constant and partly varies as F. When F = 2, E = 25, and when F = 5, E = 55.
a.  Find the formula which connects E and F.
b.  Find E when F = $2\frac{1}{2}$
c.  Find F when E = 40

# CHAPTER 12
## SIMULTANEOUS LINEAR AND QUADRATIC EQUATIONS

**Examples**

1. Solve the simultaneous equations: $2x^2 - 5y = -8$ and $3x + y = -1$

<u>Solution</u>

$\qquad 2x^2 - 5y = -8$ ..........................Equation (1)

$\qquad 3x + y = -1$ ..........................Equation (2)

From equation (2), $y = -1 - 3x$ ..................Equation (3)

Substitute $-1-3x$ for y in equation (1)

$\qquad 2x^2 - 5y = -8$ ...............Equation (1)

$\qquad 2x^2 - 5(-1 - 3x) = -8$

$\qquad 2x^2 + 5 + 15x = -8$

$\qquad 2x^2 + 15x + 5 + 8 = 0$

$\qquad 2x^2 + 15x + 13 = 0$

In order to solve this equation by factorization method, we first multiply the first and last terms. This gives:

$\qquad 2x^2 \times 13 = 26x^2$

Find two numbers in $x$ whose product is $26x^2$ and whose sum is $15x$ (i.e. the middle term in the original equation). The two numbers in $x$ are $13x$ and $2x$ (their product is $26x^2$ and their sum is $15x$). The two numbers are now used to replace the middle term in the original equation. This gives:

$\qquad 2x^2 + 13x + 2x + 13 = 0$

Factorize this equation by grouping.

$\qquad x(2x + 13) + 1(2x + 13) = 0$

$\therefore \quad (2x + 13)(x + 1) = 0.$     (Note that $(2x + 13)$ is the common factor in the expression above)

Equating each bracket above to zero and solving each of the linear equation formed, gives:

$\qquad x = -1$ or $x = -\dfrac{13}{2} = -6\dfrac{1}{2}$

Substitute $-1$ for $x$ in equation (3)

$\qquad y = -1 - 3x$ ..................Equation (3)

$\qquad = -1 - 3(-1)$

$\qquad = -1 + 3$

$\qquad y = 2$

Substitute $-\dfrac{13}{2}$ for $x$ in equation (3)

$\qquad y = -1 - 3x$ ..................Equation (3)

$\qquad = -1 - 3(-\dfrac{13}{2})$

$\qquad = -1 + \dfrac{39}{2}$

$\qquad = \dfrac{-2 + 39}{2}$

$\qquad y = \dfrac{37}{2}$

$$y = 18\frac{1}{2}$$

Therefore the solutions are $x = -1$ and $y = 2$ or $x = -6\frac{1}{2}$ and $y = 18\frac{1}{2}$. The solution can be given more neatly in $x$ and $y$ coordinates as follows:

$$(-1, 2), (-6\frac{1}{2}, 18\frac{1}{2})$$

2.  Solve the simultaneous equations: $x^2 + 3y^2 = 4$

$$x - 2y = 1$$

Solutions

$x^2 + 3y^2 = 4$ ........................Equation (1)

$x - 2y = 1$ ..........................Equation (2)

From equation (2) $x = 1 + 2y$

Substitute $1 + 2y$ for $x$ in equation (1)

$x^2 + 3y^2 = 4$

$(1 + 2y)^2 + 3y^2 = 4$

$(1 + 2y)(1 + 2y) + 3y^2 = 4$

$1 + 2y + 2y + 4y^2 + 3y^2 = 4$

$7y^2 + 4y + 1 - 4 = 0$

$7y^2 + 4y - 3 = 0$

Solving this quadratic equation by factorizing gives:

$7y^2 + 7y - 3y - 3 = 0$

$7y(y + 1) - 3(y + 1) = 0$

$(y + 1)(7y - 3) = 0$

$\therefore \quad y = -1$ or $y = \dfrac{3}{7}$

Substitute $-1$ for $y$ in equation (2). This gives:

$x - 2y = 1$ .....................Equation (2)

$x - 2(-1) = 1$

$x + 2 = 1$

$x = 1 - 2$

$x = -1$

Similarly, substitute $\dfrac{3}{7}$ for $y$ in equation (2).

$x - 2y = 1$ .....................Equation (2)

$x - 2(\dfrac{3}{7}) = 1$

$x - \dfrac{6}{7} = 1$

$x = 1 + \dfrac{6}{7}$

$\quad = \dfrac{7 + 6}{7}$

$x = \dfrac{13}{7} = 1\dfrac{6}{7}$

Therefore the solutions are $x = -1$ and $y = -1$, or $x = 1\frac{6}{7}$ and $y = \frac{3}{7}$

The solutions can also be given as: $(-1, -1)$, $(1\frac{6}{7}, \frac{3}{7})$

3.  Solve the equations $3x^2 - xy = 0$, $2y - 5x = 1$

Solution

$3x^2 - xy = 0$ ...................Equation (1)

$2y - 5x = 1$ ...................Equation (2)

From equation (1):

$x(3x - y) = 0$

$\therefore$   $x = 0$ or $3x - y = 0$

$x = 0$   or

$y = 3x$ ....................Equation (3)

Substitute 0 for $x$ in equation (2). This gives

$2y - 5(0) = 1$ ...............Equation (2)

$2y = 1$

$y = \frac{1}{2}$

Similarly, substitute $3x$ for y in equation (2).

$2y - 5x = 1$ ....................Equation (2)

$2(3x) - 5x = 1$

$6x - 5x = 1$

$x = 1$

Substitute 1 for $x$ in equation (3)

$y = 3x$ ....................Equation (3)

$y = 3(1)$

$y = 3$

$\therefore$   $x = 0$ and $y = \frac{1}{2}$   or  $x = 1$ and $y = 3$

This can also be given by: $(0, \frac{1}{2})$  or  $(1, 3)$

4.  Solve simultaneously $3x - 5y = -11$, $xy = -2$

Solution

$3x - 5y = -11$ ...................Equation (1)

$xy = -2$ ..........................Equation (2)

From equation (2), $x = \frac{-2}{y}$ ...................Equation (3)

Substitute $\frac{-2}{y}$ for $x$ in equation (1)

$3x - 5y = -11$ ...................Equation (1)

$3(\frac{-2}{y}) - 5y = -11$

$$\frac{-6}{y} - 5y = -11$$

Multiply each term by y in order to clear the fraction. This gives:

$$y\left(\frac{-6}{y}\right) - y(5y) = y(-11)$$
$$-6 - 5y^2 = -11y$$
$$0 = 5y^2 - 11y + 6$$

Or, $5y^2 - 11y + 6 = 0$

Solving this equation by factorization gives:

$$5y^2 - 5y - 6y + 6 = 0$$
$$5y(y - 1) - 6(y - 1) = 0 \quad \text{(−6 outside the bracket changes the + sign to − sign in the bracket)}$$
$$(y - 1)(5y - 6) = 0$$

Equating each bracket to zero and solving each equation gives:

$$y = 1 \text{ or } y = \frac{6}{5} = 1\frac{1}{5}$$

Substitute 1 for y in equation (3)

$$x = \frac{-2}{y} \quad .................\text{Equation (3)}$$
$$= \frac{-2}{1}$$
$$= -2$$

Substitute $\frac{6}{5}$ for y in equation (3)

$$x = \frac{-2}{y} \quad .................\text{Equation (3)}$$
$$x = \frac{-2}{\frac{6}{5}}$$
$$= -2 \times \frac{5}{6}$$
$$= \frac{-10}{6}$$
$$x = \frac{-5}{3} = -1\frac{2}{3}$$

Therefore, the solutions are $x = -2$ and $y = 1$, or $x = -1\frac{2}{3}$ and $y = 1\frac{1}{5}$. These can be given as: (−2, 1), $(-1\frac{2}{3}, 1\frac{1}{5})$.


5.  Solve the equations $9y^2 + 8x = 12$, $2x + 3y = 4$

Solution

$$9y^2 + 8x = 12 \quad .................\text{Equation (1)}$$
$$2x + 3y = 4 \quad .................\text{Equation (2)}$$

From equation (2):

$$2x = 4 - 3y$$

$\therefore \quad x = \dfrac{4 - 3y}{2}$ ..................Equation (3)

Substitute $\dfrac{4 - 3y}{2}$ for $x$ in equation (1).

$\qquad 9y^2 + 8x = 12$ .................Equation (1)

$\qquad 9y^2 + 8(\dfrac{4 - 3y}{2}) = 12$

$\qquad 9y^2 + 4(4 - 3y) = 12$     (Note that 8 divided by 2 gives 4 which is outside the bracket)

$\qquad 9y^2 + 16 - 12y = 12$

$\qquad 9y^2 - 12y + 16 - 12 = 0$

$\qquad 9y^2 - 12y + 4 = 0$

Solving this equation by factorization gives:

$\qquad 9y^2 - 6y - 6y + 4 = 0$

$\qquad 3y(3y - 2) - 2(3y - 2) = 0$

$\qquad (3y - 2)(3y - 2) = 0$

Equation each equation to zero and solving each equation gives:

$\qquad y = \dfrac{2}{3}$ in both cases.

Now substitute $\dfrac{2}{3}$ for y in equation (3)

$\qquad x = \dfrac{4 - 3y}{2}$ ..................Equation (3)

$\qquad = \dfrac{4 - 3(\frac{2}{3})}{2}$

$\qquad = \dfrac{4 - 2}{2}$     (Note that $3 \times \dfrac{2}{3} = 2$)

$\qquad x = \dfrac{2}{2} = 1$

Therefore, the solutions are $x = 1$ and $y = \dfrac{2}{3}$

6.  Solve the equations, $x^2 - y^2 = 20$, $x + y = -2$

Solution

$\qquad x^2 - y^2 = 20$ ...................Equation (1)

$\qquad x + y = -2$ .....................Equation (2)

Looking at equation (1) carefully shows that it is a difference of two squares, and that one of the factors of equation (1) is what is given in equation (2).

Recall that a difference of two squares such as $a^2 - b^2$ can be factorized as $a^2 - b^2 = (a + b)(a - b)$. Therefore the equations above can be re–written as follows:

$\qquad (x + y)(x - y) = 20$ ..................Equation (1)    (Note that $x^2 - y^2$ has been factorized)

$\qquad x + y = -2$ ...................Equation (2)

Substitute −2 for $x + y$ in equation (1)

$\qquad (x + y)(x - y) = 20$ ...................Equation (1)

$\qquad -2(x - y) = 20$     (Since $(x + y) = -2$)

Divide both sides by −2 to obtain:

$$x - y = -10 \quad \text{....................Equation (3)}$$
$$\underline{x + y = -2 \quad \text{....................Equation (2)}}$$

Equation (2) − equation (3)　　$2y = 8$　(Note that $+y - (-y) = y + y = 2y$ and $-2 -(-10) = -2 + 10$ = 8)

$$y = \frac{8}{2}$$
$$y = 4$$

Substitute 4 for y in equation (2). (Any equation can be used)

$\quad x + y = -2$ ....................Equation (2)

$\quad x + 4 = -2$

$\quad x = -2 - 4$

$\quad x = -6$

Therefore the solution are $x = -6$ and $y = 4$. This can also be given as (−6, 4).

7.  Solve the equations, $4x^2 - 9y^2 = -32$, $2x - 3y = -4$

<u>Solution</u>

$\quad 4x^2 - 9y^2 = -32$ ....................Equation (1)

$\quad 2x - 3y = -4$ ....................Equation (2)

Equation (1) is a difference of two squares. It can be expressed as:

$\quad (2x)^2 - (3y)^2 = -32$

Therefore factorizing this difference of two squares gives:

$\quad (2x - 3y)(2x + 3y) = -32$ ....................Equation (1)

$\quad 2x - 3y = -4$ ....................Equation (2)

Substitute −4 for $2x - 3y$ in equation (1)

$\quad (2x - 3y)(2x + 3y) = -32$ ....................Equation (1)

$\quad -4(2x + 3y) = -32$　(Since $(2x - 3y) = -4$)

Divide both sides by −4 to obtain equation (3) as shown below:

$$2x + 3y = 8 \quad \text{....................Equation (3)}$$
$$\underline{2x - 3y = -4 \quad \text{....................Equation (2)}}$$

Equation (3) − equation (2):　　$6y = 12$　(Note that $+3y - (-3y) = 6y$ and $8 - (-4) = 8 + 4 = 12$)

$$y = \frac{12}{6}$$
$$y = 2$$

Substitute 2 for y in equation (3). (Note that any equation can be used)

$\quad 2x + 3y = 8$ ....................Equation (3)

$\quad 2x + 3(2) = 8$

$\quad 2x + 6 = 8$

$\quad 2x = 8 - 6$

$\quad 2x = 2$

$\quad x = \frac{2}{2}$

$\quad x = 1$

Therefore the solution is $x = 1$ and $y = 2$. This can also be represented as $(1, 2)$.

8.   A boy is $x$ years old while his father is $y$ years old. The sum of their ages is equal to twice the difference of their ages. The product of their ages is equal to 675. Find their ages.

Solution
The sum of their ages is $x + y$.
The difference of their ages is $y - x$. (The father's age is larger)
Twice the difference of their ages is $2(y - x)$.
The sum of their ages is equal to twice the difference of their ages is given by:

$$x + y = 2(y - x)$$
$$x + y = 2y - 2x$$
$$x + 2x + y - 2y = 0$$
$$3x - y = 0$$

Or        $3x = y$ ....................Equation (1)

The product of their ages is 675. This is given by:

$$xy = 675 \text{ ...................Equation (2)}$$

Substitute $3x$ for $y$ in equation (2). (Note that $y = 3x$ from equation 1)

$$xy = 675 \text{ ...................Equation (2)}$$
$$x(3x) = 675$$
$$3x^2 = 675$$
$$x^2 = \frac{675}{3}$$
$$x^2 = 225$$
$$x = \sqrt{225}$$
$$x = \pm15$$
$$x = 15 \quad \text{(Since age cannot be negative)}$$

Substitute 15 for $x$ in equation (1)

$$3x = y \text{ ...................Equation (1)}$$

Or,        $y = 3x$
$$= 3(15)$$
$$y = 45$$

Therefore the boy is 15 years old (i.e. $x$), while his father is 45 years old (i.e. $y$)

9.   Twice a number added to another number is 19. The first number added to the square of the second number gives 17. Find the two numbers.

Solution
Let the first number be 'a' and the second number be 'b'.
Twice the first number = 2a
Twice the first number added to the second number is 19. This is given by:

$$2a + b = 19 \text{ ...................Equation (1)}$$

The square of the second number = $b^2$
The first number added to the square of the second number gives 17. This is represented as:

$$a + b^2 = 17 \text{ ...................Equation (2)}$$

From equation (1), $a = \dfrac{19 - b}{2}$ ..................Equation (3)

Substitute $\dfrac{19 - b}{2}$ for 'a' in equation (2)

$\qquad a + b^2 = 17$ ..................Equation (2)

$\qquad \dfrac{19 - b}{2} + b^2 = 17$

Multiply each term by 2 (L.C.M) in order to clear the fraction. This gives:

$\qquad 2\left(\dfrac{19 - b}{2}\right) + 2(b^2) = 2(17)$

$\qquad 19 - b + 2b^2 = 34$

$\qquad 2b^2 - b + 19 - 34 = 0$

$\therefore \qquad 2b^2 - b - 15 = 0$

Solving this equation by factorization method gives

$\qquad 2b^2 - 6b + 5b - 15 = 0$

$\qquad 2b(b - 3) + 5(b - 3) = 0$

Equating each bracket to zero and solving each equation gives:

$\qquad (b - 3)(2b + 5) = 0$

$\qquad b = 3 \text{ or } b = \dfrac{-5}{2} = -2\dfrac{1}{2}$

From equation (3), when b = 3, we obtain 'a' as follows:

$\qquad a = \dfrac{19 - b}{2}$ ..................Equation (3)

$\qquad = \dfrac{19 - 3}{2}$

$\qquad = \dfrac{16}{2}$

$\qquad a = 8$

Similarly, from equation (3), when $b = \dfrac{-5}{2}$, we obtain 'a' as follows:

$\qquad a = \dfrac{19 - b}{2}$ ..................Equation (3)

$\qquad = \dfrac{19 - \left(\dfrac{-5}{2}\right)}{2}$

$\qquad = \dfrac{19 + \dfrac{5}{2}}{2}$

$\qquad = \dfrac{\dfrac{38 + 5}{2}}{2}$

$\qquad = \dfrac{43}{2} \div 2$

$\qquad = \dfrac{43}{2} \times \dfrac{1}{2}$

$$= \frac{43}{4}$$

$$a = 10\frac{3}{4}$$

Therefore the solutions are $a = 8$ and $b = 3$, or $a = 10\frac{3}{4}$ and $b = -2\frac{1}{2}$. These can also be given as: $(8, 3)$ or $(10\frac{3}{4}, -2\frac{1}{2})$.

10.   The sum of the squares of the ages of two brothers is 65. If the difference between their ages is 3, find their ages.

Solution

Let the older brother be m years old, and the younger brother be n years old. Therefore the two equations that can be formed from the question are:

$m^2 + n^2 = 65$ ................Equation (1)

$m - n = 3$ ...................Equation (2)

From equation (2):

$m = 3 + n$ .................Equation (3)

Substitute $3 + n$ for m in equation (1). This gives:

$m^2 + n^2 = 65$ ................Equation (1)

$(3 + n)^2 + n^2 = 65$

$(3 + n)(3 + n) + n^2 = 65$

$9 + 3n + 3n + n^2 + n^2 = 65$

$2n^2 + 6n + 9 - 65 = 0$

$2n^2 + 6n - 56 = 0$

Solving this equation by using quadratic equation formula (you can also use factorization method) gives:

$a = 2, b = 6, c = -56$

$$\therefore \quad n = \frac{-b \pm \sqrt{b^2 - 4ac}}{2a}$$

$$= \frac{-6 \pm \sqrt{6^2 - [4 \times 2 \times (-56)]}}{2 \times 2}$$

$$= \frac{-6 \pm \sqrt{36 - (-448)}}{4}$$

$$= \frac{-6 \pm \sqrt{36 + 448}}{4}$$

$$= \frac{-6 \pm \sqrt{484}}{4}$$

$$= \frac{-6 \pm 22}{4}$$

$$n = \frac{-6 + 22}{4} \quad \text{or} \quad \frac{-6 - 22}{4}$$

$$= \frac{16}{4} \quad \text{or} \quad \frac{-28}{4}$$

$$n = 4 \quad \text{or} \quad -7$$

But, age cannot be negative. Therefore:

$$n = 4$$

Substitute 4 for n in equation (3)

$$m = 3 + n \ldots\ldots\ldots\ldots\ldots\text{Equation (3)}$$

$$= 3 + 4$$

$$m = 7$$

∴ The older brother is 7 years old while the younger brother is 4 years old.

## Exercise 12

1. Solve the following simultaneous equations:

a. $x^2 - 2y = -2$
   $5x + y = 13$

b. $x^2 + 2y^2 = 9$
   $2x - y = -4$

c. $5x^2 - 3xy = 26$, $3y - 7x = 17$

d. $6x - 5y = -2$, $2xy = 1$

e. $5y^2 + 2x = 11$, $x - 4y = 7$

f. $x^2 - y^2 = 16$, $x + y = -2$

g. $3x^2 - 2xy + y^2 = 9$, $5x - 2y = 8$

2. A boy is $x$ years old while his father is $y$ years old. The sum of their ages is equal to thrice the difference of their ages. The product of their ages is equal to 1800. Find their ages.

3. Four times a number added to twice another number is 34. Twice the first number added to twice the square of the second number gives 108. Find the two numbers.

4. The sum of the squares of the ages of two brothers is 157. If the difference between their ages is 5, find their ages.

# CHAPTER 13
## LINEAR INEQUALITY AND INTRODUCTORY LINEAR PROGRAMMING

### Linear Inequality

If a statement is expressed as $x = 10$, then this is an equation. However, if $x$ is not exactly equal to 10, but can vary in values from 10 and above or from 10 and below, then we say $x < 10$ or $x > 10$. We can also say $x \leq 10$ or $x \geq 10$. These are called inequalities.

The four symbols of inequalities and their meanings are as follows:

$<$ means less than

$>$ means greater than

$\leq$ means less than or equal to

$\geq$ means greater than or equal to

Inequalities can be solved like equations. When solving inequalities, the following points should be noted.

1.  When both sides of an inequality are divided by a negative number, the inequality sign should be reversed. This is done in order to keep the inequality true. Consider the example below:

$-3 < 5$   (This is true)

Now let us divide both sides by $-3$. This gives:

$$\frac{-3}{-3} < \frac{5}{-3}$$

$1 < -\frac{5}{3}$   (This is not true because the inequality sign has not been reversed)

Therefore the right thing to do is to reverse the inequality sign when dividing by a negative number. This means that the correct division above should be:

$-3 < 5$

$$\frac{-3}{-3} > \frac{5}{-3}$$

$1 > -\frac{5}{3}$   (This is true because the inequality sign has been reversed)

2.  When an inequality is cross multiplied, it is advisable to cross multiply from the lower right hand side to the upper left hand side first, and not the other way round first. Let us examine the example below.

$$\frac{2}{7} < \frac{3}{4}$$

The correct way to cross multiply is as follows:

$(2 \times 4) < (7 \times 3)$

$8 < 21$     (This is true)

However, if you cross multiply from the lower left to the upper right first, then it gives:

$(7 \times 3) < (2 \times 4)$

$21 < 8$     (This is not true)

In order to make this kind of cross multiplication true, you have to reverse the sign to give $21 > 8$.

Hence, in order to avoid making mistake, it is better to cross multiply from the lower right hand side to the upper left hand side first. In this case your inequality remains true, and you do not need to reverse the inequality sign.

**Examples**

1. Solve the inequality $5x - 2 > 19 - 2x$. Represent your result on a number line/line graph.

Solution

$$5x - 2 > 19 - 2x$$

We simply solve this problem just like the way we solve an equation. But we must continue to use the inequality sign.

Let us collect terms in $x$ on one side, and constant terms on the other side.

$$5x - 2 > 19 - 2x$$

$5x + 2x > 19 + 2$   (Note that the sign of any term that crosses the inequality sign must change)

$$7x > 21$$

Divide both sides by 7. This gives:

$$\frac{7x}{7} > \frac{21}{7}$$

$$x > 3$$

In representing inequalities on number lines/line graphs, we simply draw a number line and make sure the arrow starts from the solution (i.e. 3) and points to the right when it is >. It points to the left when the inequality sign is <. The beginning of the arrow line should be a small circle. When the sign is ≤ or ≥, the small circle on the number line where the arrow starts should be shaded to show that the value is included in the possible solutions.

∴   $x > 3$ is represented as shown below:

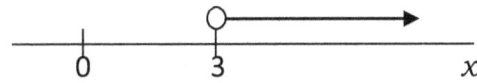

2. Solve the inequality $9x + 5 < 4x - 2(3 - x)$. Represent your result on a number line.

Solution

$$9x + 5 < 4x - 2(3 - x)$$

Let us expand the bracket on the right hand side. This gives:

$$9x + 5 < 4x - 6 + 2x$$

Let us collect terms in $x$ on one side, and constant terms on the other side.

$$9x - 4x - 2x < -6 - 5$$

$$3x < -11$$

Divide both sides by 3. This gives:

$$\frac{3x}{3} < \frac{-11}{3}$$

$$x < -\frac{11}{3}$$

151

$\therefore \quad x < -\dfrac{11}{3}$ is represented on a number line as shown below:

3. Find the range of values of $x$ for which $\dfrac{2x-1}{3} - \dfrac{1}{2}x \leq 8 + x$. Show the result on a line graph.

<u>Solution</u>

$$\dfrac{2x-1}{3} - \dfrac{1}{2}x \leq 8 + x$$

Multiply each term by 6 (i.e. the LCM of 3 and 2 which are the denominators). This is done in order to remove the fractions.

$$6(\dfrac{2x-1}{3}) - 6(\dfrac{x}{2}) \leq 6(8 + x)$$

$2(2x-1) - 3(x) \leq 48 + 6x$      (Note that 2 and 3 outside the brackets are obtained when 6 outside the brackets above divides 3 and 2 (the denominators) respectively).

$\quad 4x - 2 - 3x \leq 48 + 6x$

We now collect like terms as follows:

$\quad 4x - 3x - 6x \leq 48 + 2$

$\quad -5x \leq 50$

Divide both sides by −5 and reverse the inequality sign. This gives:

$$\dfrac{-5x}{-5} \geq \dfrac{50}{-5} \quad \text{(Take note of the reversal of the sign since we are dividing by a negative}$$

number)

$\qquad x \geq -10$

We now represent this result on a number line as follows

Note that the small circle at the beginning of the arrow is shaded to show that −10 is also part of the solution due to the sign $\geq$. And remember that greater than represents right hand direction of arrow, while less than represents left hand direction of arrow.

4. Find the range of values of $x$ for which $\dfrac{1}{5} - \dfrac{1-4x}{2} \geq \dfrac{3}{4} + 3x$. Show the result on a line graph.

<u>Solution</u>

$$\dfrac{1}{5} - \dfrac{1-4x}{2} \geq \dfrac{3}{4} + 3x$$

Multiply each term by 20 (i.e. the LCM of 5, 2 and 4 which are the denominators).

$$20(\dfrac{1}{5}) - 20(\dfrac{1-4x}{2}) \geq 20(\dfrac{3}{4}) + 20(3x)$$

$\quad 4(1) - 10(1 - 4x) \geq 5(3) + 60x$

Note that 20 has been used to divide the denominators of the fractions.

152

$$4 - 10 + 40x \geq 15 + 60x$$

We now collect like terms as follows:

$$40x - 60x \geq 15 + 10 - 4$$

$$-20x \geq 21$$

Divide both sides by −20 and reverse the sign. This gives:

$$\frac{-20x}{-20} \leq \frac{21}{-20}$$ (Take note of the reversal of the sign since we are dividing by a negative number)

$$x \leq -\frac{21}{20}$$

We now represent this result on a number line as follows

5. Given that m is an integer, find the three greatest values which satisfy the inequality:

$$2(3m + 1) < \frac{1}{2}(2m - 5)$$

Solution

$$2(3m + 1) < \frac{1}{2}(2m - 5)$$

This can also be written as follows:

$$2(3m + 1) < \frac{2m - 5}{2}$$

Cross multiply to obtain:

$$2 \times 2(3m + 1) < 2m - 5$$
$$4(3m + 1) < 2m - 5$$
$$12m + 4 < 2m - 5$$

Collect like terms. This gives:

$$12m - 2m < -5 - 4$$
$$10m < -9$$
$$\therefore \quad m < \frac{-9}{10}$$

An integer is any whole number such as, −1, 7, 0.

Hence, if $m < \frac{-9}{10}$, and m is an integer, then m can be any of the numbers below.

$$m = -1, -2, -3, -4 \ldots\ldots$$

Hence the three greatest values of m are, −1, −2, −3.

Note that the nearest whole number that is just less than $\frac{-9}{10}$ is −1 which is also the greatest whole number for 'm'.

6. Given that $x$ is an integer, find the three lowest values of $x$ which satisfy the following inequality: $2(x - \frac{1}{3}) - \frac{3}{4}(x + 5) > \frac{1}{2}(x - 4)$

Solution

$$2\left(x - \frac{1}{3}\right) - \frac{3}{4}(x + 5) > \frac{1}{2}(x - 4)$$

Expanding each of the bracket and placing each of the denominator properly gives:

$$2x - \frac{2}{3} - \frac{3x + 15}{4} > \frac{x - 4}{2}$$

Multiply each term by 12 (i.e. the LCM of the denominators). This gives:

$$12(2x) - 12\left(\frac{2}{3}\right) - 12\left(\frac{3x + 15}{4}\right) > 12\left(\frac{x - 4}{2}\right)$$

$$24x - 4(2) - 3(3x + 15) > 6(x - 4) \quad \text{(Note that 12 has been used to divide each denominator)}$$

$$24x - 8 - 9x - 45 > 6x - 24$$

Collect like term to give:

$$24x - 9x - 6x > 8 + 45 - 24$$

$$9x > 29$$

$$x > \frac{29}{9}$$

$$x > 3\frac{2}{9}$$

Since $x$ is an integer (whole number), then $x$ starts from 4 because 4 is the nearest whole number greater than $3\frac{2}{9}$

∴ The three lowest values of $x$ are 4, 5, 6.


7. Given that c is an integer, find the three greatest values which satisfy the inequality:

$$\frac{1}{3}(c - 1) > \frac{2}{5}(3c + 4)$$

Solution

$$\frac{1}{3}(c - 1) > \frac{2}{5}(3c + 4)$$

This can also be written as follows:

$$\frac{c - 1}{3} > \frac{2(3c + 4)}{5}$$

Cross multiply to obtain:

$$5(c - 1) > 3 \times 2(3c + 4)$$

$$5c - 5 > 6(3c + 4)$$

$$5c - 5 > 18c + 24$$

Collect like terms. This gives:

$$5c - 18c > 24 + 5$$

$$-13c > 29$$

∴ $\qquad c < \frac{29}{-13}$ $\qquad$ (Note that the inequality sign has been reversed due to the division by −13)

Hence, $c < -2\frac{3}{13}$

Hence, if $c < -2\frac{3}{13}$, and c is an integer, then c can be any of the numbers below.

$$c = -3, -4, -5, -6 \ ......$$

Hence the three greatest values of c are, −3, −4, −5.

8. Given that r is an integer, find the three lowest values of r which satisfy the following inequality: $\frac{3}{8} - \frac{1}{2}(3r-1) < \frac{1}{4}(2-r)$

Solution

$$\frac{3}{8} - \frac{1}{2}(3r-1) < \frac{1}{4}(2-r)$$

Multiply each term by 8 (i.e. the LCM of the denominators). This gives:

$$8(\frac{3}{8}) - \frac{8(3r-1)}{2} < \frac{8(2-r)}{4}$$

$3 - 4(3r-1) < 2(2-r)$  (Note that 8 has been used to divide each denominator)

$3 - 12r + 4 < 4 - 2r$

Collect like term to give:

$-12r + 2r < 4 - 4 - 3$

$-10r < -3$

$r > \frac{-3}{-10}$  (Take note of the reversal of the inequality sign)

$r > \frac{3}{10}$

Since r is an integer (whole number), then r starts from 1 because 1 is the nearest whole number greater than $\frac{3}{10}$

∴ The three lowest values of r are 1, 2, 3.


9. What is the range of values of y for which $4y - 7 \leq 3y$ and $3y \leq 5y + 8$ are both satisfied? Show your result on a graph.

Solution

We solve each of the inequality to obtain two values of y. This is as shown below.

$4y - 7 \leq 3y$

$4y - 3y \leq 7$

$y \leq 7$

And

$3y \leq 5y + 8$

$3y - 5y \leq 8$

$-2y \leq 8$

$y \geq \frac{8}{-2}$  (Take note of the reversal of the inequality sign due to division by negative number)

∴ $y \geq -4$

Hence, $y \leq 7$ and $y \geq -4$

Combining the two inequalities gives:

$-4 \leq y \leq 7$

When combining the solutions of inequalities, we ensure that the appropriate part of the inequality sign is facing the variable (i.e. y). In the solution, $y \leq 7$, the elbow part of the inequality sign is facing y. This was maintained in the combination above. In the solution y ≥ –4, the open part of the inequality sign is facing y. This was also maintained in the combination above.

Hence, both inequalities are satisfied if:   $-4 \le y \le 7$

This can be represented on a line graph as shown below.

10.  Express $3y - 2 < 10 + y < 2 + 5y$ in the form a < y < b, where a and b are both integers. Show the range of values on a line graph.

<u>Solution</u>

$3y - 2 < 10 + y < 2 + 5y$

We can take out two inequalities from the question above and solve. They are:

$3y - 2 < 10 + y$ ................Inequality (1)

And, $10 + y < 2 + 5y$ ................Inequality (2)

Let us solve each of them.

$3y - 2 < 10 + y$ ................Inequality (1)

$3y - y < 10 + 2$

$2y < 12$

$y < \dfrac{12}{2}$

$y < 6$

Now let us solve the second inequality.

$10 + y < 2 + 5y$ ................Inequality (2)

$y - 5y < 2 - 10$

$-4y < -8$

$y > \dfrac{-8}{-4}$

$y > 2$

Therefore the two solutions are y < 6 and y > 2

These two solutions show that y ranges from 2 to 6.

Hence the solutions can be combined to give:

$2 < y < 6$

This can be represented on a line graph as shown below.

11.  Find the range of values of y for which $2y + 1 > 21$ or $y + 5 \le 12$ is satisfied. Show your result on a line graph.

<u>Solution</u>

Let us solve the first inequality as follows:

$$2y + 1 > 21$$
$$2y > 21 - 1$$
$$2y > 20$$
$$y > \frac{20}{2}$$
$$y > 10$$

The other inequality is solved as follows:

$$y + 5 \leq 12$$
$$y \leq 12 - 5$$
$$y \leq 7$$

Hence the inequalities are satisfied by either $y > 10$ or $y \leq 7$

Therefore, the line graph is as given below.

12. Express $3m - 5 < 4m + 1 < 13 + m$ in the form $a < y < b$, where a and b are both integers.

<u>Solution</u>

$$3m - 5 < 4m + 1 < 13 + m$$

Two inequalities that can be obtained from the question above are:

$$3m - 5 < 4m + 1 \ldots\ldots\ldots\ldots\text{Inequality (1)}$$

And, $4m + 1 < 13 + m\ldots\ldots\ldots\ldots\text{Inequality (2)}$

Let us solve each of them.

$$3m - 5 < 4m + 1 \ldots\ldots\ldots\ldots\text{Inequality (1)}$$
$$3m - 4m < 1 + 5$$
$$-m < 6$$

Divide both sides by –1 and reverse the sign. This gives:

$$\frac{-m}{-1} > \frac{6}{-1}$$
$$m > -6$$

The second inequality is solved as follows:

$$4m + 1 < 13 + m \ldots\ldots\ldots\ldots\text{Inequality (2)}$$
$$4m - m < 13 - 1$$
$$3m < 12$$

Hence, $m < \dfrac{12}{3}$

$$m < 4$$

Therefore the two solutions are $m > -6$ and $m < 4$.

These two solutions show that y ranges from –6 to 4.

Hence the solutions can be combined to give:

$$-6 < m < 4$$

This can be represented on a line graph as shown below

157

## Application of Inequalities (Introduction to Linear programming)

Inequality can be applied in real life problems. This process is called linear programming. Linear programming is a mathematical method which is used to determine the best solution of a problem whose requirements have been presented using linear inequalities. A more common and direct method of solving problems involving linear inequalities (linear programming) is by graphical method.

However, we are going to use the method of solving simultaneous equations in solving linear programming.

Note that in problems involving inequalities, we use $\leq$ for 'at most' (which also means not more than), while we use $\geq$ for 'at least' (which also means not less than).

**Examples**

1. To employ workers in a baking company, a man needs at least 6 mixers and 10 bakers. He does not want to employ more than 25 workers altogether. A mixer needs 2 tables to work on and a baker needs 1 table. There are only 40 tables available.
a. Write down four inequalities that represent the restriction on the man.
b. What is the greatest number of bakers that he can employ
c. Determine the optimum number of mixers and bakers the man should employ.

<u>Solutions</u>
a. Let $x$ represent the number of mixers and y the number of bakers. This gives:

   $x \geq 6$      (This mean at least 6 mixers)
   
   $y \geq 10$     (This means at least 10 bakers)

Total number of tables required by mixers = $2x$      (Since each mixer needs 2 tables)

Total number of tables required by bakers = y        (Since each baker needs 1 tables)

Hence, total table is given by:

   $2x + y \leq 40$      (Since total table is not more than 40)

Total number of staff is given by:

   $x + y \leq 25$      (Since the man does not want to employ more than 25 workers)

Therefore, the four inequalities that are required are:

   $x \geq 6$
   
   $y \geq 10$
   
   $2x + y \leq 40$
   
   $x + y \leq 25$

b. The greatest number of bakers he can employ will be achieved if the lowest number of mixers is employed. From, $x \geq 6$, the lowest number of mixers he can employ is 6. Recall that the total number of workers is given by:

$x + y \leq 25$

Substitute $x = 6$ into the inequality above in order to obtain y (bakers). This gives:

$x + y \leq 25$

$6 + y \leq 25$

$y \leq 25 - 6$

$y \leq 19$

Therefore the greatest number of bakers he can employ is 19.

c.  In order to employ an optimum number of workers, the man must employ a number of workers (up to 25) that will use all or almost all the tables.

The values y = 19 is a maximum number of bakers. So, let us plug in $x = 6$ and y = 19 into $2x + y \leq 40$, and see if the maximum number of table will be used. This gives:

$2x + y \leq 40$

$2(6) + 19 \leq 40$

$12 + 19 \leq 40$

$31 \leq 40$

This is true (because 31 is actually less than 40), but a maximum number of tables have not been used since we still have 9 (40 − 31 = 9) tables left. Hence let us use the other extreme values for number of workers. This is obtained when y = 10 (from y $\geq$ 10). Hence when y = 10, we obtain $x$ as follows:

$x + y \leq 25$

$x + 10 \leq 25$

$x \leq 25 - 10$

$x \leq 15$

Note that extreme values in this example means one variable is at its lowest possible value while the other is at its highest possible value. Note also that y $\geq$ 10 means that the lowest possible value of y is 10.

Let us now find out if the values $x = 15$ and y = 10, will make use of a maximum number of table. Hence substitute $x = 15$ and y = 10 in the inequality for number of tables. This gives:

$2x + y \leq 40$

$2(15) + 10 \leq 40$

$30 + 10 \leq 40$

$40 \leq 40$

Hence, this is true, and the maximum number of tables is used. Hence this number of workers gives us the optimum number of workers that the man should employ.

Hence the man should employ 15 mixers and 10 bakers.


A more direct method of obtaining the optimum number of workers is to make the two inequalities to become equations and solve the two equations simultaneously. This is done as follows:

$2x + y = 40$ ................Equation(1)

$x + y = 25$................Equation (2)

Equation (1) – Equation (2):  $x$    = 15    (Note that 40 − 25 = 15)

Hence,  $x = 15$

Substitute 15 for $x$ in equation (2)

$x + y = 25$ ...................Equation (2)

$15 + y = 25$

$y = 25 - 15$

$y = 10$

Hence, $x = 15$ and $y = 10$

Therefore, the man should employ 15 mixers and 10 bakers.

2.  A shop sell large number of laptops, costing $300 each, and television, costing $400 each. The shop owner allows $30,000 to be spent in buying these electronics and needs at least twice as many laptops as televisions. There must be at least 50 laptops and 20 televisions.
a.  What is the highest number of electronics that can be bought and in what way?
b.  If it is decided to buy as many televisions as conditions allow, how many of each electronic can be bought?

Solutions

a.  Let m represent number of laptops and let n represent number of television.
Hence, total cost of laptops is 300m, while total cost of televisions is 400n. Since their combined cost should not exceed $30,000, then we have that:

$300m + 400n \leq 30000$

At least twice as many laptops as televisions can be represented as follows:

$m \geq 2n$

At least 50 laptops and 20 televisions can be represented as follows:

$m \geq 50$     (At least 50 laptops)

$n \geq 20$     (At least 20 televisions)

Therefore, the inequalities formed by this problem are:

$300m + 400n \leq 30000$

$m \geq 2n$

$m \geq 50$

$n \geq 20$

In this problem, the lowest possible value of m is 50 (from $m \geq 50$) and the lowest possible value of n is 20 (from $n \geq 20$).

Hence, when m = 50, we obtain the highest possible value of n from the total cost inequality as follows:

$300m + 400n \leq 30000$

$300(50) + 400n \leq 30000$

$15000 + 400n \leq 30000$

$400n \leq 30000 - 15000$

$400n \leq 15000$

$n \leq \dfrac{15000}{400}$

$n \leq 37.5$

$n \leq 37$    (take the whole number since number of electronics must be whole number)

Hence, when m = 50, n = 37.

Also, when n = 20 (lowest possible value of n), we obtain the highest possible value of m from the total cost inequality as follows:

$$300m + 400n \leq 30000$$
$$300m + 400(20) \leq 30000$$
$$300m + 8000 \leq 30000$$
$$300m \leq 30000 - 8000$$
$$300m \leq 22000$$
$$m \leq \frac{22000}{300}$$
$$m \leq 73.3$$
$$m \leq 73 \quad \text{(take the whole number since number of electronics must be whole number)}$$

Hence, when n = 20, m = 73

Now let us compare the two solutions obtained in order to know which one will give the highest number of electronics.

When m = 50 and n = 37, the total electronics that can be bought is = 50 + 37 = 87

When n = 20 and m = 73, the total electronics that can be bought is = 20 + 73 = 93

Therefore the highest number of electronics that can be bought is 93. This is obtained by buying 73 laptops and 20 televisions.


b.  Recall the two extreme points obtained above: When m = 50, n = 37 and when n = 20, m = 73. Let us take the values involving many televisions, i.e. m = 50 and n = 37, and see if it satisfies the two inequalities above. Let us put m = 50 and n = 37 in the inequality: m ≥ 2n. This gives:

$$m \geq 2n$$
$$50 \geq 2(37)$$
$$50 \geq 74 \quad \text{(This condition is not true)}$$

Hence we have to look for another set of maximum condition for television, i.e. the value of n. A good way of obtaining this set of values is to make the two inequalities above to become equations and solve the two equations simultaneously. This is done as follows:

$$300m + 400n = 30000 \ldots\ldots\ldots\ldots\text{Equation (1)}$$
$$m = 2n \ldots\ldots\ldots\ldots\text{Equation (2)}$$

Substitute 2n for m in equation (1).

$$300m + 400n = 30000 \ldots\ldots\ldots\ldots\text{Equation (1)}$$
$$300(2n) + 400n = 30000$$
$$600n + 400n = 30000$$
$$1000n = 30000$$
$$n = \frac{30000}{1000}$$
$$n = 30$$

Now substitute 30 for n in equation 2. This gives:

$$m = 2n$$
$$m = 2(30)$$
$$m = 60$$

Hence, m = 60 and n = 30.

Therefore, if it is decided to buy as many televisions as condition allows, then 30 televisions should be bought and 60 laptops should be bought.

3.   A man needs at least three shoes and three belts. A shoe cost $60 and a belt cost $36, and the man has $360 to spend. The man decides to spend as much as possible of his $360.

a.  How many ways can he spend his money, and how many of each item can he buy in each case?

b.  Does any of the ways leave him with any extra money? If so, how much?

Solutions

a.  Let $x$ represent shoe and let y represent belt. We now form inequalities as follows.

$x \geq 3$     (At least 3 shoes)

$y \geq 3$     (At least 3 belts)

Cost of shoes = $60x$, while cost of belts = 36y. The total cost of the items should not be greater than 360, since the man has $360 to spend. This gives:

$60x + 36y \leq 360$

Now let us substitute $x = 3$, and obtain a value of y. This gives:

$60x + 36y \leq 360$

$60(3) + 36y \leq 360$

$180 + 36y \leq 360$

$36y \leq 360 - 180$

$36y \leq 180$

$y \leq \dfrac{180}{36}$

$y \leq 5$

Hence, when $x = 3$, y = 5

Let us substitute y = 3, and obtain a value of $x$. This gives:

$60x + 36y \leq 360$

$60x + 36(3) \leq 360$

$60x + 108 \leq 360$

$60x \leq 360 - 108$

$60x \leq 252$

$x \leq \dfrac{252}{60}$

$x \leq 4.2$

$x \leq 4$     (Take only the whole number since $x$ must be a whole number)

Hence, when y = 3, $x = 4$

Therefore the man can spend his money in two ways. They are: $x = 3$ and y = 5, or $x = 4$ and y = 3.

This means that the man can buy 3 shoes and 5 belts, or 4 shoes and 3 belts

b.  The condition where $x = 4$ (This is actually 4.2) and y = 3, leaves him with extra money. This is due to the 0.2 left out of 4.2 to obtain 4.

Hence, let us obtain the cost of 4 shoes and 3 belts in order to obtain the extra money. This gives:

$60x + 36y$

$= 60(4) + 36(3)$

$= 240 + 108$

= 348

Hence the extra money that the man will have is: 360 − 348 = 12

Therefore, this condition (4 shoes and 3 belts) leaves the man with extra money of $12.


4. A patient is required to take drugs containing iron and calcium. Two types of drugs available are Fero drugs and Calco drugs. Fero contains 160mg of iron, Calco contains 40mg of iron and the required iron for a day should be at least 840mg. Also, Fero contains 4mg of calcium, Calco contains 3mg of calcium and the required calcium for a day should be at least 31mg. A daily prescription is not up to 10 or more drugs.

a. Write the inequalities to be satisfied by these conditions.

b. What prescriptions should be given for each of the drugs?

Solutions

a. Let y represents the number of Fero drugs and z represents the number of Calco drugs. Hence the total iron from Fero is 160y and the total iron from Calco is 40z. This gives a total iron of 160y + 40z. Since the total daily iron should be at least 840mg, then this gives:

$$160y + 40z \geq 840$$

Similarly, the total calcium from Fero is 4y and the total calcium from Calco is 3z. This gives a total calcium of 4y + 3z. Since the total daily calcium should be at least 31mg, then this gives:

$$4y + 3z \geq 31$$

Also, since the total daily prescription is not up to 10 or more drugs, we have:

$$y + z < 10$$

Therefore, the inequalities to be satisfied by these conditions are:

$$160y + 40z \geq 840$$
$$4y + 3z \geq 31$$
$$y + z < 10$$


b. In order to obtain the right prescriptions that should give optimum number of each of the drug, we have to make the three inequalities above to become equations and solve the right pair simultaneously. However, the pair to solve in this case is the first two equations since they have to be maximized as they contain the $\geq$ sign. This gives:

$$160y + 40z = 840 \quad\text{....................Equation (1)}$$
$$4y + 3z = 31 \quad\text{......................Equation (2)}$$

Solving equation 1 and 2 simultaneously gives:

$$160y + 40z = 840 \quad\text{....................Equation (1)}$$
$$4y + 3z = 31 \quad\text{......................Equation (2)}$$

Divide equation (1) by 40 to obtain a simplified form of the equation. This gives:

$$\frac{160}{40}y + \frac{40}{40}z = \frac{840}{40} \quad \text{(Note that we have to divide each term by 40)}$$
$$4y + z = 21 \quad\text{....................Equation (4)}$$

Solving equation 4 and 2 simultaneously gives:

$$4y + 3z = 31 \quad\text{....................Equation (2)}$$
$$\underline{4y + z = 21} \quad\text{....................Equation (4)}$$

Equation (2) – Equation (4):     $2z = 10$

$$z = \frac{10}{2}$$

$$z = 5$$

Substitute 5 for z in equation (4). This gives:

$4y + z = 21$ ...................Equation (4)

$4y + 5 = 21$

$4y = 21 - 5$

$4y = 16$

$$y = \frac{16}{4}$$

$$y = 4$$

Hence $y = 4$ and $z = 5$

Therefore, 4 Fero and 5 Calco should be prescribed.


5.  A tailoring company needs to buy a total of 10 sewing machines. A manual sewing machine cost $250, while an electric sewing machine cost $450. The company must buy at least two of each machine and it is prepared to spend no more than $3,500. The profit on each manual sewing machine is not greater than $35, while that on each electric sewing machine is not greater than $50. Find:

a.  the maximum profit the company can make in the business

b.  the number of each sewing machine that the company should buy.

Solution

a.  This is a case where profit is involved. Hence we have to maximize profit.

Let the number of manual sewing machine be $x$, and the number of electric sewing machine be y. We now form inequalities using $x$ and y as follows.

$x \geq 2$     (At least two of each machine)

$y \geq 2$     (At least two of each machine)

$x + y = 10$     (The company needs a total of 10 machines)

The cost of buying $x$ manual sewing machines at $250 each is $250x$. The cost of buying y electric sewing machines at $450 each is $450y$. Since the company is not prepared to spend more than $3,500, then we have a total cost inequality as follows:

$250x + 450y \leq 3500$

Also, since the profit on each manual sewing machine is $35, and the profit on each electric sewing machine is $50, then the total profit that the company can make is given by:

$35x + 50y = P$     (Where P is the maximum profit that the company can make)

We are not going to use this profit equation to solve for $x$ and y. We are only going to use it to find out which values of $x$ and y will give us a maximum profit.

Let us bring down our inequalities below. They are:

$x \geq 2$

$y \geq 2$

$x + y = 10$

$250x + 450y \leq 3500$

Let us obtain our extreme values first. The lowest possible value of $x$ is 2 (from $x \geq 2$). When $x$ = 2, we can obtain a value for y from $x + y = 10$. This gives y as follows:

$x + y = 10$

$2 + y = 10$

$y = 10 - 2 = 8$

Hence when $x = 2$, $y = 8$.

Similarly, the lowest possible value of y is 2. If we substitute y = 2 in $x + y = 10$, we will obtain $x$ = 8.

Hence the two solutions we have obtained so far are:

$x = 2$ and y = 8, or $x = 8$ and y = 2.

Now let us test these solutions in the cost inequality to see if they are feasible.

The cost inequality is: $250x + 450y \leq 3500$

Hence when $x = 2$ and y = 8, we have:

$250(2) + 450(8) \leq 3500$

$500 + 3600 \leq 3500$

$4100 \leq 3500$     (This is not true since 4100 is not less than 3500)

Hence $x = 2$ and y = 8 is not a possible solution.

Let us try the other solution which is $x = 8$ and y = 2. We now substitute these values as follows:

$250(8) + 450(2) \leq 3500$

$2000 + 900 \leq 3500$

$2900 \leq 3500$     (This is true)

Hence $x = 8$ and y = 2 is a possible solution

We can also obtain another solution by solving the following equations simultaneously:

$250x + 450y = 3500$ ................Equation (1)

$x + y = 10$ ..................Equation (2)

From equation (2), $x = 10 - y$ ................Equation (3)

Substitute $10 - y$ for $x$ in equation (1). This gives:

$250x + 450y = 3500$ ................Equation (1)

$250(10 - y) + 450y = 3500$

$2500 - 250y + 450y = 3500$

$200y = 3500 - 2500$

$200y = 1000$

$y = \dfrac{1000}{200}$

$y = 5$

From equation (3), we obtain $x$ as follows:

$x = 10 - y$ ................Equation (3)

$x = 10 - 5$

$x = 5$

Hence $x = 5$ and y = 5.

Therefore, the two possible solutions to this problem are:

$x = 8$ and y = 2   or $x = 5$ and y = 5

Now the solution that gives a maximum profit becomes the actual solution. The maximum profit equation is:

$35x + 50y = P$
When $x = 8$ and $y = 2$, the maximum profit is given by:

$35(8) + 50(2) = P$

$280 + 100 = P$

$P = 380$

When $x = 5$ and $y = 5$, the maximum profit is given by:

$35(5) + 50(5) = P$

$175 + 250 = P$

$P = 425$   (This is greater than the first profit)

Hence the maximum profit that the company can make is $425

b.  The company should buy 5 of each of the sewing machine.

6.   A company plans to produce at most 90 units of two products X and Y. The cost of producing each unit of X is $50, while the cost of producing each unit of Y is $100. After production, the time needed to package each unit of X is 1 hour, while the time needed to package each unit of Y is 3 hours. The profit from each unit of X and Y are $25 and $60 respectively. The company has a total budget of $5,000 and a packaging period of not more than 120 hours.

a.  What optimal quantity of X and Y should the company produce?

b.  What is the maximum profit that the company will make from this production?

Solution

a.  Let us obtain the inequalities for this problem. If X represents the total units of X produced and Y represents the total units of Y produced, then:

$X + Y \leq 90$     (The company plans to produce at most (not more than) 90 units of X and Y)

The cost of producing X at $50 each is 50X. The cost of producing Y at $100 each is 100Y. Since the company's total budget is $5,000, then we have a total cost inequality as follows:

$50X + 100Y \leq 5000$

Also, The packaging time for X at 1 hour each is X hours. The packaging time for Y at 3 hours each is 3Y. Since the company's total packaging time is not more than 120 hours, then we have a total time inequality as follows:

$X + 3Y \leq 120$

The profit on each of X is $25 and the profit on each of Y is $60. Then the total profit that the company can make is given by:

$25X + 60Y = P$     (Where P is the maximum profit that the company can make)

Hence, the three inequalities obtained from this problem are:

$X + Y \leq 90$

$50X + 100Y \leq 5000$

$X + 3Y \leq 120$

In this case, the two inequalities that we have to make to become equations and solve them simultaneously are the ones that involve the production processes. They are:

$50X + 100Y = 5000$ ....................Equation (1)

$X + 3Y = 120$ ...................Equation (2)

From equation (2), $X = 120 - 3Y$ ...................Equation (3)

166

Substitute 120 – 3Y for X in equation (1). This gives:

$\quad$ 50X + 100Y = 5000 ....................Equation (1)

$\quad$ 50(120 – 3Y) + 100Y = 5000

$\quad$ 6000 – 150Y + 100Y = 5000

$\quad$ 6000 – 5000 = 150Y – 100Y

$\quad$ 1000 = 50Y

$\quad Y = \dfrac{1000}{50}$

$\quad$ Y = 20

Substitute 20 for Y in equation (3). This gives:

$\quad$ X = 120 – 3Y ....................Equation (3)

$\quad$ X = 120 – 3(20)

$\quad\quad$ = 120 – 60

$\quad$ X = 60

Hence, X = 60 and Y = 20.

It is necessary to use the third inequality to test if this solution is true.

This gives: X + Y $\leq$ 90

$\quad$ 60 + 20 $\leq$ 90

$\quad$ 80 $\leq$ 90 $\quad$ (This is a true solution)

Therefore, the company should produce 60 units of X and 20 units of Y.


b. The maximum profit that the company will make from this production will be obtained from the profit equation given above. Let us substitute 60 for X and 20 for Y in the profit equation. This gives:

$\quad$ 25X + 60Y = P

$\quad$ 25(60) + 60(20) = P

$\quad$ 1500 + 1200 = P

$\therefore\quad$ P = 2700

Therefore, the maximum profit that the company can make is $2700


7. Solve the linear program below:

$\quad$ Maximize: 7m + 4n

$\quad$ Subject to: m + n $\leq$ 9

$\quad\quad\quad\quad\quad$ m – n $\geq$ 3

$\quad$ and $\quad$ 10m + 8n $\leq$ 66

Given that, m $\geq$ 0 and n $\geq$ 0

Solution

We have to first ignore 7m + 4n, which is the function to maximize.

In the three inequalities given in this question, our correct solution will be obtained by solving simultaneously two out of the three equations that we will convert them to. We do not know the two that will give us the solution. Hence, we take two at a time and solve them simultaneously. The solution obtained will be substituted into the third inequality to see if it is the right solution.

Let us bring down the three inequalities as follows.

$$m + n \leq 9 \quad ...................Inequality\ (1)$$
$$m - n \geq 3 \quad ...................Inequality\ (2)$$
$$10m + 8n \leq 66 \quad ...................Inequality\ (3)$$

Converting them to equations gives:

$$m + n = 9 \quad ...................Equation\ (1)$$
$$m - n = 3 \quad ................... Equation\ (2)$$
$$10m + 8n = 66 \quad ................... Equation\ (3)$$

Let us solve equations (1) and (2) simultaneously as follows:

$$m + n = 9 \quad ...................Equation\ (1)$$
$$m - n = 3 \quad ................... Equation\ (2)$$

Equation (1) – (2):     $2n = 6$     (Note that $n - (-n) = n + n = 2n$, and $9 - 3 = 6$)

$$n = \frac{6}{2}$$
$$n = 3$$

Substitute 3 for n in equation (1). This gives:

$$m + n = 9 \quad ...................Equation\ (1)$$
$$m + 3 = 9$$
$$m = 9 - 3$$
$$m = 6$$

Hence, $m = 6$ and $n = 3$.

Let us now substitute these values in inequality (3), to see if this solution is correct

$$10m + 8n \leq 66 \quad ...................Inequality\ (3)$$
$$10(6) + 8(3) \leq 66$$
$$60 + 24 \leq 66$$
$$84 \leq 66 \quad \text{(This is not correct)}$$

Hence, $m = 6$ and $n = 3$ are not correct solutions to this problem.

Let us solve equation (1) and (3) simultaneously as follows:

$$m + n = 9 \quad ...................Equation\ (1)$$
$$10m + 8n = 66 \quad ................... Equation\ (3)$$

From equation (1), $m = 9 - n$.

Substitute $9 - n$ for m in equation (3). This gives:

$$10m + 8n = 66 \quad ................... Equation\ (3)$$
$$10(9 - n) + 8n = 66$$
$$90 - 10n + 8n = 66$$
$$90 - 2n = 66$$
$$90 - 66 = 2n$$
$$24 = 2n$$
$$n = \frac{24}{2}$$
$$n = 12$$

Substitute 12 for n in equation (1). This gives:

$$m + n = 9 \quad ...................Equation\ (1)$$
$$m + 12 = 9$$

m = 9 − 12

m = −3

But m cannot be a negative value since $m \geq 0$

Hence, m = −3 and n = 12 is not a correct solution.

Finally, let us solve equations (2) and (3) simultaneously as follows:

   m − n = 3 ................... Equation (2)

   10m + 8n = 66 ................... Equation (3)

From equation (2), m = 3 + n.

Substitute 3 + n for m in equation (3). This gives:

   10m + 8n = 66 ................... Equation (3)

   10(3 + n) + 8n = 66

   30 + 10n + 8n = 66

   30 + 18n = 66

   18n = 66 − 30

   $n = \dfrac{36}{18}$

   n = 2

Substitute 2 for n in equation (2). This gives:

   m − n = 3 ...................Equation (2)

   m − 2 = 3

   m = 3 + 2

   m = 5

Hence m = 5 and n = 2

Let us now substitute these values in inequality (1), to see if this solution is correct

   m + n ≤ 9 ...................Inequality (1)

   5 + 2 ≤ 9

   7 ≤ 9     (This is true)

Hence, m = 5 and n = 2 is the correct solution.

Therefore we now maximize 7m + 4n as follows:

   7m + 4n = 7(5) + 4(2)

   = 35 + 8

   = 43


8.  Solve the linear program:

Minimize:  8x + 5y

Subject to: x + y ≥ 11

           x − y ≤ 5

        7x + 12y ≥ 35

        x ≥ 0 and y ≥ 0

Solution

We have to first ignore 8x + 5y, which is the function to minimize.

In the three inequalities, our correct solution will be obtained by solving simultaneously two out of the three equations that will convert them to. We do not know the two that will give us

169

the solution. Hence, we take two at a time and solve them simultaneously. The solution obtained will be substituted into the third inequality to see if it is the right solution.

Let us bring down the three inequalities as follows.

$x + y \geq 11$ ....................Inequality (1)

$x - y \leq 5$ ....................Inequality (2)

$7x + 12y \geq 35$ ....................Inequality (3)

Converting them to equations gives:

$x + y = 11$ ....................Equation (1)

$x - y = 5$ ....................Equation (2)

$7x + 12y = 35$ ....................Equation (3)

Let us solve equations (1) and (2) simultaneously as follows:

$x + y = 11$ ....................Equation (1)

$\underline{x - y = 5}$ .................... Equation (2)

Equation (1) – (2):     $2y = 6$

$y = \dfrac{6}{2}$

$y = 3$

Substitute 3 for y in equation (1). This gives:

$x + y = 11$ ....................Equation (1)

$x + 3 = 11$

$x = 11 - 3$

$x = 8$

Hence, x = 8 and y = 3.

Let us now substitute these values in inequality (3), to see if this solution is correct

$7x + 12y \geq 35$ ....................Inequality (3)

$7(8) + 12(3) \geq 35$

$56 + 36 \geq 35$

$92 \geq 35$     (This is correct)

Hence, x = 8 and y = 3 are the correct solutions to this problem.

Note that if we take other pairs of equations and solve simultaneously, we will not obtain feasible solutions.

Therefore the only correct solution is x = 8 and y = 3.

Hence, we now minimize 8x + 5y as follows:

$8x + 5y = 8(8) + 5(3)$

$= 64 + 15$

$= 79$

## Exercise 13

1.  Solve the inequality $9x - 12 > 2 + 2x$. Represent your result on a number line/line graph.

2.  Solve the inequality $5x + 8 < 5x + 2(3 - x)$. Represent your result on a number line.

3. Find the range of values of $x$ for which $\dfrac{3x+1}{4} - \dfrac{3}{5}x \le 11 - 2x$. Show the result on a line graph.

4. Find the range of values of $x$ for which $\dfrac{3}{4}x - \dfrac{2-3x}{2} \ge \dfrac{1}{2} + \dfrac{21}{8}x$. Show the result on a line

5. Given that m is an integer, find the three greatest values which satisfy the inequality:

$$5(7m + 3) < \dfrac{2}{5}(9m - 2)$$

6. Given that $x$ is an integer, find the three lowest values of $x$ which satisfy the following inequality: $3(2x - \dfrac{3}{4}) - \dfrac{1}{2}(x + 3) > \dfrac{1}{3}(5x - 2)$

7. Given that c is an integer, find the three greatest values which satisfy the inequality:

$$\dfrac{3}{10}(3c - 4) < \dfrac{4}{5}(c + 7)$$

8. Given that r is an integer, find the three lowest values of r which satisfy the following inequality:

$$\dfrac{5}{6} - \dfrac{3}{2}(4r - 9) < \dfrac{5}{12}(3 - r)$$

9. What is the range of values of y for which $5y - 4 \le 3y$ and $9y \le 15y + 6$ are both satisfied? Show your result on a graph.

10. Express $7y - 4 < 11 + 2y < 5y + 17$ in the form $a < y < b$, where a and b are both integers. Show the range of values on a line graph.

11. Find the range of values of y for which $4y - 1 > -5$ or $5 \ge y + 12$ are both satisfied. Show your result on a line graph.

12. Express $5m - 9 < 7m - 17 < 23 - m$ in the form $a < y < b$, where a and b are both integers.

13. To employ workers in a farm, a man needs at least 5 drivers and 2 supervisors. He does not want to employ more than 10 workers altogether. A driver needs 2 cars to drive and a supervisor needs 1 car. There are only 15 cars available.
a. Write down four inequalities that represent the restriction on the man.
b. What is the greatest number of supervisors that he can employ
c. Determine the optimum number of drivers and supervisors the man should employ.

14. A shop sells bicycles, costing $200 each, and tricycles, costing $1300 each. The shop owner allows $21,000 to be spent in buying these items and needs at least three times as many bicycles as tricycles. There must be at least 30 bicycles and 10 tricycles.
a. What is the highest number of items that can be bought and in what way?
b. If it is decided to buy as many tricycles as conditions allow, how many of each item can be bought?

15. A boy needs at least five shirts and three trousers. A shirt cost $120 and a trouser cost $150, and the boy has $1600 to spend. The boy decides to spend as much as possible of his $1600.
a. In what way can he spend the maximum of his money?
b. How much is his balance from (a) above?

16.  A paint seller produces two types of tins of paints each containing indigo and cyan. The two types of tins of paints that he makes are Fux tin and Clusion tin. Fux tin contains $320cm^3$ of indigo, Clusion tin contains $80cm^3$ of indigo and the required indigo for a day production should be at least $1680cm^3$. Also, Fux tin contains $8cm^3$ of cyan, Clusion tin contains $6cm^3$ of cyan and the required cyan for a day production should be at least $62cm^3$. A day's production is not more than 20 tins
a.  Write the inequalities to be satisfied by these conditions.
b.  What production quantity should be made for each of the tin?

17.  In a mixing process, there is need to buy a total of 15 mixing machines. A manual mixing machine cost $8000, while an electric mixing machine cost $12000. The process must use at least three of each machine and the money to be spent is no more than $80,000. The profit on each manual mixing machine is not greater than $600, while that on each electric sewing machine is not greater than $500. Find:
a.  the maximum profit the process can make in the business
b.  the number of each mixing machine that should be bought for the process.

18.  A company plans to produce at most 120 units of two products A and B. The cost of producing each unit of A is $60, while the cost of producing each unit of B is $90. After production, the time needed to package each unit of A is 2 hour, while the time needed to package each unit of B is 5 hours. The profit from each unit of A and B are $20 and $40 respectively. The company has a total budget of $8,700 and a packaging period of not more than 400 hours.
a.  What optimal quantity of A and B should the company produce?
b.  What is the maximum profit that the company will make from this production?

19.  Solve the linear program below:
Maximize: $2a + 3b$
Subject to:  $a + b \leq 25$
$\qquad\quad a - b \geq 8$
and    $8a + 6b \leq 180$
Given that, $a \geq 0$ and $b \geq 0$

20.  Solve the linear program:
Minimize $9m + 4n$
Subject to: $m + n \geq 14$
$\qquad\quad m - n \leq 1$
$\qquad 11m + 7n \geq 130$
$\qquad m \geq 0$ and $n \geq 0$

# CHAPTER 14
## QUADRATIC INEQUALITY

If we have a factorized quadratic inequality such as $(x + 2)(x - 1) > 0$, then the product $(x + 2)(x - 1)$ has to be positive ($> 0$). This means that either $(x + 2)$ and $(x - 1)$ are both positive or both negative.

If $(x + 2)(x - 1) < 0$, then the product $(x + 2)(x - 1)$ has to be negative ($< 0$). This means that only $(x + 2)$ is negative or only $(x - 1)$ is negative.

Therefore it is necessary to test the solutions of a quadratic inequality in order to know the correct solution.

However, there is a general rule that applies to quadratic inequalities (and even other inequalities having two solutions). If the coefficient of $x^2$ in a quadratic inequality is a positive value, then of the two values of the solutions obtained, the larger value will take the original sign ($<$ or $>$) of the inequality, while the smaller value will take the reverse sign of the inequality. The following examples illustrate this rule.

**Examples**
1.  Solve the inequality: $x^2 - 2x - 8 < 0$

Solution
$$x^2 - 2x - 8 < 0$$
Let us equate this inequality to zero and solve by factorization.
$$\therefore \quad x^2 - 2x - 8 = 0$$
$$(x - 4)(x + 2) = 0$$
$$\therefore \quad x = 4 \text{ or } x = -2$$
In $x^2 - 2x - 8 < 0$, the coefficient of $x^2$ is positive. Therefore, between the solutions $x = 4$ and $x = -2$, the larger value i.e. 4 will take the sign of the inequality in the question, i.e. $<$. The lower solution i.e. $-2$ will take $>$ (i.e. the reverse sign).
Therefore, the solutions is $x < 4$ or $x > -2$. These can be combined to give:
$$-2 < x < 4$$
Hence, $x^2 - 2x - 8 < 0$ is satisfied when $-2 < x < 4$
We can test these solutions to see if our rule and values obtained are correct. When $x < 4$, a possible value is 3 since $3 < 4$. We now substitute $x = 3$ in the original equation as follows:
$$x^2 - 2x - 8 < 0$$
$$(3)^2 - 2(3) - 8 < 0$$
$$9 - 6 - 8 < 0$$
$$-5 < 0 \quad \text{(This is true)}$$
Hence, $x < 4$ is a correct solution to the inequality.
Similarly, when $x > -2$, a possible value is $-1$ since $-1 > -2$. We now substitute $x = -1$ in the original inequality as follows:
$$x^2 - 2x - 8 < 0$$
$$(-1)^2 - 2(-1) - 8 < 0$$
$$1 + 2 - 8 < 0$$

$-5 < 0$    (This is true)

Hence, $x > -2$ is also a correct solution of the inequality.

2. Solve the inequality $2x^2 + 3x - 5 \geq 0$

<u>Solution</u>

$\qquad 2x^2 + 3x - 5 \geq 0$

Equating to zero and solving by factorization gives:

$\qquad 2x^2 + 3x - 5 = 0$

$\qquad 2x^2 + 5x - 2x - 5 = 0$

$\qquad x(2x + 5) - 1(2x + 5) = 0$

$\therefore \qquad (2x + 5)(x - 1) = 0$

This gives: $x = -\dfrac{5}{2}$ or $x = 1$

Of these two values, the larger solution i.e. 1 will take the inequality sign in the question, while $-\dfrac{5}{2}$ will take the reverse sign. Therefore the solution is $x \geq 1$ or $x \leq -\dfrac{5}{2}$

Hence, $2x^2 + 3x - 5 \geq 0$ is satisfied when $x \geq 1$ or $x \leq -\dfrac{5}{2}$

3. Solve the inequality: $3x^2 - 9x + 4 \leq 0$

<u>Solution</u>

$\qquad 3x^2 - 9x + 4 = 0$    (Equating to zero first)

This equation cannot be factorized. So let's solve it by using quadratic formula as follows:

$\qquad 3x^2 - 9x + 4 = 0$

$\qquad a = 3, b = -9, c = 4$

$\therefore \qquad x = \dfrac{-b \pm \sqrt{b^2 - 4ac}}{2a}$

$\qquad = \dfrac{-(-9) \pm \sqrt{(-9)^2 - (4 \times 3 \times 4)}}{2 \times 3}$

$\qquad = \dfrac{9 \pm \sqrt{81 - 48}}{6}$

$\qquad = \dfrac{9 \pm \sqrt{33}}{6}$

$\qquad x = \dfrac{9 + 5.74}{6}$ or $x = \dfrac{9 - 5.74}{6}$

$\qquad x = \dfrac{14.74}{6}$ or $x = \dfrac{3.26}{6}$

$\therefore \qquad x = 2.46 \quad x = 0.54$

Therefore, $x \leq 2.46$ or $x \geq 0.54$   (Note that the larger value takes the original inequality sign)

Hence, $3x^2 - 9x + 4 \leq 0$ is satisfied when $0.54 \leq x \leq 2.46$

4. Solve the inequality $7 + 6x - x^2 > 0$

### Solution

$7 + 6x - x^2 > 0$     (Note that the coefficient of $x^2$ is negative, i.e. $-1$)

$7 + 6x - x^2 = 0$     (This can be factorized)

$(7 - x)(1 + x) = 0$

$\therefore \quad x = 7 \text{ or } x = -1$

Since the coefficient of $x^2$ in this inequality is negative, then the smaller value of the solution i.e. $-1$ will take the sign of the inequality in the question, while the larger value i.e. 7 will take the reverse sign. This is exactly the opposite of what we do when the coefficient of $x^2$ is positive like in examples 1 to 3.

$\therefore \quad x > -1 \text{ or } x < 7$

Hence, $7 + 6x - x^2 > 0$ is satisfied when $-1 < x < 7$

5. Solve the inequality $5 + 8x - 2x^2 \geq 0$

### Solution

$5 + 8x - 2x^2 \geq 0$

Equating to zero and solving by using quadratic formula gives:

$5 + 8x - 2x^2 = 0$

$a = -2, b = 8, c = 5$

$$\therefore \quad x = \frac{-b \pm \sqrt{b^2 - 4ac}}{2a}$$

$$= \frac{-8 \pm \sqrt{8^2 - (4 \times -2 \times 5)}}{2(-2)}$$

$$= \frac{-8 \pm \sqrt{64 + 40}}{-4}$$

$$= \frac{-8 \pm \sqrt{104}}{-4}$$

$$x = \frac{-8 + 10.2}{-4} \quad \text{or} \quad x = \frac{-8 - 10.2}{-4}$$

$$x = -0.55 \quad \text{or} \quad x = 4.55$$

Since the coefficient of $x^2$ in this inequality is negative, then the smaller value will take the inequality sign in the question.

$\therefore \quad x \geq -0.55 \text{ or } x \leq 4.55$

Hence $5 + 8x - 2x^2 \geq 0$ is satisfied when $-0.55 \leq x \leq 4.55$

### Exercise 14

1. Solve the inequality: $x^2 - x - 30 < 0$
2. Solve the inequality $3x^2 + 10x - 8 \geq 0$
3. Solve the inequality: $5x^2 - 9x + 4 \leq 0$
4. Solve the inequality $9 + 9x - 10x^2 > 0$
5. Solve the inequality $11 + 2x - 2x^2 \geq 0$

6. Solve the inequality: $2x^2 - 5x - 18 < 0$
7. Solve the inequality $7x^2 + x - 6 \geq 0$
8. Solve the inequality: $5x^2 + 17x + 14 \leq 0$
9. Solve the inequality $15 + 2x - x^2 > 0$
10. Solve the inequality $22 + 9x - 4x^2 \geq 0$

# CHAPTER 15
# INTRODUCTORY VECTOR ALGEBRA

## Representation of a Vector

A vector is a quantity which has magnitude and direction. A scalar quantity has only magnitude. A vector can be represented by a line such as AB. This line usually has a corresponding $x$ and y values in the Cartesian plane.

A column vector can be written as $\begin{pmatrix} x \\ y \end{pmatrix}$.

## Modulus or Magnitude of a Vector

The modulus or magnitude of a vector can be determined by using Pythagoras theorem as follows:

$|AB| = \sqrt{x^2 + y^2}$     (where AB is the vector).

A zero or null vector represented as $\begin{pmatrix} 0 \\ 0 \end{pmatrix}$ has a magnitude of zero.

Two vectors are parallel if they have the same direction. They are equal if they have the same magnitude.

**Examples**

1. What is the magnitude of the vector AB = $\begin{pmatrix} 5 \\ 12 \end{pmatrix}$

$$|AB| = \sqrt{5^2 + 12^2}$$
$$= \sqrt{25 + 144}$$
$$= \sqrt{169}$$

$\therefore$ $|AB|$ = 13 units

2. If the magnitude of the vector $\begin{pmatrix} x \\ 8 \end{pmatrix}$ is 17, find the value of $x$.

<u>Solution</u>

$$\text{Magnitude} = \sqrt{x^2 + 8^2}$$
$$17 = \sqrt{x^2 + 64}$$

Square both sides in order to remove the square root sign. This gives:

$$17^2 = (\sqrt{x^2 + 64})^2$$
$$\therefore 289 = x^2 + 64$$
$$x^2 = 289 - 64$$
$$x^2 = 225$$

Take the square root of both sides. This gives:

$$\sqrt{x^2} = \sqrt{225}$$
$$x = 15$$

## Vector Algebra

**Examples**

(1) Given that $p = \begin{pmatrix} -3 \\ 4 \end{pmatrix}$, $q = \begin{pmatrix} -2 \\ -5 \end{pmatrix}$, $r = \begin{pmatrix} 4 \\ 1 \end{pmatrix}$, find:

a. $p - q + r$

b. $2p + q - 3r$

c. $\frac{1}{5}(p + q) - \frac{1}{2}r$

d. $|3p + q - 2r|$

<u>Solution</u>

a. $p - q + r = \begin{pmatrix} -3 \\ 4 \end{pmatrix} - \begin{pmatrix} -2 \\ -5 \end{pmatrix} + \begin{pmatrix} 4 \\ 1 \end{pmatrix}$

$\quad = \begin{pmatrix} -3 - (-2) + 4 \\ 4 - (-5) + 1 \end{pmatrix}$

$\quad = \begin{pmatrix} -3 + 2 + 4 \\ 4 + 5 + 1 \end{pmatrix}$

$\quad = \begin{pmatrix} 3 \\ 10 \end{pmatrix}$

In solving this problem, simply combine the upper numbers separately and the lower numbers separately. Make sure the signs are well organized. For example, where $4 - (-5)$ should be written, should not be written as $4 - 5$.

`

b. $2p + q - 3r = 2\begin{pmatrix} -3 \\ 4 \end{pmatrix} + \begin{pmatrix} -2 \\ -5 \end{pmatrix} - 3\begin{pmatrix} 4 \\ 1 \end{pmatrix}$

$\quad = \begin{pmatrix} -6 \\ 8 \end{pmatrix} + \begin{pmatrix} -2 \\ -5 \end{pmatrix} - \begin{pmatrix} 12 \\ 3 \end{pmatrix}$

$\quad = \begin{pmatrix} -6 - 2 - 12 \\ 8 - 5 - 3 \end{pmatrix}$

$\quad = \begin{pmatrix} -20 \\ 0 \end{pmatrix}$

Note that a number multiplying a bracket is used to multiply each term in the bracket.

c. $\frac{1}{5}(p + q) - \frac{1}{2}r = \frac{1}{5}\left[\begin{pmatrix} -3 \\ 4 \end{pmatrix} + \begin{pmatrix} -2 \\ -5 \end{pmatrix}\right] - \frac{1}{2}\begin{pmatrix} 4 \\ 1 \end{pmatrix}$

$\quad = \frac{1}{5}\begin{pmatrix} -3 - 2 \\ 4 - 5 \end{pmatrix} - \begin{pmatrix} 2 \\ \frac{1}{2} \end{pmatrix}$

$\quad = \frac{1}{5}\begin{pmatrix} -5 \\ -1 \end{pmatrix} - \begin{pmatrix} 2 \\ \frac{1}{2} \end{pmatrix}$

$\quad = \begin{pmatrix} -1 \\ -\frac{1}{5} \end{pmatrix} - \begin{pmatrix} 2 \\ \frac{1}{2} \end{pmatrix}$

$\quad = \begin{pmatrix} -1 - 2 \\ -\frac{1}{5} - \frac{1}{2} \end{pmatrix}$

$\quad = \begin{pmatrix} -3 \\ -\frac{7}{10} \end{pmatrix}$ (Note that $-\frac{1}{5} - \frac{1}{2} = -\frac{7}{10}$)

d. In order to determine $|3p + q - 2r|$, we have to first determine $3p + q - 2r$

$$\therefore \quad 3p + q - 2r = 3\begin{pmatrix} -3 \\ 4 \end{pmatrix} + \begin{pmatrix} -2 \\ -5 \end{pmatrix} - 2\begin{pmatrix} 4 \\ 1 \end{pmatrix}$$

$$= \begin{pmatrix} -9 \\ 12 \end{pmatrix} + \begin{pmatrix} -2 \\ -5 \end{pmatrix} - \begin{pmatrix} 8 \\ 2 \end{pmatrix}$$

$$= \begin{pmatrix} -9 - 2 - 8 \\ 12 - 5 - 2 \end{pmatrix}$$

$$3p + q - 2r = \begin{pmatrix} -19 \\ 5 \end{pmatrix}$$

$$\therefore \quad |3p + q - 2r| = \sqrt{(-19)^2 + 5^2} \quad \text{(Note that we are now calculating the magnitude of } \begin{pmatrix} -19 \\ 5 \end{pmatrix})$$

$$= \sqrt{361 + 25}$$

$$= \sqrt{386}$$

$$= 19.6 \text{ units}$$

2. If $a = \begin{pmatrix} 3 \\ 1 \end{pmatrix}$, $b = \begin{pmatrix} 2 \\ -1 \end{pmatrix}$, $c = \begin{pmatrix} 8 \\ 6 \end{pmatrix}$, find the values of:

a. $m$ and $n$ such that $ma + nb = c$

b. $|d|$ if $d = 2b - a$

<u>Solution</u>

(a) $ma + nb = c$, means that:

$$m\begin{pmatrix} 3 \\ 1 \end{pmatrix} + n\begin{pmatrix} 2 \\ -1 \end{pmatrix} = \begin{pmatrix} 8 \\ 6 \end{pmatrix}$$

$$\begin{pmatrix} 3m \\ m \end{pmatrix} + \begin{pmatrix} 2n \\ -n \end{pmatrix} = \begin{pmatrix} 8 \\ 6 \end{pmatrix}$$

$$\begin{pmatrix} 3m + 2n \\ m - n \end{pmatrix} = \begin{pmatrix} 8 \\ 6 \end{pmatrix}$$

By comparing the upper part of the vectors, and then the lower part, two equations can be obtained as shown below:

$3m + 2n = 8$ .....................Equation (1)

$m - n = 6$ ........................Equation (2)

Equation (2) x 2 gives:

$2m - 2n = 12$ ...................Equation (3)

Bringing equations (1) and (3) together in order to solve them simultaneously gives:

$3m + 2n = 8$ ..................Equation (1)

$\underline{2m - 2n = 12}$ ................Equation (3)

Equation (1) + Equation (3): $5m \quad = 20$ \quad (n has been eliminated, since $+2n + - 2n = 0$

$$m = \frac{20}{5}$$

$$m = 4$$

Substitute 4 for $m$ in equation (2)

$m - n = 6$ ..................Equation (2)

$4 - n = 6$

$4 - 6 = n$

$n = -2$

$\therefore \quad n = -2$ and $m = 4$

b. $d = 2b - a$

$$= 2\begin{pmatrix} 2 \\ -1 \end{pmatrix} - \begin{pmatrix} 3 \\ 1 \end{pmatrix}$$

$$= \begin{pmatrix} 4 \\ -2 \end{pmatrix} - \begin{pmatrix} 3 \\ 1 \end{pmatrix}$$

$$= \begin{pmatrix} 4 - 3 \\ -2 - 1 \end{pmatrix}$$

$$d = \begin{pmatrix} 1 \\ -3 \end{pmatrix}$$

$\therefore$   $|d| = \sqrt{1^2 + (-3)^2}$

$$= \sqrt{1 + 9}$$

$|d| = \sqrt{10}$ units

## Direction of a Vector

If the position vector of a point A is $(x_1 \ y_1)$, and that of point B is $(x_2 \ y_2)$, then the component of the vector AB is given by:

AB $= (x_2 - x_1, \ y_2 - y_1)$

This can be written in column vector as:

$$AB = \begin{pmatrix} x_2 - x_1 \\ y_2 - y_1 \end{pmatrix}$$

If a vector is expressed as CD $= \begin{pmatrix} x \\ y \end{pmatrix}$, then the angle that CD makes with the $x$-axis (horizontal axis) is given by:

Tan $\theta = \dfrac{y}{x}$

$\therefore$   $\theta = \tan^{-1} \dfrac{y}{x}$

### Examples

1. If u $= \begin{pmatrix} 3 \\ 2 \end{pmatrix}$ and v $= \begin{pmatrix} -2 \\ 5 \end{pmatrix}$

a. Find u − 2v

b. Express u in the form (k, $\theta°$) where k is magnitude of u and $\theta$ is the angle u makes with the $x$-axis.

### Solution

a.  u − 2v $= \begin{pmatrix} 3 \\ 2 \end{pmatrix} - 2\begin{pmatrix} -2 \\ 5 \end{pmatrix}$

$$= \begin{pmatrix} 3 \\ 2 \end{pmatrix} - \begin{pmatrix} -4 \\ 10 \end{pmatrix}$$

$$= \begin{pmatrix} 3 - -4 \\ 2 - 10 \end{pmatrix}$$

$$= \begin{pmatrix} 3 + 4 \\ 2 - 10 \end{pmatrix}$$

$$= \begin{pmatrix} 7 \\ -8 \end{pmatrix}$$

b.  u $= \begin{pmatrix} 3 \\ 2 \end{pmatrix}$

The angle that u makes with the $x$ axis is given by:

Tan $\theta = \dfrac{y}{x}$

$Tan\ \theta = \dfrac{2}{3}$    (Note that the upper number of the vector is $x$ while the lower number is y)

$Tan\ \theta = 0.6667$

$\therefore\qquad \theta = tan^{-1}\ 0.6667$

$\theta = 33.7^0$    (From the use of calculator)

Also, $|u| = \sqrt{3^2 + 2^2}$    [Note that u = $\begin{pmatrix} 3 \\ 2 \end{pmatrix}$]

$= \sqrt{9 + 4}$

$= \sqrt{13}$ units

$\therefore$  u in the form (k, $\theta$) is given by: $(\sqrt{13},\ 33.7^0)$

2.  If b = $\begin{pmatrix} -2 \\ 5 \end{pmatrix}$, find the angle $\theta$ that b makes with the horizontal axis. Hence express b in the form (u, $\theta^0$), where u is the magnitude of b.

Solution

b = $\begin{pmatrix} -2 \\ 5 \end{pmatrix}$,

The angle that b makes with the horizontal is given by:

$Tan\ \theta = \dfrac{y}{x}$

$Tan\ \theta = \dfrac{5}{-2}$    (Note that in the vector$\begin{pmatrix} -2 \\ 5 \end{pmatrix}$, –2 represents $x$, while 5 represents y)

$Tan\ \theta = \dfrac{5}{2}$    (Ignore the negative sign of –2)

$\theta = tan^{-1}\ 2.5$

$\therefore\qquad \theta = 68.2^0$

Also, $|b| = u = \sqrt{(-2)^2 + 5^2}$    [Since b = $\begin{pmatrix} -2 \\ 5 \end{pmatrix}$]

$= \sqrt{4 + 25}$

$u = \sqrt{29}$

$\therefore$  Expressing b in the form (u, $\theta^0$), gives: $(\sqrt{29},\ 68.2^0)$

3.  The vertices of a quadrilateral WXYZ are W(2, –1), X(–1, 3), Y(–2, 1), Z(–3, –2). Determine the components and lengths of:

a.  WX b.  WZ c.  –2XY d.  YW

Solutions

a.  With respect to the origin, each of the points can be expressed as:

W = $\begin{pmatrix} 2 \\ -1 \end{pmatrix}$, X = $\begin{pmatrix} -1 \\ 3 \end{pmatrix}$, Y = $\begin{pmatrix} -2 \\ 1 \end{pmatrix}$, Z = $\begin{pmatrix} -3 \\ -2 \end{pmatrix}$

$\therefore$  WX = X – W    (Take note of the arrangement of the letters, as the second letter comes first)

$= \begin{pmatrix} -1 \\ 3 \end{pmatrix} - \begin{pmatrix} 2 \\ -1 \end{pmatrix}$

$$= \begin{pmatrix} -1 - 2 \\ 3 - -1 \end{pmatrix}$$

$$= \begin{pmatrix} -1 - 2 \\ 3 + 1 \end{pmatrix}$$

$\therefore$ WX = $\begin{pmatrix} -3 \\ 4 \end{pmatrix}$

Also, the length of WX is the magnitude of WX. It is given by:

$$|WX| = \sqrt{(-3)^2 + 4^2}$$
$$= \sqrt{9 + 16}$$
$$= \sqrt{25}$$
$$= 5$$

$\therefore$ |WX| = 5 units

b. Similarly, WZ = Z – W    (Take note of the arrangement of the letters, as the second letter comes first in the subtraction)

$$WZ = \begin{pmatrix} -3 \\ -2 \end{pmatrix} - \begin{pmatrix} 2 \\ -1 \end{pmatrix}$$

$$= \begin{pmatrix} -3 - 2 \\ -2 - -1 \end{pmatrix}$$

$$= \begin{pmatrix} -3 - 2 \\ -2 + 1 \end{pmatrix}$$

$\therefore$ WZ = $\begin{pmatrix} -5 \\ -1 \end{pmatrix}$

Also, the length of WZ is the magnitude of WZ. It is given by:

$$|WZ| = \sqrt{(-5)^2 + (-1)^2}$$
$$= \sqrt{25 + 1}$$

|WZ| = $\sqrt{26}$ units

c.  −2XY = −2( Y − X)

$$= -2\left[\begin{pmatrix} -2 \\ 1 \end{pmatrix} - \begin{pmatrix} -1 \\ 3 \end{pmatrix}\right]$$
$$= -2\left[\begin{pmatrix} -2 - -1 \\ 1 - 3 \end{pmatrix}\right]$$
$$= -2\begin{pmatrix} -2 + 1 \\ -2 \end{pmatrix}$$
$$= -2\begin{pmatrix} -1 \\ -2 \end{pmatrix}$$
$$= \begin{pmatrix} 2 \\ 4 \end{pmatrix} \quad \text{(Take note of the change in sign due to the multiplication by −2)}$$

$\therefore$   −2XY = $\begin{pmatrix} 2 \\ 4 \end{pmatrix}$

Also, the length of −2XY is its magnitude. It is given by:

$$\therefore |{-2XY}| = \sqrt{2^2 + 4^2}$$
$$= \sqrt{4 + 16}$$
$$= \sqrt{20}$$

$\therefore$ $|-2XY| = \sqrt{20}$ units

d. $YW = W - Y$

$$YW = \begin{pmatrix} 2 \\ -1 \end{pmatrix} - \begin{pmatrix} -2 \\ 1 \end{pmatrix}$$

$$= \begin{pmatrix} 2 - -2 \\ -1 - 1 \end{pmatrix}$$

$$= \begin{pmatrix} 2 + 2 \\ -1 - 1 \end{pmatrix}$$

$\therefore$ $YW = \begin{pmatrix} 4 \\ -2 \end{pmatrix}$

Also, $|YW| = \sqrt{4^2 + (-2)^2}$

$$= \sqrt{16 + 4}$$

$\therefore$ $|YW| = \sqrt{20}$ units

4. A(4, 7) is the vertex of a triangle ABC. $BA = \begin{pmatrix} 5 \\ 3 \end{pmatrix}$ and $AC = \begin{pmatrix} 4 \\ -3 \end{pmatrix}$.

a. Find the coordinates of B and C

b. If M is the midpoint of the line BC, find AM.

Solution

a. With respect to the origin, $A = \begin{pmatrix} 4 \\ 7 \end{pmatrix}$

Let the column vector B be, $B = \begin{pmatrix} d \\ e \end{pmatrix}$

$\therefore$ The component, $BA = A - B$.   Substitute each value into this equation. This gives:

$$BA = A - B$$

$$\begin{pmatrix} 5 \\ 3 \end{pmatrix} = \begin{pmatrix} 4 \\ 7 \end{pmatrix} - \begin{pmatrix} d \\ e \end{pmatrix}$$

$$\begin{pmatrix} 5 \\ 3 \end{pmatrix} = \begin{pmatrix} 4 - d \\ 7 - e \end{pmatrix}$$

$\therefore$ $\quad 5 = 4 - d$   (By equating the upper parts)

$\quad d = 4 - 5$

$\quad d = -1$

Similarly, $3 = 7 - e$   (By equating the lower parts)

$\quad e = 7 - 3$

$\quad e = 4$

Hence, $B = \begin{pmatrix} -1 \\ 4 \end{pmatrix}$

$\therefore$ The coordinates of B are (−1, 4)

Also, let the column vector C be, $C = \begin{pmatrix} f \\ g \end{pmatrix}$

$\therefore$ The component:

$$AC = C - A$$

$$\begin{pmatrix} 4 \\ -3 \end{pmatrix} = \begin{pmatrix} f \\ g \end{pmatrix} - \begin{pmatrix} 4 \\ 7 \end{pmatrix}$$

$$\begin{pmatrix} 4 \\ -3 \end{pmatrix} = \begin{pmatrix} f - 4 \\ g - 7 \end{pmatrix}$$

$\therefore$ $\quad 4 = f - 4$   (By equating the upper parts)

183

$$4 + 4 = f$$
$$f = 8$$

Similarly, $-3 = g - 7$   (By equating the lower parts)
$$-3 + 7 = g$$
$$g = 4$$

Hence, $C = \begin{pmatrix} 8 \\ 4 \end{pmatrix}$

$\therefore$ The coordinates of C are (8, 4)

b. The midpoint of BC is $M = \frac{1}{2}(B + C)$

$$M = \frac{1}{2}\left[\begin{pmatrix} -1 \\ 4 \end{pmatrix} + \begin{pmatrix} 8 \\ 4 \end{pmatrix}\right]$$
$$= \frac{1}{2}\left[\begin{pmatrix} -1 + 8 \\ 4 + 4 \end{pmatrix}\right]$$
$$= \frac{1}{2}\begin{pmatrix} 7 \\ 8 \end{pmatrix}$$
$$M = \begin{pmatrix} \frac{7}{2} \\ 4 \end{pmatrix}$$

$\therefore$   $AM = M - A$

$$= \begin{pmatrix} \frac{7}{2} \\ 4 \end{pmatrix} - \begin{pmatrix} 4 \\ 7 \end{pmatrix}$$
$$= \begin{pmatrix} \frac{7}{2} - 4 \\ 4 - 7 \end{pmatrix}$$
$$AM = \begin{pmatrix} -\frac{1}{2} \\ -3 \end{pmatrix}$$

## Exercise 15

1. What is the magnitude of the vector $AB = \begin{pmatrix} 8 \\ 6 \end{pmatrix}$

2. If the magnitude of the vector $\begin{pmatrix} x \\ 5 \end{pmatrix}$ is 13, find the value of $x$.

3. Given that $p = \begin{pmatrix} -6 \\ 10 \end{pmatrix}$, $q = \begin{pmatrix} -3 \\ 1 \end{pmatrix}$, $r = \begin{pmatrix} 2 \\ -8 \end{pmatrix}$, find:

a. $p + q - 2r$

b. $p - q + r$

c. $\frac{2}{3}(p - q) - \frac{1}{4}r$

d. $|2p + 5q - 3r|$

4. If $a = \begin{pmatrix} 5 \\ 2 \end{pmatrix}$, $b = \begin{pmatrix} 3 \\ -8 \end{pmatrix}$, $c = \begin{pmatrix} -4 \\ -3 \end{pmatrix}$, find the values of:

a. p and q such that $pa - qb = 2c$

b. $|m|$ if $m = 5a - 3b$

184

5. If u = $\begin{pmatrix} 1 \\ 7 \end{pmatrix}$ and v = $\begin{pmatrix} 3 \\ -2 \end{pmatrix}$

a. Find 4u − 5v

b. Express u in the form (K, $\theta°$) where k is magnitude of u and $\theta$ is the angle u makes with the x− axis.

6. If n = $\begin{pmatrix} 6 \\ -4 \end{pmatrix}$, find the angle $\theta$ that n makes with the horizontal axis. Hence express n in the form (u, $\theta°$), where u is the magnitude of n.

7. The vertices of a quadrilateral ABCD are A(3, −2), B(1, 2), C(−1, 3), D(−3, −4). Determine the components and lengths of:

a. AB  b. AD  c. −3BC  d. BD

8. P(2, 5) is the vertex of a triangle PQR. QP = $\begin{pmatrix} 3 \\ -1 \end{pmatrix}$ and PR = $\begin{pmatrix} 6 \\ 4 \end{pmatrix}$.

a. Find the coordinates of Q and R

b. If M is the midpoint of the line QR, find PM.

9. What is the magnitude of the vector AB = $\begin{pmatrix} 24 \\ 17 \end{pmatrix}$

10. If the magnitude of the vector $\begin{pmatrix} x \\ 14 \end{pmatrix}$ is 20, find the value of x.

11. Given that p = $\begin{pmatrix} -1 \\ 5 \end{pmatrix}$, q = $\begin{pmatrix} -4 \\ 7 \end{pmatrix}$, r = $\begin{pmatrix} 2 \\ 3 \end{pmatrix}$, find:

a. 2p − 3q + r

b. p + q − r

c. $\frac{1}{2}$( p + 2q) − $\frac{1}{4}$r

d. |5p + 3q − 4r|

12. If a = $\begin{pmatrix} 2 \\ 5 \end{pmatrix}$, b = $\begin{pmatrix} 3 \\ -3 \end{pmatrix}$, c = $\begin{pmatrix} 5 \\ 6 \end{pmatrix}$, find the values of:

a. p and q such that pa + qb = 2c

b. |d| if d = 2a − 3b

13. The vertices of a quadrilateral PQRS are P(3, −2), Q(−5, 1), R(−1, 3), S(−2, −4). Determine the components and lengths of:

a. PR  b. RQ  c. −5QS  d. 2PQ

# CHAPTER 16
## LIMIT OF A FUNCTION

The concept of limits is very important in differential calculus. A function has a limiting value when its variable approaches a certain value. For example, if the limiting value of f(x) as $x$ approaches 4 is 16, it is written as:

$$\lim_{x \to 4} f(x) = 16$$

In evaluating limit, some problems are simply solved by just putting in the values of the variables, while some problems are solved by applying certain rules.

**Some important limits**

1. $\lim\limits_{x \to 0} \dfrac{\sin x}{x} = 1$ and $\lim\limits_{x \to 0} \dfrac{\tan x}{x} = 1$ (Or $\lim\limits_{x \to 0} \dfrac{\sin ax}{ax} = 1$ and $\lim\limits_{x \to 0} \dfrac{\tan ax}{ax} = 1$)

2. $\lim\limits_{x \to 0} \dfrac{1 - \cos x}{x} = 0$ and $\lim\limits_{x \to 0} \dfrac{1 - \cos a x}{x} = 0$

3. $\lim\limits_{x \to \infty} \left(1 + \dfrac{1}{x}\right)^x = e$

4. $\lim\limits_{x \to 0} \dfrac{a^x - 1}{a^x} = 1$

5. $\lim\limits_{x \to 0} \dfrac{e^x - 1}{x} = 1$

6. $\lim\limits_{x \to 0} \dfrac{1 + x}{x} = 1$

7. $\lim\limits_{x \to a} \dfrac{x^n - a^n}{x - a} = na^{n-1}$

8. $\lim\limits_{x \to 0} \dfrac{a^x - 1}{x} = \ln a$

9. $\lim\limits_{x \to \infty} \dfrac{\ln x}{x} = 0$

10. $\lim\limits_{x \to \infty} x^{\frac{1}{x}} = 0$

11. $\lim\limits_{x \to 0} (1 + x)^{\frac{1}{x}} = e$

12. $\lim\limits_{x \to 0} (1 + \sin x)^{\frac{1}{x}} = e$

13. $\lim\limits_{x \to a} c = c$

**Examples**

1. Evaluate $\lim_{x \to 0} 3x^3 - 5x^2 + 2x + 6$

Solution

$\lim_{x \to 0} 3x^3 - 5x^2 + 2x + 6$

Simply substitute zero for $x$ in the given expression. This gives:

$= 3(0^3) - 5(0^2) + 2(0) + 6$

$= 0 - 0 + 0 + 6$

$= 6$

2. Evaluate $\lim_{x \to 0} \dfrac{x^2 + 3x - 11}{4x^2 - 5x - 5}$

Solution

$\lim_{x \to 0} \dfrac{x^2 + 3x - 11}{4x^2 - 5x - 5}$

Substituting 0 for $x$ into the expression gives:

$= \dfrac{0^2 + 3(0) - 11}{4(0)^2 - 5(0) - 5}$

$= \dfrac{0 + 0 - 11}{0 - 0 - 5}$

$= \dfrac{-11}{-5}$

$= \dfrac{11}{5}$

$= 2\dfrac{1}{5}$

3. Evaluate $\lim_{x \to 4} \dfrac{x^2 - 2x - 8}{x - 4}$

Solution

A close look at the expression shows that if 4 is substituted for $x$, it will give $\dfrac{0}{0}$. This has no value as it is indeterminate. Therefore, in order to solve a limit such as this, we have to factorize the denominator and then simplify the expression. This is done as follows:

$\lim_{x \to 4} \dfrac{x^2 - 2x - 8}{x - 4} = \lim_{x \to 4} \dfrac{(x+2)(x-4)}{x - 4}$

Cancelling out $(x - 4)$ gives:

$\lim_{x \to 4} (x + 2)$

We now substitute 4 for $x$ to obtain our final answer as follows:

$\lim_{x \to 4} (x + 2) = 4 + 2$

$= 6$

Therefore, $\lim_{x \to 4} \dfrac{x^2 + x - 8}{x - 4} = 6$

4. Evaluate $\displaystyle\lim_{x\to2} \frac{x^2-4}{x-2}$

<u>Solution</u>

This is similar to example 3 above since substituting 2 for $x$ in the expression will $\frac{0}{0}$ which has no value. Hence we factorize the numerator and simplify as follows:

$\displaystyle\lim_{x\to2} \frac{x^2-4}{x-2} = \lim_{x\to2} \frac{(x-2)(x+2)}{x-2}$   [Note that $x^2-4 = x^2-2^2$ and recall that $a^2-b^2 = (a+b)(a-b)$]

Cancelling out $(x-2)$ gives:

$\displaystyle\lim_{x\to2} (x+2)$

We now substitute 2 for $x$ to obtain our final answer as follows:

$\displaystyle\lim_{x\to2} (x+2) = 2+2$

$= 4$

Therefore, $\displaystyle\lim_{x\to2} \frac{x^2-4}{x-2} = 4$

5. Evaluate $\displaystyle\lim_{x\to0} (2x-5)(x+2)(3x-1)$

<u>Solution</u>

$\displaystyle\lim_{x\to0} (2x-5)(x+2)(3x-1)$

Substituting 0 for $x$ in the expression gives:

$\displaystyle\lim_{x\to0} (2x-5)(x+2)(3x-1) = [2(0)-5][(0)+2][3(0)-1]$

$= (-5)(2)(-1)$

$= 10$

6. Evaluate $\displaystyle\lim_{x\to\infty} \frac{x^3+2x^2+5x+3}{x^3+x^2+3x+7}$

<u>Solution</u>

In this case of $x$ approaching infinity, we evaluate it by first dividing each term in the numerator and denominator by the variable having the highest exponent (power). Hence we divide each term by $x^3$. This is done as follows:

$\displaystyle\lim_{x\to\infty} \frac{x^3+2x^2+5x+3}{x^3+x^2+3x+7} = \lim_{x\to\infty} \frac{\frac{x^3}{x^3}+\frac{2x^2}{x^3}+\frac{5x}{x^3}+\frac{3}{x^3}}{\frac{x^3}{x^3}+\frac{x^2}{x^3}+\frac{3x}{x^3}+\frac{7}{x^3}}$

$= \displaystyle\lim_{x\to\infty} \frac{1+\frac{2}{x}+\frac{5}{x^2}+\frac{3}{x^3}}{1+\frac{1}{x}+\frac{3}{x^2}+\frac{7}{x^3}}$

$= \dfrac{1+\frac{2}{\infty}+\frac{5}{\infty^2}+\frac{3}{\infty^3}}{1+\frac{1}{\infty}+\frac{3}{\infty^2}+\frac{7}{\infty^3}}$

$$= \frac{1+0+0+0}{1+0+0+0}$$  (Note that a number divided by $\infty$ gives 0)

$$= \frac{1}{1}$$

$$= 1$$

7. Evaluate $\displaystyle\lim_{x \to \infty} \frac{2x^4 - 3x^2 + 8x - 1}{x^4 - 5x + 3}$

<u>Solution</u>

The variable in its highest exponent (power) is $x^4$. Hence we divide each term by $x^4$ as follows:

$$\lim_{x \to \infty} \frac{2x^4 - 3x^2 + 8x - 1}{x^4 - 5x + 3} = \lim_{x \to \infty} \frac{\frac{2x^4}{x^4} - \frac{3x^2}{x^4} + \frac{8x}{x^4} - \frac{1}{x^4}}{\frac{x^4}{x^4} - \frac{5x}{x^4} + \frac{3}{x^4}}$$

$$= \lim_{x \to \infty} \frac{2 - \frac{3}{x^2} + \frac{8}{x^3} - \frac{1}{x^4}}{1 - \frac{5}{x^3} + \frac{3}{x^4}}$$

$$= \frac{2 - \frac{3}{\infty} + \frac{8}{\infty} - \frac{1}{\infty}}{1 - \frac{5}{\infty} + \frac{3}{\infty}}$$

$$= \frac{2 - 0 + 0 - 0}{1 - 0 + 0}$$  (Note that a number divided by $\infty$ gives 0)

$$= \frac{2}{1}$$

$$= 2$$

8. Evaluate $\displaystyle\lim_{x \to 4} \frac{2x^2 - 7x - 4}{x^2 - 3x - 4}$

<u>Solution</u>

If 4 is substituted for $x$ in the function above, it will give $\frac{0}{0}$ which is an indeterminate value.

Hence we factorize the expression and simplify as follows:

$$\lim_{x \to 4} \frac{2x^2 - 7x - 4}{x^2 - 3x - 4} = \lim_{x \to 4} \frac{(2x+1)(x-4)}{(x+1)(x-4)}$$

Cancelling out $(x - 4)$ gives:

$$\lim_{x \to 4} \frac{2x + 1}{x + 1}$$

We now substitute 4 for $x$ to obtain our final answer as follows:

$$\lim_{x \to 4} \frac{2x + 1}{x + 1} = \frac{2(4) + 1}{(4) + 1}$$

$$= \frac{8+1}{4+1}$$

$$= \frac{9}{5}$$

$$= 1\frac{4}{5}$$

Therefore, $\lim\limits_{x\to 4} \dfrac{2x^2 - 7x - 4}{x^2 - 3x - 4} = 1\frac{4}{5}$

9. Evaluate $\lim\limits_{x\to 3} \dfrac{x^2 - 9}{x^2 + 6x + 9}$

Solution

Substituting 3 for $x$ in the function can give us the final answer as follows:

$$\lim\limits_{x\to 3} \frac{x^2 - 9}{x^2 + 6x + 9} = \frac{3^2 - 9}{3^2 + 6(3) + 9}$$

$$= \frac{9 - 9}{9 + 18 + 9}$$

$$= \frac{0}{36}$$

$$= 0$$

Hence, $\lim\limits_{x\to 3} \dfrac{x^2 - 9}{x^2 + 6x + 9} = 0$

Note that even if this expression had been factorized and simplified, the final answer would still give zero.

10. Determine the limiting value of $\dfrac{x + 2}{x^2 + 6x + 8}$ as $x$ tends to $-2$ .

Solution

The question can also be written as $\lim\limits_{x\to -2} \dfrac{x + 2}{x^2 + 6x + 8}$

If $-2$ is substituted for $x$ in the function above, it will give $\dfrac{0}{0}$ . Hence we factorize the expression and simplify as follows:

$$\lim\limits_{x\to -2} \frac{x + 2}{x^2 + 6x + 8} = \lim\limits_{x\to -2} \frac{x + 2}{(x + 2)(x + 4)}$$

Cancelling out $(x + 2)$ gives:

$$\lim\limits_{x\to -2} \frac{1}{x + 4}$$

We now substitute $-2$ for $x$ as follows:

$$\lim\limits_{x\to -2} \frac{1}{x + 4} = \frac{1}{-2 + 4}$$

$$= \frac{1}{2}$$

11. Find the limit of $\dfrac{x^3 - 125}{x - 5}$ as $x \to 5$

Solution

The question can also be written as $\displaystyle\lim_{x \to 5} \dfrac{x^3 - 125}{x - 5}$

We factorize the expression and simplify as follows:

$$\lim_{x \to 5} \dfrac{x^3 - 125}{x - 5} = \lim_{x \to 5} \dfrac{x^3 - 5^3}{x - 5}$$

$$= \lim_{x \to 5} \dfrac{(x - 5)(x^2 + 5x + 5^2)}{x - 5} \quad [\text{Recall the identity } a^3 - b^3 = (a - b)(a^2 + ab + b^2)]$$

Cancelling out $(x - 5)$ gives:

$$\lim_{x \to 5} (x^2 + 5x + 5^2)$$

We now substitute 5 for $x$ as follows:

$$\lim_{x \to 5} (x^2 + 5x + 5^2) = (5^2 + 5(5) + 25)$$

$$= 25 + 25 + 25$$

$$= 75$$

12. Find the limit of $\dfrac{x^3 + 64}{x + 4}$ as $x \to -4$

Solution

The question can also be written as $\displaystyle\lim_{x \to -4} \dfrac{x^3 + 4^3}{x + 4}$

We factorize the expression and simplify as follows:

$$\lim_{x \to -4} \dfrac{x^3 + 64}{x + 4} = \lim_{x \to -4} \dfrac{x^3 + 4^3}{x + 4}$$

$$= \lim_{x \to -4} \dfrac{(x + 4)(x^2 - 4x + 4^2)}{x + 4} \quad [\text{Recall the identity } a^3 + b^3 = (a + b)(a^2 - ab + b^2)]$$

Cancelling out $(x + 4)$ gives:

$$\lim_{x \to -4} (x^2 - 4x + 4^2)$$

We now substitute $-4$ for $x$ as follows:

$$\lim_{x \to -4} (x^2 - 4x + 4^2) = (-4^2 - 4(-4) + 16)$$

$$= 16 + 16 + 16$$

$$= 48$$

13. Evaluate $\displaystyle\lim_{x \to 9} \dfrac{3 - \sqrt{x}}{9 - x}$

Solution

If 9 is substituted for $x$ in the expression above, it will give $\dfrac{0}{0}$ which is not the desired answer.

Hence a way to go around this problem is to multiply the top and bottom by the conjugate of the surd at the top. The conjugate of $3 - \sqrt{x}$ is $3 + \sqrt{x}$ (only a difference in their middle signs).

Hence we multiply top and bottom by $3 + \sqrt{x}$ and simplify as follows:

$$\lim_{x \to 9} \frac{3 - \sqrt{x}}{9 - x} = \lim_{x \to 9} \frac{(3 - \sqrt{x})(3 + \sqrt{x})}{(9 - x)(3 + \sqrt{x})}$$

$$= \lim_{x \to 9} \frac{3^2 - (\sqrt{x})^2}{(9 - x)(3 + \sqrt{x})} \quad \text{(Note that the top was simplified by using } (a - b)(a + b) = a^2 - b^2)$$

$$= \lim_{x \to 9} \frac{9 - x}{(9 - x)(3 + \sqrt{x})}$$

$$= \lim_{x \to 9} \frac{1}{(3 + \sqrt{x})} \quad \text{(Since } 9 - x \text{ cancels out)}$$

We now substitute 9 for $x$ as follows:

$$\lim_{x \to 9} \frac{1}{(3 + \sqrt{x})} = \frac{1}{(3 + \sqrt{9})}$$

$$= \frac{1}{3 + 3}$$

$$= \frac{1}{6}$$

14. Evaluate $\lim_{x \to 1} \dfrac{x^2 - 1}{x^5 - 1}$

Solution

If 2 is substituted for $x$ in the expression above, it will give $\dfrac{0}{0}$ which is not a good answer. Hence a way to go around this problem is to divide the top and bottom by $x - 1$ (i.e. the difference between the variable and 1). This is done as follows:

$$\lim_{x \to 1} \frac{x^2 - 1}{x^5 - 1} = \lim_{x \to 1} \frac{\dfrac{x^2 - 1}{x - 1}}{\dfrac{x^5 - 1}{x - 1}} \quad \text{(Note that this has not changed the original fraction)}$$

Note that $\lim_{x \to a} \dfrac{x^n - a^n}{x - a} = na^{n-1}$. With this rule we now simplify the expression above as follows:

$$\lim_{x \to 1} \frac{\dfrac{x^2 - 1}{x - 1}}{\dfrac{x^5 - 1}{x - 1}} = \frac{2 \times 1^{(2-1)}}{5 \times 1^{(5-1)}}$$

$$= \frac{2 \times 1^1}{5 \times 1^5}$$

$$= \frac{2}{5}$$

**Continuity of a function**

A function is said to be continuous if the three conditions below are satisfied.

1. $f(a)$ exists

2. $\underset{x \to a}{lim}$ f(x) exists

3. $\underset{x \to a}{lim}$ f(x) = f(a)    (i.e. if the values of conditions 1 and 2 above are equal)

**Examples**

1. Determine if the function f(x) = $2x^2 - 8x + 5$ is continuous at the point $x = -1$

Solution

When $x = -1$, the f(-1) is obtained as follows:

$$f(x) = 2x^2 - 8x + 5$$
$$f(-1) = 2(-1)^2 - 8(-1) + 5$$
$$= 2 + 8 + 5$$
$$= 15$$

Since f(-1) = 15, it means that f(-1) exists.

The next step is to find the value of $\underset{x \to -1}{lim}$ f(x) as follows:

$$\underset{x \to -1}{lim} \ 2x^2 - 8x + 5 = 2(-1)^2 - 8(-1) + 5$$
$$= 2 + 8 + 5$$
$$= 15$$

Hence f(-1) = $\underset{x \to -1}{lim}$ f(x) = 15

Therefore, the function is continuous.


2. Determine if the function f(x) = $\dfrac{2x + 1}{x^2 - 3x + 7}$ is continuous at $x = 2$

Solution

When $x = 2$, the f(2) is obtained as follows:

$$f(x) = \frac{2x + 1}{x^2 - 3x + 7}$$
$$f(2) = \frac{2(2) + 1}{2^2 - 3(2) + 7}$$
$$= \frac{4 + 1}{4 - 6 + 7}$$
$$= \frac{5}{5}$$
$$= 1$$

Since f(2) = 1, it means that f(2) exists.

The next step is to find the value of $\underset{x \to 2}{lim}$ f(x) as follows:

$$\underset{x \to 2}{lim} \ \frac{2x + 1}{x^2 - 3x + 7} = \frac{2(2) + 1}{2^2 - 3(2) + 7}$$
$$= \frac{4 + 1}{4 - 6 + 7}$$
$$= \frac{5}{5}$$

$$= 1$$

Hence $f(2) = \underset{x \to 2}{lim} f(x) = 1$

Therefore, the function is continuous.

3. Determine if the function $f(x) = \dfrac{2x^2 - 8}{x + 4}$ is continuous at $x = -4$.

When $x = -4$, the $f(-4)$ is obtained as follows:

$$f(x) = \frac{2x^2 - 8}{x + 4}$$

$$f(-4) = \frac{2(-4^2) - 8}{-4 + 4}$$

$$= \frac{2(16) - 8}{0}$$

$$= \frac{24}{0} \quad \text{(Undefined)}$$

Hence $f(-4)$ is undefined, and does not exist. This shows that the function is not continuous. It is discontinuous as $x = -4$.

4. Determine if the function $f(x) = \dfrac{x^2 - 9}{x - 3}$ is continuous at $x = 3$.

When $x = 3$, the $f(3)$ is obtained as follows:

$$f(x) = \frac{x^2 - 9}{x - 3}$$

$$f(3) = \frac{3^2 - 9}{3 - 3}$$

$$= \frac{0}{0} \quad \text{(This has no value)}$$

Hence $f(3)$ does not exist.

Let us determine the value of $\underset{x \to 3}{lim} f(x)$ as follows:

$$\underset{x \to 3}{lim} \frac{x^2 - 9}{x - 3} = \underset{x \to 3}{lim} \frac{(x - 3)(x + 3)}{x - 3}$$

$$= \underset{x \to 3}{lim} (x + 3) \quad \text{(Since } x - 3 \text{ cancels out)}$$

$$= 3 + 3 \quad \text{(When 3 is substituted for } x\text{)}$$

$$= 6$$

Hence, $f(3)$ has no value while $\underset{x \to 3}{lim} f(x) = 6$

This shows that $f(3) \neq \underset{x \to 3}{lim} f(x)$

Therefore the function discontinuous.

5. Test for the continuity of the function: $f(x) = \begin{cases} -5x + 7 & \text{if } x > 1 \\ x^2 + 2 & \text{if } x \leq 1 \end{cases}$

Solution

$$f(x) = \begin{cases} -5x + 7 & \text{if } x > 1 \\ x^2 + 2 & \text{if } x \leq 1 \end{cases}$$

194

When $x = 1$, we use the expression $x^2 + 2$ to evaluate f(1). Note that $x \leq 1$ means $x < 1$ or $x = 1$.

Hence, $f(x) = x^2 + 2$    (for $x = 1$)

$\quad\quad f(1) = 1^2 + 2$

$\quad\quad\quad\quad = 3$

Since f(1) = 3, it means that f(1) exist.

Let us now determine the limiting value of f(x).

If $x$ approaches 1 from the left, which is $x \leq 1$ (left hand elbow) we write it as $\lim\limits_{x \to 1-} f(x)$, and we pick the expression given by:

$\quad f(x) = x^2 + 2$

Hence, $\lim\limits_{x \to 1} f(x) = 1^2 + 3$

$\quad\quad\quad\quad = 3$

If $x$ approaches 1 from the right, which is $x > 1$ (right hand elbow), we write it as $\lim\limits_{x \to 1+} f(x)$, and we pick the expression given by:

$\quad f(x) = -5x + 7$

Hence, $\lim\limits_{x \to 1} f(x) = -5(1) + 7$

$\quad\quad\quad\quad = 2$

Hence, $\lim\limits_{x \to 1-} f(x) = 3$ while $\lim\limits_{x \to 1+} f(x) = 2$. They are not equal. This shows that $\lim\limits_{x \to 1} f(x)$ does not exist. Therefore, the function is not continuous.


6. Determine the continuity of the function: $f(x) = \begin{cases} x^2 & \text{if } x \geq -2 \\ x+6 & \text{if } x < -2 \end{cases}$

Solution

$\quad f(x) = \begin{cases} x^2 & \text{if } x \geq -2 \\ x+6 & \text{if } x < -2 \end{cases}$

A direct way of determining the continuity of functions such as these is to find the limits of the two expressions in the function as $x$ tends to the values given in the question.

Therefore, when $x$ tends to $-2$ in the function $f(x) = x^2$, we have:

$\quad f(x) = x^2$

Hence, $\lim\limits_{x \to -2} f(x) = (-2)^2$

$\quad\quad\quad\quad = 4$

Also, when $x$ tends to $-2$ in the function $f(x) = x + 6$, we have:

$\quad f(x) = x + 6$

Hence, $\lim\limits_{x \to -2} f(x) = -2 + 6$

$\quad\quad\quad\quad = 4$

Hence, $\lim\limits_{x \to -2} x^2 = \lim\limits_{x \to -2} x + 6 = 4$. Their limits give the same value. This shows that $\lim\limits_{x \to -2} f(x)$ exist. Therefore, the function is continuous.

7. Determine if the function below is continuous or not.

$$f(x) = \begin{cases} \dfrac{x^2 - 16}{x + 4} & \text{if } x \geq -4 \\[2mm] \dfrac{x + 4}{2x - 3} & \text{if } x < -4 \end{cases}$$

Solution

As $x$ tends to $-4$ in the function $f(x) = \dfrac{x^2 - 16}{x + 4}$ we simplify as follows:

$$f(x) = \frac{(x + 4)(x-4)}{x + 4}$$

Hence, $\displaystyle\lim_{x \to -4} f(x) = (x - 4)$     (Note that $(x + 4)$ cancels out)

$$= -4 - 4$$
$$= -8$$

Also, when $x$ tends to $-4$ in the function $f(x) = \dfrac{x + 4}{2x - 3}$, we have:

$$\lim_{x \to -4} f(x) = \frac{-4 + 4}{2(-4) - 3}$$

$$= \frac{-4 + 4}{-8 - 3}$$

$$= \frac{0}{-11}$$

$$= 0$$

Their limits give different values. This shows that $\displaystyle\lim_{x \to -4} f(x)$ does not exist. Therefore, the function is not continuous.

8. If $f(x) = \begin{cases} \dfrac{x^2 - 4}{x - 2} & \\[2mm] 4 & \text{if } x = 2 \end{cases}$     test for the continuity of the function at $x = 2$.

Solution

$$f(x) = \begin{cases} \dfrac{x^2 - 4}{x - 2} & \\[2mm] 4 & \text{if } x = 2 \end{cases}$$

Let us find the limits of the two expressions in the function as $x$ tends 2.

From the first expression, as $x$ tends to 2 we have:

$$f(x) = \frac{x^2 - 4}{x - 2}$$

Hence, $\displaystyle\lim_{x \to 2} f(x) = \dfrac{(x-2)(x+2)}{x - 2}$     (When the numerator is factorized)

$$\lim_{x \to 2} f(x) = (x + 2)     (x - 2 \text{ has cancelled out})$$

$$\lim_{x \to 2} f(x) = 2 + 2$$

$$= 4$$

Also, when $x$ tends to 2 in the function $f(x) = 4$, we have:

$$f(x) = 4$$

Hence, $\lim\limits_{x\to 2} 4 = 4$    (Since $\lim\limits_{x\to a} c = c$)

The question also tells us that when $x = 2$, f($x$) = 4

Hence, $\lim\limits_{x\to 2} 4 = \lim\limits_{x\to 2} \dfrac{x^2-4}{x-2} = 4$. Their limits give the same value of 4. This shows that f($x$) exist,

and that f(2) = $\lim\limits_{x\to 2}$ f($x$). Therefore, the function is continuous.


## Examples on Limits of Trigonometric Functions

1. Evaluate $\lim\limits_{x\to 0} \dfrac{\sin 2x}{5x}$

Solution

$$\lim\limits_{x\to 0} \dfrac{\sin 2x}{5x}$$

Substituting zero directly into the expression above will give $\dfrac{0}{0}$. Hence we have to apply the

appropriate rule of limit. Let us first make an adjustment to the expression as follows.

Multiply the expression by $\dfrac{2x}{5x}$ and change the denominator to $2x$ just like the numerator, as

follows:

$$\lim\limits_{x\to 0} \dfrac{\sin 2x}{5x} = \lim\limits_{x\to 0} \left(\dfrac{\sin 2x}{2x} \times \dfrac{2x}{5x}\right)$$

Note that when $2x$ cancels out, the original expression remains the same.

Recall that $\lim\limits_{x\to 0} \dfrac{\sin ax}{ax} = 1$

Hence, $\lim\limits_{x\to 0} \dfrac{\sin 2x}{2x} = 1$

Substituting 1 for $\lim\limits_{x\to 0} \dfrac{\sin 2x}{2x}$ in the simplification given above to gives:

$$\lim\limits_{x\to 0} \left(\dfrac{\sin 2x}{2x} \times \dfrac{2x}{5x}\right) = \lim\limits_{x\to 0} \dfrac{\sin 2x}{2x} \times \lim\limits_{x\to 0} \dfrac{2x}{5x}$$

$$= 1 \times \lim\limits_{x\to 0} \dfrac{2}{5} \quad (x \text{ cancels out})$$

$$= 1 \times \dfrac{2}{5} \quad (\text{Since } \lim\limits_{x\to a} c = c)$$

$$= \dfrac{2}{5}$$


2. 1. Evaluate $\lim\limits_{x\to 0} \dfrac{\sin 5x}{7x}$

Solution

$$\lim\limits_{x\to 0} \dfrac{\sin 5x}{7x}$$

Multiply the expression by $\dfrac{5x}{7x}$ and change the denominator to $5x$ just like the numerator. This

gives:

$$\lim_{x \to 0} \frac{\sin 5x}{7x} = \lim_{x \to 0} \left( \frac{\sin 5x}{5x} \times \frac{5x}{7x} \right)$$

Hence, $\lim_{x \to 0} \frac{\sin 5x}{5x} = 1$

Substituting 1 for $\lim_{x \to 0} \frac{\sin 5x}{5x}$ in the simplification given above to gives:

$$\lim_{x \to 0} \left( \frac{\sin 5x}{5x} \times \frac{5x}{7x} \right) = \lim_{x \to 0} \frac{\sin 5x}{5x} \times \lim_{x \to 0} \frac{5x}{7x}$$

$$= 1 \times \lim_{x \to 0} \frac{5}{7} \quad (x \text{ cancels out})$$

$$= 1 \times \frac{5}{7} \quad (\text{Since } \lim_{x \to a} c = c)$$

$$= \frac{5}{7}$$

The two examples above show that $\lim_{x \to 0} \frac{\sin a\,x}{x} = a$ or $\lim_{x \to 0} \frac{\sin a\,x}{bx} = \frac{a}{b}$. This rule also applies to tangent as $\lim_{x \to 0} \frac{\tan a\,x}{x} = a$ or $\lim_{x \to 0} \frac{\tan a\,x}{bx} = \frac{a}{b}$.

3. Find the value of $\lim_{x \to \frac{\pi}{2}} \frac{\cos 2x}{\sin 3x}$

Solution

$$\lim_{x \to \frac{\pi}{2}} \frac{\cos 2x}{\sin 3x}$$

We can simply substitute in the value of x and obtain our answer.

$$\lim_{x \to \frac{\pi}{2}} \frac{\cos 2x}{\sin 3x} = \frac{\cos 2\left(\frac{\pi}{2}\right)}{\sin 3\left(\frac{\pi}{2}\right)}$$

$$= \frac{\cos \pi}{\sin \frac{3\pi}{2}}$$

$$= \frac{\cos 180}{\sin 270} \quad (\text{Note that } \pi \text{ radians} = 180^{\circ})$$

$$= \frac{-1}{-1}$$

$$= 1$$

This problem can also be solved by using the angles directly in radians, but I prefer to work in degrees. Note that $\cos \pi$ in radians $= -1$, and $\sin\frac{3\pi}{2}$ in radians $= -1$ as obtained above.

4. Evaluate $\lim_{x \to 0} \frac{\cos x}{\sin x - 3}$

Solution

$$\lim_{x \to 0} \frac{\cos x}{\sin x - 3} = \frac{\cos 0}{\sin 0 - 3}$$

198

$$= \frac{1}{0-3}$$

$$= -\frac{1}{3}$$

5. Evaluate $\lim_{x \to 0} \dfrac{\sin 3x}{\sin 2x}$

Solution

Multiply by $\dfrac{x}{\sin 2x}$ and change the denominator to $x$ as follows.

$\lim_{x \to 0} \dfrac{\sin 3x}{\sin 2x} = \lim_{x \to 0} \left( \dfrac{\sin 3x}{x} \text{ X } \dfrac{x}{\sin 2x} \right)$    (When $x$ cancels out it gives the original expression)

$= \lim_{x \to 0} \left( \dfrac{\sin 3x}{x} \text{ X } \left(\dfrac{\sin 2x}{x}\right)^{-1} \right)$    (Note that the inverse of $\dfrac{x}{\sin 2x}$ has been taken)

$= \lim_{x \to 0} \dfrac{\sin 3x}{x} \text{ X } \left( \lim_{x \to 0} \dfrac{\sin 2x}{x}\right)^{-1}$

$= 3 \text{ x } 2^{-1}$    (Note that $\lim_{x \to 0} \dfrac{\sin 3x}{x} = 3$ and $\lim_{x \to 0} \dfrac{\sin 2x}{x} = 2$)

$= 3 \text{ x } \dfrac{1}{2}$

$= \dfrac{3}{2}$

This example shows that: $\lim_{x \to 0} \dfrac{\sin a\, x}{\sin b\, x} = \dfrac{a}{b}$

6. Evaluate $\lim_{x \to 0} \dfrac{\sin 2x}{\sin 5x}$

Solution

$\lim_{x \to 0} \dfrac{\sin 2x}{\sin 5x}$

From the rule established in example 5 above, we can see that the solution to this problem is $\dfrac{2}{5}$.

Hence, $\lim_{x \to 0} \dfrac{\sin 2x}{\sin 5x} = \dfrac{2}{5}$    (Since $\lim_{x \to 0} \dfrac{\sin a\, x}{\sin b\, x} = \dfrac{a}{b}$ )

**Exercise 16**

1. Evaluate $\lim_{x \to 0} 2x^3 - 4x^2 + x + 9$

2. Evaluate $\lim_{x \to 0} \dfrac{5x^2 + x - 8}{x^2 - 2x + 5}$

3. Evaluate $\lim_{x \to 5} \dfrac{x^2 - x - 20}{x - 5}$

4. Evaluate $\lim_{x \to 7} \dfrac{x^2 - 49}{x - 7}$

5. Evaluate $\lim\limits_{x \to 0} (5x - 1)(3x + 7)(3x + 2)$

6. Evaluate $\lim\limits_{x \to \infty} \dfrac{2x^3 + x^2 - 9x - 5}{5x^3 + 2x^2 - 7x + 2}$

7. Evaluate $\lim\limits_{x \to \infty} \dfrac{x^4 - 5x^2 - 2x - 6}{3x^4 + 2x + 3}$

8. Evaluate $\lim\limits_{x \to 4} \dfrac{x^2 - 25}{x^2 + 2x - 15}$

9. Evaluate $\lim\limits_{x \to 8} \dfrac{2x^2 - 17x + 8}{8 - x}$

10. Evaluate $\lim\limits_{m \to 0} \dfrac{(5 + m)^2 - 25}{m}$

11. Determine the limiting value of $\dfrac{\sqrt{x} - 2}{x - 4}$ as $x$ tends to 4.

12. Find the limit of $\dfrac{x^3 - 27}{x - 3}$ as $x \to 3$

13. Find the limit of $\dfrac{x^3 + 8}{x + 2}$ as $x \to -2$

14. Evaluate $\lim\limits_{x \to 25} \dfrac{5 - \sqrt{x}}{25 - x}$

15. Evaluate $\lim\limits_{x \to 1} \dfrac{x^3 - 1}{x^7 - 1}$

16. Determine if the function $f(x) = x^2 + 3x - 1$ is continuous at the point $x = 2$

17. Determine if the function $f(x) = \dfrac{5x - 2}{3x^2 - x - 2}$ is continuous at $x = -1$

18. Determine if the function $f(x) = \dfrac{2x^2 - 18}{x - 3}$ is continuous at $x = 3$.

19. Determine if the function $f(x) = \dfrac{x^3 + 64}{x + 4}$ is continuous at $x = -4$.

20. Test for the continuity of the function: $f(x) = \begin{cases} -x + 5 & \text{if } x > 2 \\ x^3 + 10 & \text{if } x \leq 2 \end{cases}$

21. Determine the continuity of the function: $f(x) = \begin{cases} 3x^2 & \text{if } x \geq -3 \\ x + 30 & \text{if } x < -3 \end{cases}$

22. Determine if the function below is continuous or not.

$$f(x) = \begin{cases} \dfrac{x^2 - 25}{x + 5} & \text{if } x \geq -5 \\ \dfrac{x + 5}{2x + 10} & \text{if } x < -5 \end{cases}$$

23. If $f(x) = \begin{cases} \dfrac{x^3 - 27}{x - 3} \\ 1 \quad \text{if } x = 3 \end{cases}$     test for the continuity of the function at $x = 3$.

24. If $f(x) = \begin{cases} 2 - 3x & x < 1 \\ x^3 + 4 & x \geq 1 \end{cases}$     evaluate the following limits, if they exist.

(a) $\lim\limits_{x \to -2} f(x)$

(b) $\lim\limits_{x \to 1} f(x)$

25. Evaluate $\lim\limits_{x \to 2} (5 + |x - 2|)$ if it exists.

26. Evaluate $\lim\limits_{x \to 0} \dfrac{\sin 7x}{2x}$

27. Evaluate $\lim\limits_{x \to 0} \dfrac{\sin x}{3x}$

28. Find the value of $\lim\limits_{x \to \pi} \dfrac{\cos 2x}{\sin \frac{x}{2}}$

29. Evaluate $\lim\limits_{x \to 0} \dfrac{\cos 5x}{\sin 3x - 2}$

30. Evaluate $\lim\limits_{x \to 0} \dfrac{\sin 9x}{\sin 4x}$

# CHAPTER 17
## DIFFERENTIATION FROM FIRST PRINCIPLE

If y = f($x$), then the gradient function of y or f($x$) is given by:

$$g(x) = \frac{f(x + \Delta x) - f(x)}{\Delta x}$$

where $\Delta x$ is the increment in $x$.

Applying limits has shown that as $\Delta x$ tends to zero,

$$f'(x) = \lim_{\Delta x \to 0} \frac{f(x + \Delta x) - f(x)}{\Delta x}$$

$$\frac{dy}{dx} = \lim_{\Delta x \to 0} \frac{f(x + \Delta x) - f(x)}{\Delta x}$$

$\dfrac{dy}{dx}$ reads dee y dee $x$.

Note that f'($x$) and $\dfrac{dy}{dx}$ can be interpreted as instantaneous rate of change of y with respect to $x$.

The technique of finding the derivative of a function by considering the limiting value is called differentiation from first principle.

**Examples**

1. Find the derivative of $x$ from first principle.

Solution

Let, y = $x$

Increasing $x$ by $\Delta x$ will give an increase in y by $\Delta$y. This gives:

$$y + \Delta y = x + \Delta x$$
$$\Delta y = x + \Delta x - y$$
$$\Delta y = x + \Delta x - x \quad \text{(substitute } x \text{ for y since y = } x \text{ as given above from the question)}$$
$$\Delta y = \Delta x \quad (x \text{ cancels out})$$

Dividing both sides by $\Delta x$ in order to get the derivative gives:

$$\frac{\Delta y}{\Delta x} = \frac{\Delta x}{\Delta x}$$
$$\frac{\Delta y}{\Delta x} = 1$$

Taking limits as $\Delta x$ tends to zero gives:

$$\lim_{\Delta x \to 0} \frac{\Delta y}{\Delta x} = 1 \quad \left[\text{Recall that } \lim_{x \to 0} c = c, \text{ hence } \lim_{\Delta x \to 0} f(\Delta x) = 1 \text{ gives 1, since } \frac{\Delta y}{\Delta x} \text{ is regarded as } f(\Delta x)\right]$$

We now write $\displaystyle\lim_{\Delta x \to 0} \frac{\Delta y}{\Delta x}$ as $\dfrac{dy}{dx}$ . This gives:

$$\frac{dy}{dx} = 1$$

2. Find from first principle, the derivative of $2x^3$.

<u>Solution</u>

Let, $y = 2x^3$

Increasing $x$ by $\Delta x$ will give an increase in y by $\Delta y$. This gives:

$$y + \Delta y = 2(x + \Delta x)^3$$
$$y + \Delta y = 2[x^3 + 3x^2\Delta x + 3x(\Delta x)^2 + (\Delta x)^3]$$
$$= 2x^3 + 6x^2\Delta x + 6x(\Delta x)^2 + 2(\Delta x)^3$$
$$\Delta y = 2x^3 + 6x^2\Delta x + 6x(\Delta x)^2 + 2(\Delta x)^3 - y$$
$$= 2x^3 + 6x^2\Delta x + 6x(\Delta x)^2 + 2(\Delta x)^3 - 2x^3 \quad \text{(Since } y = 2x^3\text{)}$$
$$\Delta y = 6x^2\Delta x + 6x(\Delta x)^2 + 2(\Delta x)^3 \quad \text{(Since } 2x^3 \text{ cancels out)}$$

Dividing both sides by $\Delta x$ in order to get the derivative gives:

$$\frac{\Delta y}{\Delta x} = \frac{6x^2\Delta x}{\Delta x} + \frac{6x(\Delta x)^2}{\Delta x} + \frac{2(\Delta x)^3}{\Delta x}$$
$$\frac{\Delta y}{\Delta x} = 6x^2 + 6x\Delta x + 2(\Delta x)^2$$

Taking limits as $\Delta x$ tends to zero gives:

$$\lim_{\Delta x \to 0} \frac{\Delta y}{\Delta x} = 6x^2 + 6x(0) + 2(0)^2 \quad \text{(Note that zero replaces } \Delta x\text{)}$$
$$\lim_{\Delta x \to 0} \frac{\Delta y}{\Delta x} = 6x^2$$

We now write $\lim_{\Delta x \to 0} \dfrac{\Delta y}{\Delta x}$ as $\dfrac{dy}{dx}$. This gives:

$$\frac{dy}{dx} = 6x^2$$

3. Find the derivative of $f(x) = \dfrac{1}{x^2}$ from first principle.

<u>Solution</u>

$$f(x) = \Delta x$$

Increasing $x$ by $\Delta x$ gives:

$$f(x + \Delta x) = \frac{1}{(x + \Delta x)^2}$$
$$f(x + \Delta x) = \frac{1}{x^2 + 2x\Delta x + (\Delta x)^2}$$

Subtract $f(x)$ from both sides of the equation

$$f(x + \Delta x) - f(x) = \frac{1}{x^2 + 2x\Delta x + (\Delta x)^2} - f(x)$$
$$f(x + \Delta x) - f(x) = \frac{1}{x^2 + 2x\Delta x + (\Delta x)^2} - \frac{1}{x^2} \quad \text{(Note that } f(x) = \frac{1}{x^2}\text{)}$$
$$= \frac{x^2 - (x^2 + 2x\Delta x + (\Delta x)^2)}{x^2(x^2 + 2x\Delta x + (\Delta x)^2)}$$
$$= \frac{x^2 - x^2 - 2x\Delta x - (\Delta x)^2}{x^2(x^2 + 2x\Delta x + (\Delta x)^2)}$$

$$= \frac{-2x\Delta x - (\Delta x)^2}{x^2(x^2 + 2x\Delta x + (\Delta x)^2)}$$

$$f(x + \Delta x) - f(x) = \frac{\Delta x(-2x - \Delta x)}{x^2(x^2 + 2x\Delta x + (\Delta x)^2)}$$

Dividing both sides by $\Delta x$ gives:

$$\frac{f(x + \Delta x) - f(x)}{\Delta x} = \frac{-2x - \Delta x}{x^2(x^2 + 2x\Delta x + (\Delta x)^2)}$$  ($\Delta x$ has cancelled out from the right hand side)

Taking limits as $\Delta x$ tends to zero gives:

$$\lim_{\Delta x \to 0} \frac{f(x + \Delta x) - f(x)}{\Delta x} = \frac{-2x - 0}{x^2(x^2 + 2x(0) + (0)^2)}$$

Replacing $\lim_{\Delta x \to 0} \frac{f(x + \Delta x) - f(x)}{\Delta x}$ with f'(x) gives the derivative of f(x) as:

$$f'(x) = \frac{-2x}{x^2(x^2)}$$

$$f'(x) = \frac{-2}{x^3}$$

4. If $f(x) = 3x^2$

(a) write down and simplify the expression $\dfrac{f(x + h) - f(x)}{h}$  (h ≠ 0)

(b) find $\lim_{\Delta x \to 0} \dfrac{f(x + h) - f(x)}{h}$

Solution

(a) Let, $f(x) = 3x^2$

Increasing $x$ by h gives:

$$f(x + h) = 3(x + h)^2$$
$$= 3[x^2 + 2xh + h^2]$$
$$= 3x^2 + 6xh + 3h^2$$

Subtracting f(x) from both sides gives:

$$f(x + h) - f(x) = 3x^2 + 6xh + 3h^2 - 3x^2$$  (Note that $f(x) = 3x^2$ from the question)
$$= 6xh + 3h^2$$

Dividing both side by h gives:

$$\frac{f(x + h) - f(x)}{h} = \frac{6xh}{h} + \frac{3h^2}{h}$$

$$\frac{f(x + h) - f(x)}{h} = 6x + 3h$$

(b)  $\dfrac{f(x + h) - f(x)}{h} = 6x + 3h$

Taking limits as h tends to zero gives:

$$\lim_{h \to 0} \frac{f(x + h) - f(x)}{h} = 6x + 3(0)$$

$$\lim_{h \to 0} \frac{f(x + h) - f(x)}{h} = 6x$$

5. If $f(x) = 4x^3 - 5$,

(a) evaluate $\dfrac{f(x+h)-f(x)}{h}$ , where $h \neq 0$

(b) From your result in (5a) above , find the derivatives of $f(x)$ with respect to $x$.

Solution

(a) Let, $f(x) = 4x^3 - 5$

Increasing $x$ by h gives:

$\qquad f(x + h) = 4(x + h)^3 - 5$
$\qquad\qquad = 4[x^3 + 3x^2h + 3xh^2 + h^3] - 5$
$\qquad\qquad = 4x^3 + 12x^2h + 12xh^2 + 4h^3 - 5$

Subtracting $f(x)$ from both sides gives:

$\qquad f(x + h) - f(x) = 4x^3 + 12x^2h + 12xh^2 + 4h^3 - 5 - (4x^3 - 5)$ $\qquad$ $(f(x) = 4x^3 - 5$ from the question)
$\qquad\qquad = 4x^3 + 12x^2h + 12xh^2 + 4h^3 - 5 - 4x^3 + 5$
$\qquad\qquad = 12x^2h + 12xh^2 + 4h^3$

Dividing both side by h gives:

$\dfrac{f(x+h)-f(x)}{h} = 12x^2 + 12xh + 4h^2$

(b) $\dfrac{f(x+h)-f(x)}{h} = 12x^2 + 12xh + 4h^2$

Taking limits as h tends to zero gives:

$\displaystyle\lim_{h\to 0} \dfrac{f(x+h)-f(x)}{h} = 12x^2 + 12x(0)\ 4(0)^2$

$\displaystyle\lim_{h\to 0} \dfrac{f(x+h)-f(x)}{h} = 12x^2$

6. If $f(x) = \dfrac{x^2+1}{2x}$

(a) write down and simplify the expression $\dfrac{f(x+\Delta x)-f(\Delta x)}{\Delta x}$ , where $\Delta x \neq 0$

(b) Find the derivatives of $y = \dfrac{x^2+1}{2x}$

Solution

$\qquad f(x) = \dfrac{x^2+1}{2x}$

$\qquad f(x + \Delta x) = \dfrac{(x+\Delta x)^2+1}{2(x+\Delta x)}$

$\qquad f(x + \Delta x) = \dfrac{x^2+2x\Delta x+(\Delta x)^2+1}{2x+2\Delta x}$

Subtract $f(x)$ from both sides of the equation

$\qquad f(x + \Delta x) - f(x) = \dfrac{x^2+2x\Delta x+(\Delta x)^2+1}{2x+2\Delta x} - \dfrac{x^2+1}{2x}$

205

$$f(x + \Delta x) - f(x) = \frac{2x[x^2 + 2x\Delta x + (\Delta x)^2 + 1] - (x^2 + 1)(2x + 2\Delta x)}{2x(2x + 2\Delta x)}$$

$$= \frac{2x^3 + 4x^2\Delta x + 2x(\Delta x)^2 + 2x - (2x^3 + 2x^2\Delta x + 2x + 2\Delta x)}{2x(2x + 2\Delta x)}$$

$$= \frac{2x^3 + 4x^2\Delta x + 2x(\Delta x)^2 + 2x - 2x^3 - 2x^2\Delta x - 2x - 2\Delta x)}{2x(2x + 2\Delta x)}$$

$$= \frac{2x^2\Delta x + 2x(\Delta x)^2 - 2\Delta x)}{2x(2x + 2\Delta x)}$$

$$f(x + \Delta x) - f(x) = \frac{\Delta x(2x^2 + 2x\Delta x - 2)}{2x(2x + 2\Delta x)} \qquad \text{(After factorizing the numerator)}$$

Dividing both sides by $\Delta x$ gives:

$$\frac{f(x + \Delta x) - f(x)}{\Delta x} = \frac{2x^2 + 2x\Delta x - 2}{2x(2x + 2\Delta x)}$$

(b)Taking limits as $\Delta x$ tends to zero gives:

$$\lim_{\Delta x \to 0} \frac{f(x + \Delta x) - f(x)}{\Delta x} = \frac{2x^2 + 2x(0) - 2}{2x(2x + 2(0))}$$

$$= \frac{2x^2 - 2}{2x(2x)}$$

$$= \frac{2(x^2 - 1)}{4x^2}$$

Replacing $\lim_{\Delta x \to 0} \frac{f(x + \Delta x) - f(x)}{\Delta x}$ with f'(x) gives the derivative of f(x) as:

$$f'(x) = \frac{x^2 - 1}{2x^2}$$

Separating into fractions gives:

$$f'(x) = \frac{x^2}{2x^2} - \frac{1}{2x^2}$$

$$f'(x) = \frac{1}{2} - \frac{1}{2x^2}$$

7. Find from first principle the derivatives with respect to $x$ of $y = 5x^3 - x + 7$

Solution

$$y = 5x^3 - x + 7$$

$$y + \Delta y = 5(x + \Delta x)^3 - (x + \Delta x) + 7$$

$$y + \Delta y = 5[x^3 + 3x^2\Delta x + 3x(\Delta x)^2 + (\Delta x)^3] - x - \Delta x + 7$$

$$= 5x^3 + 15x^2\Delta x + 15x(\Delta x)^2 + 5(\Delta x)^3 - x - \Delta x + 7$$

$$\Delta y = 5x^3 + 15x^2\Delta x + 15x(\Delta x)^2 + 5(\Delta x)^3 - x - \Delta x + 7 - y$$

$$= 5x^3 + 15x^2\Delta x + 15x(\Delta x)^2 + 5(\Delta x)^3 - x - \Delta x + 7 - (5x^3 - x + 7) \quad \text{(Since } y = 5x^3 - x + 7)$$

$$\Delta y = 5x^3 + 15x^2\Delta x + 15x(\Delta x)^2 + 5(\Delta x)^3 - x - \Delta x + 7 - 5x^3 + x - 7$$

$$= 15x^2\Delta x + 15x(\Delta x)^2 + 5(\Delta x)^3 - \Delta x$$

Dividing both sides by $\Delta x$ gives:

$$\frac{\Delta y}{\Delta x} = \frac{15x^2 \Delta x}{\Delta x} + \frac{15x(\Delta x)^2}{\Delta x} + \frac{5(\Delta x)^3}{\Delta x} - \frac{\Delta x}{\Delta x}$$

$$\frac{\Delta y}{\Delta x} = 15x^2 + 15x\Delta x + 5(\Delta x)^2 - 1$$

Taking limits as $\Delta x$ tends to zero gives:

$$\lim_{\Delta x \to 0} \frac{\Delta y}{\Delta x} = 15x^2 + 15x(0) + 5(0)^2 - 1$$

$$\lim_{\Delta x \to 0} \frac{\Delta y}{\Delta x} = 15x^2 - 1$$

We now replace $\lim_{\Delta x \to 0} \frac{\Delta y}{\Delta x}$ with $\frac{dy}{dx}$. This gives:

$$\frac{dy}{dx} = 15x^2 - 1$$

8. Differentiate $3x - \frac{1}{2x^2}$ from first principle.

<u>Solution</u>

$$y = 3x - \frac{1}{x^2}$$

$$y + \Delta y = 3(x + \Delta x) - \frac{1}{2(x + \Delta x)^2}$$

$$= 3x + 3\Delta x - \frac{1}{2[x^2 + 2x\Delta x + (\Delta x)^2]}$$

Subtract y from both sides of the equation

$$\Delta y = 3x + 3\Delta x - \frac{1}{2x^2 + 4x\Delta x + 2(\Delta x)^2} - y$$

$$= 3x + 3\Delta x - \frac{1}{2x^2 + 4x\Delta x + 2(\Delta x)^2} - (3x - \frac{1}{2x^2})$$

$$= 3x + 3\Delta x - \frac{1}{2x^2 + 4x\Delta x + 2(\Delta x)^2} - 3x + \frac{1}{2x^2}$$

$$= 3\Delta x - \frac{1}{2x^2 + 4x\Delta x + 2(\Delta x)^2} + \frac{1}{2x^2}$$

Combining them into one fraction gives:

$$\Delta y = \frac{3\Delta x[2x^2(2x^2 + 4x\Delta x + 2(\Delta x)^2] - 2x^2 + 2x^2 + 4x\Delta x + 2(\Delta x)^2}{2x^2[2x^2 + 4x\Delta x + 2(\Delta x)^2]}$$

$$= \frac{3\Delta x[4x^4 + 8x^3\Delta x + 4x^2(\Delta x)^2] + 4x\Delta x + 2(\Delta x)^2}{2x^2[2x^2 + 4x\Delta x + 2(\Delta x)^2]} \quad (-2x^2 + 2x^2 \text{ has cancelled out})$$

$$= \frac{12x^4\Delta x + 24x^3(\Delta x)^2 + 12x^2(\Delta x)^3 + 4x\Delta x + 2(\Delta x)^2}{2x^2[2x^2 + 4x\Delta x + 2(\Delta x)^2]}$$

Factorizing the numerator gives:

$$= \frac{\Delta x[12x^4 + 24x^3\Delta x + 12x^2(\Delta x)^2 + 4x + 2\Delta x]}{2x^2[2x^2 + 4x\Delta x + 2(\Delta x)^2]}$$

Dividing both sides by $\Delta x$ gives:

$$\frac{\Delta y}{\Delta x} = \frac{12x^4 + 24x^3\Delta x + 12x^2(\Delta x)^2 + 4x + 2\Delta x}{2x^2[2x^2 + 4x\Delta x + 2(\Delta x)^2]}$$  ($\Delta x$ that is outside the bracket cancels out)

Taking limits as $\Delta x$ tends to zero gives:

$$\lim_{\Delta x \to 0} \frac{\Delta y}{\Delta x} = \frac{12x^4 + 24x^3(0) + 12x^2(0)^2 + 4x + 2(0)}{2x^2[2x^2 + 4x(0) + 2(0)^2]}$$

$$= \frac{12x^4 + 4x}{2x^2[2x^2]}$$

$$\lim_{\Delta x \to 0} \frac{\Delta y}{\Delta x} = \frac{12x^4 + 4x}{4x^4}$$

$$= \frac{12x^4}{4x^4} + \frac{4x}{4x^4}$$

$$= 3 + \frac{1}{x^3}$$

Replacing $\lim_{\Delta x \to 0} \frac{\Delta y}{\Delta x}$ with $\frac{dy}{dx}$ gives the derivative as:

$$\frac{dy}{dx} = 3 + \frac{1}{x^3}$$

## Exercise 17

1. Find the derivative of $2x$ from first principle.

2. Find from first principle, the derivative of $x^2$.

3. Find the derivative of $f(x) = \frac{1}{x^3}$ from first principle.

4. If $f(x) = 5x^2$

(a) write down and simplify the expression $\frac{f(x+h) - f(x)}{h}$   ($h \neq 0$)

(b) find $\lim_{h \to 0} \frac{f(x+h) - f(x)}{h}$

5. If $f(x) = 9x^3$,

(a) evaluate $\frac{f(x+h) - f(x)}{h}$, where $h \neq 0$

(b) From your result above, find the derivatives of $f(x)$ with respect to $x$.

6. If $f(x) = \frac{x^3 - 2}{x}$

(a) write down and simplify the expression $\frac{f(x + \Delta x) - f(\Delta x)}{\Delta x}$, where $\Delta x \neq 0$

(b) Find the derivatives of $y = \frac{x^3 - 2}{x}$

7. Find from first principle the derivatives with respect to $x$ of $y = 3x^2 - 10x$

8. Differentiate $x + \dfrac{3}{x}$ from first principle.

9. From first principle, find the derivatives with respect to $x$ of $y = 5x - 3x^2$

10. Differentiate $2x + \dfrac{x}{5}$ from first principle.

# CHAPTER 18
## GENERAL RULE OF DIFFERENTIATION AND COMPOSITE FUNCTIONS

The general rule for the derivative/differentiation of a function is as given below.

If $y = x^n$

then, $\dfrac{dy}{dx} = nx^{n-1}$

The rule for differentiating a composite function (or function of a function) is given as follows:

If $y = (2x + 5)^4$

then we write, $u = 2x + 5$

and express y as:

$y = u^4$

Therefore, $\dfrac{dy}{dx} = \dfrac{dy}{du} \times \dfrac{du}{dx}$

This rule is called the chain rule.

**Examples**

1. Find the derivatives of the following:

(a) $y = 2x^7$

(b) $y = \dfrac{3}{4}x^8$

(c) $y = 3\sqrt{x}$

(d) $y = \dfrac{10}{\sqrt[5]{x^3}}$

(e) $y = \dfrac{1}{x^{\frac{1}{4}}}$

Solution

(a) $y = 2x^7$

Solutions

(a) $y = 2x^7$

$\dfrac{dy}{dx} = 7 \times 2x^{7-1}$

This is done by multiplying the exponent (power) by the term and subtracting 1 from the exponent (power). Hence, the answer is:

$\dfrac{dy}{dx} = 14x^6$

(b) $y = \dfrac{3}{4}x^8$

Multiply the term by the exponent (power) (i.e. 8) and subtract 1 from the exponent. This gives:

$\dfrac{dy}{dx} = 8 \times \dfrac{3}{4}x^{8-1}$

$$= \frac{24}{4}x^7$$

$$\frac{dy}{dx} = 6x^7$$

(c) $y = 3\sqrt{x}$

Since $\sqrt{x} = x^{\frac{1}{2}}$, we can rewrite the expression as:

$$y = 3x^{\frac{1}{2}}$$

$$\frac{dy}{dx} = \frac{1}{2} \times 3x^{\frac{1}{2}-1}$$

$$= \frac{3}{2}x^{-\frac{1}{2}}$$

$$= \frac{3}{2} \times \frac{1}{x^{\frac{1}{2}}} \qquad \text{(Note that } x^{-\frac{1}{2}} = \frac{1}{x^{\frac{1}{2}}} \text{ from indices)}$$

$$\frac{dy}{dx} = \frac{3}{2x^{\frac{1}{2}}}$$

$$\frac{dy}{dx} = \frac{3}{2\sqrt{x}} \qquad \text{(Since } x^{\frac{1}{2}} = \sqrt{x} \text{ )}$$

(d) $y = \frac{10}{\sqrt[5]{x^3}}$

Expressing the root in fractional form gives:

$$y = \frac{10}{(x^3)^{\frac{1}{5}}}$$

$$y = \frac{10}{x^{\frac{3}{5}}} \qquad \text{(The two exponents (powers) 3 and } \frac{1}{5} \text{ have been multiplied)}$$

Taking the denominator to the numerator changes the sign of its exponent as follows:

$$y = 10x^{-\frac{3}{5}}$$

Hence, $\dfrac{dy}{dx} = -\dfrac{3}{5} \times 10x^{-\frac{3}{5}-1}$

$$= -\frac{30}{5}x^{-\frac{8}{5}}$$

$$= -6x^{-\frac{8}{5}}$$

$$= -6 \times \frac{1}{x^{\frac{8}{5}}} \qquad \text{(Note that the inverse of a term changes the sign of its exponent)}$$

$$\frac{dy}{dx} = \frac{-6}{x^{\frac{8}{5}}}$$

Or, $\dfrac{dy}{dx} = \dfrac{-6}{\sqrt[5]{x^8}}$

(e) $y = \dfrac{1}{x^{\frac{1}{4}}}$

This can be expressed as:

$y = x^{-\frac{1}{4}}$

$\dfrac{dy}{dx} = -\dfrac{1}{4}x^{-\frac{1}{4}-1}$

$\qquad = -\dfrac{1}{4}x^{-\frac{5}{4}}$

$\qquad = -\dfrac{1}{4} \times \dfrac{1}{x^{\frac{5}{4}}}$

$\dfrac{dy}{dx} = -\dfrac{1}{4x^{\frac{5}{4}}}$      (Note that the inverse of a term changes the sign of its exponent)

Or,    $\dfrac{dy}{dx} = \dfrac{1}{4\sqrt[4]{x^5}}$

2. Find the derivative of each of the following:

(a) $5x^3 - 7x^2 - 3x + 8$

(b) $\dfrac{3}{5}x^5 + 2x^3 - x$

(c) $\dfrac{2x^4 - 5x^3 - 4x^2 + 3}{x^2}$

(d) $\sqrt{x} + \dfrac{1}{\sqrt{x}}$

Solutions

(a) Let the expression be $y = 5x^3 - 7x^2 - 3x + 8$

Hence, $\dfrac{dy}{dx} = \dfrac{d(5x^3)}{dx} - \dfrac{d(7x^2)}{dx} - \dfrac{d(3x)}{dx} + \dfrac{d(8)}{dx}$

This means that each part should be differentiated separately.

Hence, $\dfrac{dy}{dx} = (3 \times 5x^{3-1}) - (2 \times 7x^{2-1}) - (1 \times 3x^{1-1}) + 0$

$\dfrac{dy}{dx} = 15x^2 - 14x - 3$    (Note that $x^0 = 1$)

Note that the derivative of a constant is zero as shown by the derivative of 8

(b) $y = \dfrac{3}{5}x^5 + 2x^3 - x$

$\qquad \dfrac{dy}{dx} = (5 \times \dfrac{3}{5}x^{5-1}) + (3 \times 2x^{3-1}) - (1 \times x^{1-1})$

$\qquad\qquad = 3x^4 + 6x^2 - 1$

(c) $y = \dfrac{2x^4 - 5x^3 - 4x^2 + 3}{x^2}$

Dividing each term in the numerator by the denominator in order to separate the expression into it different fractions gives:

$\qquad y = \dfrac{2x^4}{x^2} - \dfrac{5x^3}{x^2} - \dfrac{4x^2}{x^2} + \dfrac{3}{x^2}$

$\qquad\qquad = 2x^2 - 5x - 4 + \dfrac{3}{x^2}$

$\qquad y = 2x^2 - 5x - 4 + 3x^{-2}$

$\qquad \dfrac{dy}{dx} = (2 \times 2x) - 5 - 0 + (-2 \times 3x^{-2-1})$

$\qquad\qquad = 4x - 5 - 6x^{-3}$

$\qquad \dfrac{dy}{dx} = 4x - 5 - \dfrac{6}{x^2}$

(d) $y = \sqrt{x} + \dfrac{1}{\sqrt{x}}$

$\qquad = x^{\frac{1}{2}} + \dfrac{1}{x^{\frac{1}{2}}}$

$\qquad y = x^{\frac{1}{2}} + x^{-\frac{1}{2}}$

$\qquad \dfrac{dy}{dx} = \dfrac{1}{2}x^{-\frac{1}{2}} - \dfrac{1}{2}x^{-\frac{3}{2}}$ $\qquad$ (Note that $-\dfrac{1}{2} - 1 = -\dfrac{3}{2}$ )

$\qquad\qquad = \dfrac{1}{2} \times \dfrac{1}{x^{\frac{1}{2}}} - \dfrac{1}{2} \times \dfrac{1}{x^{\frac{3}{2}}}$

$\qquad\qquad = \dfrac{1}{2x^{\frac{1}{2}}} - \dfrac{1}{2x^{\frac{3}{2}}}$

$\qquad \dfrac{dy}{dx} = \dfrac{1}{2\sqrt{x}} - \dfrac{1}{2\sqrt{x^3}}$

3. If $y = (5x - 2)^3$, find $\dfrac{dy}{dx}$

Solution

$\qquad y = (5x - 2)^3$ $\qquad$ (This is a composite function)

213

Let us take u = $5x - 2$

If $5x - 2$ is replaced with U, then the question (i.e. y = $(5x - 2)^3$) becomes:

$$y = u^3$$

Hence, $\frac{dy}{du} = 3u^2$

Since u = $5x - 2$

then $\frac{du}{dx} = 5$

Therefore, $\frac{dy}{dx} = \frac{dy}{du} \times \frac{du}{dx}$     (Chain rule)

$$= 3u^2 \times 5$$

$$= 15u^2$$

Now substitute $5x - 2$ for u to obtain $\frac{dy}{dx}$ as follows:

$$\frac{dy}{dx} = 15(5x - 2)^2$$

4. If $y = \dfrac{1}{(5x^2 - 1)^4}$ find $\dfrac{dy}{dx}$

Solution

$$y = \frac{1}{(5x^2 - 1)^4}$$     (This is a composite function)

It can also be represented as follows:

$$y = (5x^2 - 1)^{-4}$$     (Its inverse changes the sign of its exponent)

Now, let us take u = $5x^2 - 1$

Hence, $y = u^{-4}$   (When $5x^2 - 1$ is replaced with u in the original question)

Therefore, $\frac{dy}{du} = -4u^{-5}$

Since  u = $5x^2 - 1$

then, $\frac{du}{dx} = 10x$

Therefore, $\frac{dy}{dx} = \frac{dy}{du} \times \frac{du}{dx}$   (Chain rule)

$$= -4u^{-5} \times 10x$$

$$\frac{dy}{dx} = -40xu^{-5}$$

Now substitute $5x^2 - 1$ for u. This gives:

$$\frac{dy}{dx} = -40x(5x^2 - 1)^{-5}$$

Or, $\frac{dy}{dx} = \dfrac{-40x}{(5x^2 - 1)^5}$     (Note the change in the sign of the exponent as it becomes

denominator)

5. If $y = (2x^3 + 7x)^{\frac{1}{2}}$ find $\dfrac{dy}{dx}$

Solution

$$y = (2x^3 + 7x)^{\frac{1}{2}}$$

Let $u = 2x^3 + 7x$

Hence, $y = u^{\frac{1}{2}}$

$$\frac{dy}{du} = \frac{1}{2}u^{-\frac{1}{2}}$$

Also, $u = 2x^3 + 7x$

$$\frac{du}{dx} = 6x^2 + 7$$

Therefore, $\dfrac{dy}{dx} = \dfrac{dy}{du} \times \dfrac{du}{dx}$

$$= \frac{1}{2}u^{-\frac{1}{2}} \times 6x^2 + 7$$

$$= \frac{6x^2 + 7}{2}u^{-\frac{1}{2}}$$

$$= \frac{6x^2 + 7}{2} \times \frac{1}{u^{1/2}}$$

$$\frac{dy}{dx} = \frac{6x^2 + 7}{2u^{1/2}}$$

Now, replace u with $2x^3 + 7x$. This gives:

$$\frac{dy}{dx} = \frac{6x^2 + 7}{2(2x^3 + 7x)^{1/2}}$$

Or, $\dfrac{dy}{dx} = \dfrac{6x^2 + 7}{2\sqrt{2x^3 + 7x}}$     (Note that $(2x^3 + 7x)^{\frac{1}{2}} = \sqrt{2x^3 + 7x}$)

6. Find the derivative of $3x^2 - x + 9)^4$

Solution

$$y = (3x^2 - x + 9)^4$$

Let $u = 3x^2 - x + 9$

Hence, $y = u^4$

$$\frac{dy}{du} = 4u^3$$

$$\frac{du}{dx} = 6x - 1$$

Therefore, $\dfrac{dy}{dx} = \dfrac{dy}{du} \times \dfrac{du}{dx}$

$$= 4u^3 \times 6x - 1$$

$$= 4(6x - 1)u^3$$

215

$$= (24x - 4)u^3$$

$$\frac{dy}{dx} = (24x - 4)(3x^2 - x + 9)^3 \quad \text{(When u is replaced with } 3x^2 - x + 9)$$

**7. Find the derivative of** $\left(x - \dfrac{5}{x}\right)^4$

Solution

$$y = \left(x - \frac{5}{x}\right)^4$$

Let $u = x - \dfrac{5}{x}$

Therefore, $y = u^4$

$$\frac{dy}{du} = 4u^3$$

$$\frac{du}{dx} = \frac{d(x)}{dx} - \frac{d(5x^{-1})}{dx} \quad \text{(Note that } = \frac{5}{x} = 5x^{-1})$$

$$= 1 - (-1)5x^{-2}$$

$$= 1 + 5x^{-2}$$

$$\frac{du}{dx} = 1 + \frac{5}{x^2}$$

Therefore, $\dfrac{dy}{dx} = \dfrac{dy}{du} \times \dfrac{du}{dx}$

$$= 4u^3 \times \left(1 + \frac{5}{x^2}\right)$$

$$= \left(4 + \frac{20}{x^2}\right)u^3$$

$$\frac{dy}{dx} = \left(4 + \frac{20}{x^2}\right)\left(x - \frac{5}{x}\right)^3 \quad \text{(When u is replaced with } \left(x - \frac{5}{x}\right))$$

**8. Differentiate with respect to** $x$: $\dfrac{1}{2x^5 - 3x + 1}$

Solution

Let $y = \dfrac{1}{2x^5 - 3x + 1}$

Or, $y = (2x^5 - 3x + 1)^{-1} \quad \text{(Recall from indices that } \dfrac{1}{a} = a^{-1})$

Let us take $u = 2x^5 - 3x + 1$

Therefore, $y = u^{-1}$

$$\frac{dy}{du} = -1u^{-2}$$

$$\frac{du}{dx} = 10x^4 - 3$$

Therefore, $\dfrac{dy}{dx} = \dfrac{dy}{du} \times \dfrac{du}{dx}$

$$= -1u^{-2} \times 10x^4 - 3$$

$$= -1(10x^4 - 3)u^{-2}$$

$$= (-10x^4 + 3)u^{-2}$$

$$= \frac{3 - 10x^4}{u^2}$$

$$\frac{dy}{dx} = \frac{3 - 10x^4}{(2x^5 - 3x + 1)^2}$$

9. Differentiate with respect to $x$: $\sqrt{7 - 5x^3}$

Solution

Let $y = \sqrt{7 - 5x^3}$

Or, $y = (7 - 5x^3)^{\frac{1}{2}}$     (Recall from indices that $\sqrt{a} = a^{\frac{1}{2}}$)

Let $u = 7 - 5x^3$

Hence, $y = u^{\frac{1}{2}}$

$$\frac{dy}{du} = \frac{1}{2} u^{-\frac{1}{2}}$$

$$\frac{du}{dx} = -15x^2$$

Therefore, $\dfrac{dy}{dx} = \dfrac{dy}{du} \times \dfrac{du}{dx}$

$$= \frac{1}{2} u^{-\frac{1}{2}} \times (-15x^2)$$

$$= -\frac{15}{2} x^2 \left( u^{-\frac{1}{2}} \right)$$

$$= -\frac{15x^2}{2u^{\frac{1}{2}}}$$

$$= -\frac{15x^2}{2\sqrt{u}}$$

$$\frac{dy}{dx} = -\frac{15x^2}{2\sqrt{7 - 5x^3}}$$

10. Find $\dfrac{dy}{dx}$ if $y = \dfrac{1}{\sqrt{2x^3 - 5}}$

Solution

$$y = \frac{1}{\sqrt{2x^3 - 5}}$$

Let $u = 2x^3 - 5$

hence, $y = \dfrac{1}{\sqrt{u}}$

$$y = \frac{1}{u^{\frac{1}{2}}}$$

$$y = u^{-\frac{1}{2}}$$

$$\frac{dy}{du} = -\frac{1}{2} u^{-\frac{3}{2}} \quad \text{(Note that } -\frac{1}{2} - 1 = -\frac{3}{2}\text{)}$$

217

$$\frac{du}{dx} = 6x^2$$

Therefore, $\dfrac{dy}{dx} = \dfrac{dy}{du} \times \dfrac{du}{dx}$

$$= -\frac{1}{2}u^{-\frac{3}{2}} \times (6x^2)$$

$$= -\frac{6}{2}x^2\,(u^{-\frac{3}{2}})$$

$$= -3x^2(u^{-\frac{3}{2}})$$

$$= -\frac{3x^2}{u^{\frac{3}{2}}}$$

$$= -\frac{3x^2}{\sqrt{u^3}}$$

$$\frac{dy}{dx} = -\frac{3x^2}{\sqrt{(2x^3-5)^3}}$$

## Exercise 18

1. Find the derivatives of the following:

(a) $y = 8x^5$

(b) $y = \dfrac{2}{5}x^5$

(c) $y = \sqrt[3]{x}$

(d) $y = 7\sqrt[7]{x}$

(e) $y = \dfrac{1}{\sqrt[8]{x^5}}$

(f) $y = \dfrac{2}{x^{\frac{5}{2}}}$

2. Find the derivative of each of the following:

(a) $2x^5 - 3x^4 - 4x^3 + 5x^2 - 6x + 7$

(b) $x^7 + 2x^4 - \dfrac{3}{x}$

(c) $\dfrac{3x^9 - x^7 - 5x^4 + 2x^2 - 1}{x^3}$

(d) $5(\sqrt[4]{x}) + \dfrac{5}{\sqrt[3]{2x}}$

3. If $y = (2x - 5)^4$, find $\dfrac{dy}{dx}$

4. If $y = \dfrac{3}{(x^3 - 7)^2}$ find $\dfrac{dy}{dx}$

5. If $y = (2x^3 + 7x)^{\frac{1}{2}}$ find $\dfrac{dy}{dx}$

6. Find the derivative of $(7x^3 - x^2 + 3)^5$

7. Find the derivative of $\left(3x - \dfrac{2}{3x}\right)^3$

8. Differentiate with respect to $x$:  $-\dfrac{9}{3x^2 - x - 10}$

9. Differentiate with respect to $x$:  $\sqrt{1 - 2x^4}$

10. Find $\dfrac{dy}{dx}$ if $y = \dfrac{1}{\sqrt[3]{5x^3 - 1}}$

11. Find the derivative of $(x^5 - 3)^9$

12. Find the derivative of $\left(x - \dfrac{1}{5x}\right)^2$

13. Differentiate with respect to $x$:  $\dfrac{2}{x^3 - x - \frac{1}{x}}$

14. Differentiate with respect to $x$:  $\sqrt{5x - x^2}$

15. Find $\dfrac{dy}{dx}$ if $y = \dfrac{1}{\sqrt[5]{x^3 + 2}}$

# CHAPTER 19
## PRODUCT RULE OF DERIVATIVE

If $y = uv$ where u and v are functions of $x$, then:

$$\frac{dy}{dx} = u\frac{dv}{dx} + v\frac{du}{dx}$$

This is called the product rule of differentiation.

Similarly, If $y = uvw$ where u, v and w are functions of $x$, then:

$$\frac{dy}{dx} = \frac{du}{dx}vw + \frac{dv}{dx}uw + \frac{dw}{dx}uv$$

**Examples**

1. If $f(x) = (x - 3)(x + 4)$, find $f'(x)$

Solution

$$f(x) = (x - 3)(x + 4)$$

This is a product of two functions of $x$.

Let $u = x - 3$

and $v = x + 4$

Hence, $\frac{du}{dx} = 1$

$$\frac{dv}{dx} = 1$$

Therefore, the derivative of $f(x)$ is:

$$f'(x) = u\frac{dv}{dx} + v\frac{du}{dx}$$

$$= (x - 3) \times 1 + (x + 4) \times 1$$

$$= x - 3 + x + 4$$

$$f'(x) = 2x + 1$$

Note that another way to differentiate the function in example 1 above is to expand the bracket and differentiate directly.

2. Find the derivative of $y = (4x^2 + 1)(x^2 - 3)$

Solution

$$y = (4x^2 + 1)(x^2 - 3)$$

Let $u = 4x^2 + 1$

and $v = (x^2 - 3)$

Hence, $\frac{du}{dx} = 8x$

$$\frac{dv}{dx} = 2x$$

$$\frac{dy}{dx} = u\frac{dv}{dx} + v\frac{du}{dx}$$
$$= (4x^2 + 1)2x + (x^2 - 3)8x$$
$$= 8x^3 + 2x + 8x^3 - 24x$$
$$\frac{dy}{dx} = 16x^3 - 22x$$

3. If $y = x^2(1 + 2x)^{\frac{1}{2}}$, find $\frac{dy}{dx}$

<u>Solution</u>

$$y = x^2(1 + 2x)^{\frac{1}{2}}$$
$$u = x^2$$

and $v = (1 + 2x)^{\frac{1}{2}}$

Hence, $\dfrac{du}{dx} = 2x$

$$\frac{dv}{dx} = \frac{1}{2} \times 2 \times (1 + 2x)^{\frac{1}{2} - 1} \qquad \text{(Use of chain rule)}$$

$$= (1 + 2x)^{-\frac{1}{2}}$$

$$\frac{dv}{dx} = \frac{1}{(1+2x)^{\frac{1}{2}}}$$

Hence, $\dfrac{dy}{dx} = u\dfrac{dv}{dx} + v\dfrac{du}{dx}$

$$= x^2 \frac{1}{(1+2x)^{\frac{1}{2}}} + (1 + 2x)^{\frac{1}{2}} \times 2x$$

$$= \frac{x^2}{(1+2x)^{\frac{1}{2}}} + 2x(1 + 2x)^{\frac{1}{2}}$$

let us now simplify by using $(1 + 2x)^{\frac{1}{2}}$ as the LCM as follows:

$$= \frac{x^2 + 2x\ (1+2x)^{\frac{1}{2}}(1+2x)^{\frac{1}{2}}}{(1+2x)^{\frac{1}{2}}}$$

$$= \frac{x^2 + 2x\ (1+2x)}{(1+2x)^{\frac{1}{2}}} \qquad \text{(Note that } (1 + 2x)^{\frac{1}{2}} \times (1 + 2x)^{\frac{1}{2}} = (1 + 2x)^{\frac{1}{2} + \frac{1}{2}} = 1 + 2x\text{)}$$

$$= \frac{x^2 + 2x + 4x^2}{(1+2x)^{\frac{1}{2}}}$$

$$\frac{dy}{dx} = \frac{5x^2 + 2x}{\sqrt{1+2x}}$$

4. Find the derivative of $(2x + 3)^3(4x^2 - 1)^2$

<u>Solution</u>

$$y = (2x + 3)^3(4x^2 - 1)^2$$

Let $u = (2x + 3)^3$

and $v = (4x^2 - 1)^2$

$\quad \dfrac{du}{dx} = 3(2x + 3)^{3-1} \times 2 \quad$ (Use of chain rule. Also note that 2 is from the derivative of $2x + 3$)

$\quad\quad = 6(2x + 3)^2$

$\quad \dfrac{dv}{dx} = 2(4x^2 - 1)^{2-1} \times 8x \quad$ (Note that $8x$ is from the derivative of $4x^2 - 1$)

$\quad\quad = 16x(4x^2 - 1)$

Hence, $\dfrac{dy}{dx} = u\dfrac{dv}{dx} + v\dfrac{du}{dx}$

$\quad\quad = (2x + 3)^3 \times 16x(4x^2 - 1) + (4x^2 - 1)^2 \times 6(2x + 3)^2$

Let us factorize the expression by taking out $(2x + 3)^2$ and $(4x^2 - 1)$ which are the common terms as follows:

$\quad \dfrac{dy}{dx} = (2x + 3)^2(4x^2 - 1)[(2x + 3)16x + (4x^2 - 1)6]$

$\quad\quad = (2x + 3)^2(4x^2 - 1)(32x^2 + 48x + 24x^2 - 6)$

$\quad\quad = (2x + 3)^2(4x^2 - 1)(56x^2 + 48x - 6)$

$\quad\quad = (2x + 3)^2(4x^2 - 1)\, 2(28x^2 + 24x - 3)$

$\quad \dfrac{dy}{dx} = 2(2x + 3)^2(4x^2 - 1)(28x^2 + 24x - 3)$

5. Differentiate: $y = x(x + 1)(x^2 - 4)$

Solution

$\quad y = x(x + 1)(x^2 - 4)$

Expanding the first two brackets gives:

$\quad y = (x^2 + x)(x^2 - 4)$

Hence, $u = (x^2 + x)$

and $v = (x^2 - 4)$

Hence, $\dfrac{du}{dx} = 2x + 1$

$\quad \dfrac{dv}{dx} = 2x$

Therefore, $\dfrac{dy}{dx} = u\dfrac{dv}{dx} + v\dfrac{du}{dx}$

$\quad\quad = (x^2 + x)2x + (x^2 - 4)(2x + 1)$

$\quad\quad = 2x^3 + 2x^2 + 2x^3 + x^2 - 8x - 4$

$\quad \dfrac{dy}{dx} = 4x^3 + 3x^2 - 8x - 4$

6. Find the derivative of $(x^2 + 3x - 2)^2 \sqrt{x}$

Solution

$\quad y = (x^2 + 3x - 2)^2 \sqrt{x}$

$$u = (x^2 + 3x - 2)^2$$

$$v = \sqrt{x}$$

Or, $v = x^{\frac{1}{2}}$

$$\frac{du}{dx} = 2(x^2 + 3x - 2)^{2-1} \times 2x + 3 \quad \text{(Note that } 2x + 3 \text{ is from the derivative of } x^2 + 3x - 2)$$

$$= (4x + 6)(x^2 + 3x - 2)$$

$$\frac{dv}{dx} = \frac{1}{2} x^{-\frac{1}{2}} \quad \text{(Note that } 8x \text{ is from the derivative of } 4x^2 - 1)$$

$$= \frac{1}{2x^{\frac{1}{2}}}$$

$$\frac{dv}{dx} = \frac{1}{2\sqrt{x}}$$

Hence, $\dfrac{dy}{dx} = u\dfrac{dv}{dx} + v\dfrac{du}{dx}$

$$= (x^2 + 3x - 2)^2 \frac{1}{2\sqrt{x}} + \sqrt{x}(4x + 6)(x^2 + 3x - 2)$$

$$= (x^2 + 3x - 2)^2 \frac{1}{2\sqrt{x}} + 2\sqrt{x}(2x + 3)(x^2 + 3x - 2)$$

Factorize the expression by taking out ($x^2 + 3x - 2$) which is the common factor. This gives:

$$\frac{dy}{dx} = (x^2 + 3x - 2)\left[\frac{x^2 + 3x - 2}{2\sqrt{x}} + 2\sqrt{x}(2x + 3)\right]$$

Simplifying the part in the bracket by taking $2\sqrt{x}$ as the LCM gives:

$$\frac{dy}{dx} = (x^2 + 3x - 2)\left[\frac{x^2 + 3x - 2 + 4x(2x + 3)}{2\sqrt{x}}\right] \quad \text{(Note that } 2\sqrt{x} \times 2\sqrt{x} = 4x)$$

$$= (x^2 + 3x - 2)\left[\frac{x^2 + 3x - 2 + 8x^2 + 12x}{2\sqrt{x}}\right]$$

$$\frac{dy}{dx} = \frac{(x^2 + 3x - 2)(9x^2 + 15x - 2)}{2\sqrt{x}}$$

7. Find the derivative of $(1 + x)(5x - 2)^{\frac{3}{2}}$

<u>Solution</u>

$$y = (1 + x)(5x - 2)^{\frac{3}{2}}$$

$$u = (1 + x)$$

$$v = (5x - 2)^{\frac{3}{2}}$$

$$\frac{du}{dx} = 1$$

$$\frac{dv}{dx} = \frac{3}{2}(5x - 2)^{\frac{3}{2} - 1} \times 5 \quad \text{(Note that the derivative of } 5x - 2 \text{ is 5)}$$

$$= \frac{15}{2}(5x - 2)^{\frac{1}{2}}$$

Hence, $\dfrac{dy}{dx} = u\dfrac{dv}{dx} + v\dfrac{du}{dx}$

223

$$= (1 + x) \times \frac{15}{2} (5x - 2)^{\frac{1}{2}} + (5x - 2)^{\frac{3}{2}} \times 1$$

$$= \frac{15}{2} (1 + x)(5x - 2)^{\frac{1}{2}} + (5x - 2)^{\frac{3}{2}}$$

Factorize by taking out $(5x - 2)^{\frac{1}{2}}$ (i.e. the lower exponent) which is the common factor gives:

$$\frac{dy}{dx} = (5x - 2)^{\frac{1}{2}} \left[\frac{15}{2} (1 + x) + 5x - 2\right] \quad \left(\text{Note that } \frac{(5x-2)^{\frac{3}{2}}}{(5x-2)^{\frac{1}{2}}} = (5x - 2)^{\frac{3}{2} - \frac{1}{2}} = 5x - 2\right)$$

$$= (5x - 2)^{\frac{1}{2}} \left[\frac{15}{2} + \frac{15x}{2} + 5x - 2\right]$$

$$= (5x - 2)^{\frac{1}{2}} \left[\frac{25x}{2} + \frac{11}{2}\right]$$

$$\frac{dy}{dx} = \sqrt{5x - 2} \left[\frac{25x + 11}{2}\right]$$

8. If y = (1 + x)(2 − 3x)(2x − 1), find $\frac{dy}{dx}$ by using product rule.

Solution

$\quad$ y = (1 + x)(2 − 3x)(2x − 1)

This is a product of three expressions, u, v and w.

Hence, u = (1 + x)

$\quad\quad\quad$ v = (2 − 3x)

and $\quad$ w = (2x − 1)

Therefore, $\frac{du}{dx} = 1$

$\quad$ $\frac{dv}{dx} = -3$

$\quad$ $\frac{dw}{dx} = 2$

Hence the formula for product rule of three terms is given by:

$$\frac{dy}{dx} = \frac{du}{dx} vw + \frac{dv}{dx} uw + \frac{dw}{dx} uv$$

$$= 1(2 - 3x)(2x - 1) + (-3)(1 + x)(2x - 1) + 2(1 + x)(2 - 3x)$$

$$= 4x - 2 - 6x^2 + 3x + (-3 - 3x)(2x - 1) + (2 + 2x)(2 - 3x)$$

$$= 7x - 2 - 6x^2 - 6x + 3 - 6x^2 + 3x + 4 - 6x + 4x - 6x^2$$

$$= -2 + 3 + 4 + 7x - 6x + 3x - 6x + 4x - 6x^2 - 6x^2 - 6x^2$$

$$\frac{dy}{dx} = 5 + 2x - 18x^2$$

9. Differentiate with respect to $x$: $(x^2 - 3x + 5)(2x - 7)$

Solution

$\quad$ y = $(x^2 - 3x + 5)(2x - 7)$

Let us differentiate this product by applying product rule but without the use u and v. This is done as follows:

$$\frac{dy}{dx} = (x^2 - 3x + 5)\frac{d(2x-7)}{dx} + (2x-7)\frac{d(x^2 - 3x + 5)}{dx}$$

$$= (x^2 - 3x + 5)2 + (2x - 7)(2x - 3)$$

$$= 2x^2 - 6x + 10 + 4x^2 - 6x - 14x + 21$$

$$\frac{dy}{dx} = 6x^2 - 26x + 31$$

10. If $y = (5x^2 - 3)(2 + \frac{3}{x})$, find $\frac{dy}{dx}$

Solution

$$y = (5x^2 - 3)(2 + \frac{3}{x})$$

$$u = (5x^2 - 3)$$

$$v = (2 + \frac{3}{x})$$

$$\frac{du}{dx} = 10x$$

$$\frac{dv}{dx} = \frac{d(3x^{-1})}{dx} \quad \text{(Note that } \frac{3}{x} = 3x^{-1})$$

$$= -3x^{-2}$$

$$\frac{dv}{dx} = \frac{-3}{x^2}$$

Hence, $\frac{dy}{dx} = u\frac{dv}{dx} + v\frac{du}{dx}$

$$= (5x^2 - 3)\left(\frac{-3}{x^2}\right) + (2 + \frac{3}{x})10x$$

$$= -15 + \frac{9}{x^2} + 20x + 30$$

$$\frac{dy}{dx} = 15 + 20x + \frac{9}{x^2}$$

Exercise 19

1. If $f(x) = (2x - 1)(3x + 1)$, find $f'(x)$

2. Find the derivative of $y = (3x^2 - 5)(x^2 + 10)$

3. If $y = 5x(3 + x)^{\frac{1}{2}}$ find $\frac{dy}{dx}$

4. Find the derivative of $(x + 5)(x^2 - 7)^3$

5. Differentiate: $y = 2x(3x + 2)(2x^2 - 5)$

6. Find the derivative of $(3x^2 - 1)^3 \sqrt{2x}$

7. Find the derivative of $(9 - x)(x + 3)^{\frac{3}{4}}$

8. If $y = (2 - x)(5 - 3x^2)(x + 3)$, find $\dfrac{dy}{dx}$ by using product rule.

9. Differentiate with respect to $x$: $(3x^4 - x^2 + 2x)(x - 1)$

10. If $y = (3x^2 - x)(1 + \dfrac{1}{2x})$, find $\dfrac{dy}{dx}$

11. Find the derivative of $x^2(\sqrt{x^5})$

12. Find the derivative of $x^4(2x - 11)^{\frac{2}{3}}$

13. If $y = (7 + x)(1 - x^2)(5x^3 + 1)$, find $\dfrac{dy}{dx}$.

14. Differentiate with respect to $x$: $(3x^4 - x^3 + x^2 + 2x - 3)(5x + 4)$

15. If $y = (x^3 - 3x + 5)\left(\dfrac{1}{x^5}\right)$, find $\dfrac{dy}{dx}$

# CHAPTER 20
## QUOTIENT RULE OF DERIVATIVE

If $y = \dfrac{u}{v}$ where u and v are functions of $x$, then:

$$\frac{dy}{dx} = \frac{v\frac{du}{dx} - u\frac{dv}{dx}}{v^2}$$

This is called the quotient rule of differentiation.

**Examples**

1. If $y = \dfrac{3x^2 - 8x + 5}{5x - 2}$ find $\dfrac{dy}{dx}$

<u>Solution</u>

$$y = \frac{3x^2 - 8x + 5}{5x - 2}$$

This is of the form $y = \dfrac{u}{v}$. Therefore, we are going to apply product rule.

Let $u = 3x^2 - 8x + 5$

and $v = 5x - 2$

Hence, $\dfrac{du}{dx} = 6x - 8$

$\dfrac{dv}{dx} = 5$

Therefore, $\dfrac{dy}{dx} = \dfrac{v\frac{du}{dx} - u\frac{dv}{dx}}{v^2}$

$$= \frac{(5x - 2)(6x - 8) - (3x^2 - 8x + 5)(5)}{(5x - 2)^2}$$

$$= \frac{30x^2 - 40x - 12x + 16 - 15x^2 - 40x - 25}{(5x - 2)^2}$$

$$\frac{dy}{dx} = \frac{15x^2 - 12x - 9}{(5x - 2)^2}$$

2. Differentiate with respect to $x$, the function: $\dfrac{3x^2 - 2x}{x + 5}$

<u>Solution</u>

$$y = \frac{3x^2 - 2x}{x + 5}$$

Let $u = 3x^2 - 2x$

and $v = x + 5$

Hence, $\dfrac{du}{dx} = 6x - 2$

$\dfrac{dv}{dx} = 1$

Therefore, $\dfrac{dy}{dx} = \dfrac{v\frac{du}{dx} - u\frac{dv}{dx}}{v^2}$

$$= \dfrac{(x+5)(6x-2) - (3x^2 - 2x)(1)}{(x+5)^2}$$

$$= \dfrac{6x^2 - 2x + 30x - 10 - 3x^2 + 2x}{(x+5)^2}$$

$$\dfrac{dy}{dx} = \dfrac{3x^2 + 30x - 10}{(x+5)^2}$$

3. Find the differential coefficient of $y = \dfrac{-3}{x^2 + 5}$

Solution

$$y = \dfrac{-3}{x^2 + 5}$$

$$u = -3$$

and $v = x^2 + 5$

Hence, $\dfrac{du}{dx} = 0$  (The derivative of a constant is zero)

$$\dfrac{dv}{dx} = 2x$$

Therefore, $\dfrac{dy}{dx} = \dfrac{v\frac{du}{dx} - u\frac{dv}{dx}}{v^2}$

$$= \dfrac{(x^2 + 5) \times 0 - (-3)2x}{(x^2 + 5)^2}$$

$$= \dfrac{0 - 6x}{(x^2 + 5)^2}$$

$$\dfrac{dy}{dx} = \dfrac{-6x}{(x^2 + 5)^2}$$

4. Differentiate: $y = \dfrac{\sqrt{3-x}}{\sqrt{3+x}}$

Solution

$$y = \dfrac{\sqrt{3-x}}{\sqrt{3+x}}$$

$$u = \sqrt{3-x}$$

$$= (3-x)^{\frac{1}{2}}$$

and $v = \sqrt{3+x}$

$$= (3+x)^{\frac{1}{2}}$$

Hence, $\dfrac{du}{dx} = \dfrac{1}{2}(3-x)^{\frac{1}{2}-1} \times -1$  (Note that the derivative of $3-x$ is $-1$)

$$= -\dfrac{1}{2}(3-x)^{-\frac{1}{2}}$$

228

$$\frac{du}{dx} = \frac{-(3-x)^{-\frac{1}{2}}}{2}$$

$$\frac{dv}{dx} = \frac{1}{2}(3+x)^{\frac{1}{2}-1} \times 1 \quad \text{(Note that the derivative of } 3-x \text{ is } -1\text{)}$$

$$= \frac{1}{2}(3+x)^{-\frac{1}{2}}$$

$$\frac{dv}{dx} = \frac{(3+x)^{-\frac{1}{2}}}{2}$$

Therefore, $\dfrac{dy}{dx} = \dfrac{v\frac{du}{dx} - u\frac{dv}{dx}}{v^2}$

$$= \frac{(3+x)^{\frac{1}{2}}\left(\frac{-(3-x)^{-\frac{1}{2}}}{2}\right) - (3-x)^{\frac{1}{2}}\left(\frac{(3+x)^{-\frac{1}{2}}}{2}\right)}{(3+x)^{\frac{1}{2}})^2}$$

Taking out the common terms which are terms with positive exponent in order to factorize the expression gives:

$$\frac{dy}{dx} = \frac{\left((3+x)^{\frac{1}{2}}\right)\left((3-x)^{\frac{1}{2}}\right)\left[\frac{-(3-x)^{-1}}{2} - \frac{(3+x)^{-1}}{2}\right]}{3+x}$$

Note that in order to obtain the terms in the square bracket, we subtracted the exponents of the factors from the exponents of the original term. For example, $(3+x)^{\frac{1}{2}}\left(\frac{-(3-x)^{-\frac{1}{2}}}{2}\right)$ divided by $\left((3+x)^{\frac{1}{2}}\right)\left((3-x)^{\frac{1}{2}}\right)$ gave $\frac{-(3-x)^{-1}}{2}$ since the equal terms canceled out and $-\frac{1}{2} - \frac{1}{2} = -1$, which gave the exponent of $-1$. Similarly, $(3-x)^{\frac{1}{2}}\left(\frac{(3+x)^{-\frac{1}{2}}}{2}\right)$ divided by $\left((3+x)^{\frac{1}{2}}\right)\left((3-x)^{\frac{1}{2}}\right)$ gave $\frac{(3+x)^{-1}}{2}$ since the equal terms canceled out and $-\frac{1}{2} - \frac{1}{2} = -1$, which gave the exponent of $-1$ as the terms in the square bracket.

Let us now continue with the solution by simplifying further as follows:

$$\frac{dy}{dx} = \frac{\left((3+x)^{\frac{1}{2}}\right)\left((3-x)^{\frac{1}{2}}\right)\left[\frac{-(3-x)^{-1}}{2} - \frac{(3+x)^{-1}}{2}\right]}{3+x}$$

$$= \frac{\left((3+x)^{\frac{1}{2}}\right)\left((3-x)^{\frac{1}{2}}\right)\left[\frac{-1}{2(3-x)} - \frac{1}{2(3+x)}\right]}{3+x}$$

$$= \frac{\left((3+x)^{\frac{1}{2}}\right)\left((3-x)^{\frac{1}{2}}\right)\left[\frac{-(3+x)-(3-x)}{2(3-x)(3+x)}\right]}{3+x}$$

$$= \frac{\left((3+x)^{\frac{1}{2}}\right)\left((3-x)^{\frac{1}{2}}\right)\left[\frac{-3-x-3+x}{2(3-x)(3+x)}\right]}{3+x}$$

$$= \frac{\left((3+x)^{\frac{1}{2}}\right)\left((3-x)^{\frac{1}{2}}\right)\left[\frac{-6}{2(3-x)(3+x)}\right]}{3+x}$$

$$= \frac{\left((3+x)^{\frac{1}{2}}\right)\left((3-x)^{\frac{1}{2}}\right)\left[\frac{-3}{(3-x)(3+x)}\right]}{3+x}$$

$$= \frac{-3\left((3+x)^{\frac{1}{2}}\right)\left((3-x)^{\frac{1}{2}}\right)}{(3-x)(3+x)(3+x)}$$

$$= \frac{-3\left((3+x)^{\frac{1}{2}}\right)\left((3-x)^{\frac{1}{2}}\right)}{(3+x)^2(3-x)}$$

$$= -3(3+x)^{\frac{1}{2}-2}(3-x)^{\frac{1}{2}-1} \quad \text{(Subtraction of exponents due to the division above)}$$

$$= -3(3+x)^{-\frac{3}{2}}(3-x)^{-\frac{1}{2}}$$

$$= \frac{-3}{(3+x)^{\frac{3}{2}}(3-x)^{\frac{1}{2}}}$$

$$= \frac{-3}{\sqrt{(3+x)^3(3-x)}}$$

5. Differentiate with respect to $x$:  $y = \dfrac{(2x^2-3)^3}{x}$

<u>Solution</u>

$$y = \frac{(2x^2-3)^3}{x}$$

$$u = (2x^2-3)^3$$

and  $v = x$

Hence,  $\dfrac{du}{dx} = 3(2x^2-3)^{3-1} \times 4x$

$$= 12x(2x^2-3)^2$$

$$\frac{dv}{dx} = 1$$

Therefore,  $\dfrac{dy}{dx} = \dfrac{v\frac{du}{dx} - u\frac{dv}{dx}}{v^2}$

$$= \frac{x[12x(2x^2-3)^2] - (2x^2-3)^3 \times 1}{x^2}$$

$$= \frac{12x^2(2x^2-3)^2 - (2x^2-3)^3}{x^2}$$

Factorize the expression by taking out $(2x^2-3)^2$ which has the lower exponent. This gives:

230

$$\frac{dy}{dx} = \frac{(2x^2-3)^2[12x^2-(2x^2-3)]}{x^2}$$

$$= \frac{(2x^2-3)^2(12x^2-2x^2+3)}{x^2}$$

$$\frac{dy}{dx} = \frac{(2x^2-3)^2(10x^2+3)}{x^2}$$

6. Find the derivative of $\dfrac{\sqrt{(1+2x^2)^3}}{x}$

Solution

$$y = \frac{\sqrt{(1+2x^2)^3}}{x}$$

$$u = \sqrt{(1+2x^2)^3}$$

$$= [(1+2x^2)^3]^{\frac{1}{2}} \qquad \text{(When the square root sign is removed, we use an exponent of } \tfrac{1}{2})$$

$$u = (1+2x^2)^{\frac{3}{2}} \qquad \text{(After multiplying the exponents)}$$

$$v = x$$

Hence, $\dfrac{du}{dx} = \dfrac{3}{2}(1+2x^2)^{\frac{3}{2}-1} \text{ x } 4x \quad$ (Note that the derivative of $3-x$ is $-1$)

$$= 6x(1+2x^2)^{\frac{1}{2}}$$

$$\frac{dv}{dx} = 1$$

Therefore, $\dfrac{dy}{dx} = \dfrac{v\frac{du}{dx} - u\frac{dv}{dx}}{v^2}$

$$= \frac{x\left(6x(1+2x^2)^{\frac{1}{2}}\right) - (1+2x^2)^{\frac{3}{2}} \text{ x } 1}{x^2}$$

Take out $(1+2x^2)^{\frac{1}{2}}$ which has the lower exponent and factorize the expression. This gives:

$$\frac{dy}{dx} = \frac{(1+2x^2)^{\frac{1}{2}}[6x^2-(1+2x^2)]}{x^2} \qquad \text{(Note that } \frac{(1+2x^2)^{\frac{3}{2}}}{(1+2x^2)^{\frac{1}{2}}} \text{ gives } (1+2x^2) \text{ by subtracting exponents.}$$

$$= \frac{(1+2x^2)^{\frac{1}{2}}(6x^2-1-2x^2)}{x^2}$$

$$= \frac{(1+2x^2)^{\frac{1}{2}}(4x^2-1)}{x^2}$$

$$\frac{dy}{dx} = \frac{(\sqrt{1+2x^2})(4x^2-1)}{x^2}$$

7. If $y = \dfrac{(4x^3 - 3x^2 + x + 1)^{\frac{1}{2}}}{(x+1)^2}$

Solution

$$y = \dfrac{(4x^3 - 3x^2 + x + 1)^{\frac{1}{2}}}{(x+1)^2}$$

$u = (4x^3 - 3x^2 + x + 1)^{\frac{1}{2}}$     (After multiplying the exponents)

$v = (x + 1)^2$

Hence, $\dfrac{du}{dx} = \dfrac{1}{2}(4x^3 - 3x^2 + x + 1)^{\frac{1}{2} - 1} \times (12x^2 - 6x + 1)$

$$= \dfrac{(12x^2 - 6x + 1)(4x^3 - 3x^2 + x + 1)^{-\frac{1}{2}}}{2}$$

$\dfrac{dv}{dx} = 2(x + 1)^{2-1}$

$\qquad = 2(x + 1)$

Therefore, $\dfrac{dy}{dx} = \dfrac{v\dfrac{du}{dx} - u\dfrac{dv}{dx}}{v^2}$

$$= \dfrac{\dfrac{(x+1)^2(12x^2 - 6x + 1)(4x^3 - 3x^2 + x + 1)^{-\frac{1}{2}} - (4x^3 - 3x^2 + x + 1)^{\frac{1}{2}}[2(x+1)]}{2}}{[(x+1)^2]^2}$$

Take out the common terms with lower exponents [i.e. $(x + 1)$ and $(4x^3 - 3x^2 + x + 1)^{-\frac{1}{2}}$] and factorize the expression. This gives:

$$\dfrac{dy}{dx} = \dfrac{\dfrac{(x+1)(4x^3 - 3x^2 + x + 1)^{-\frac{1}{2}}[(x+1)(12x^2 - 6x + 1) - (4x^3 - 3x^2 + x + 1)2]}{2}}{[(x+1)^2]^2}$$

Remember to subtract the exponents of the factors from the exponents of the original expression when simplifying. Simplifying further, the expression above gives:

$$\dfrac{dy}{dx} = \dfrac{(x+1)(4x^3 - 3x^2 + x + 1)^{-\frac{1}{2}}[12x^3 - 6x^2 + x + 12x^2 - 6x + 1 - (8x^3 - 6x^2 + 2x + 2)]}{2(x+1)^4}$$

$$= \dfrac{(x+1)(4x^3 - 3x^2 + x + 1)^{-\frac{1}{2}}(12x^3 - 6x^2 + x + 12x^2 - 6x + 1 - 8x^3 + 6x^2 - 2x - 2)}{2(x+1)^4}$$

$$= \dfrac{(x+1)(4x^3 - 3x^2 + x + 1)^{-\frac{1}{2}}(4x^3 + 12x^2 - 7x - 1)}{2(x+1)^4}$$

$$= \dfrac{(4x^3 - 3x^2 + x + 1)^{-\frac{1}{2}}(4x^3 + 12x^2 - 7x - 1)}{2(x+1)^3}$$

232

Note that $(x + 1)$ cancels out from the numerator and denominator.

Therefore, $\dfrac{dy}{dx} = \dfrac{4x^3 + 12x^2 - 7x - 1}{2(4x^3 - 3x^2 + x + 1)^{\frac{1}{2}}(x+1)^3}$

8. Determine $\dfrac{d}{dx}\left(\dfrac{3 + 2x - x^2}{\sqrt{1+x}}\right)$

Solution

Let $y = \dfrac{3 + 2x - x^2}{\sqrt{1+x}}$

Hence, $u = 3 + 2x - x^2$

$\quad \dfrac{du}{dx} = 2 - 2x$

$\quad v = \sqrt{1 + x}$

$\quad = (1 + x)^{\frac{1}{2}}$

$\quad \dfrac{du}{dx} = \dfrac{1}{2}(1 + x)^{\frac{1}{2} - 1} \times 1$

$\quad = \dfrac{1}{2}(1 + x)^{-\frac{1}{2}}$

Therefore, $\dfrac{dy}{dx} = \dfrac{v\frac{du}{dx} - u\frac{dv}{dx}}{v^2}$

$= \dfrac{(1 + x)^{\frac{1}{2}}(2 - 2x) - (3 + 2x - x^2)\frac{1}{2}(1 + x)^{-\frac{1}{2}}}{[(1 + x)^{\frac{1}{2}}]^2}$

Take out $(1 + x)^{-\frac{1}{2}}$ as the common factor since it has the lower exponent and factorize the expression. This gives:

$\dfrac{dy}{dx} = \dfrac{(1 + x)^{-\frac{1}{2}}\left[(1+x)(2 - 2x) - (3 + 2x - x^2)\frac{1}{2}\right]}{1+x}$

$= \dfrac{(1 + x)^{-\frac{1}{2}}\left[\frac{2(1+x)(2 - 2x) - (3 + 2x - x^2)}{2}\right]}{1+x}$

$= \dfrac{(1 + x)^{-\frac{1}{2}}[(2 + 2x)(2 - 2x) - 3 - 2x + x^2]}{2(1+x)}$

$= \dfrac{(1 + x)^{-\frac{1}{2}}(4 - 4x + 4x - 4x^2 - 3 - 2x + x^2)}{2(1+x)}$

$$= \frac{1 - 2x - 3x^2}{2(1 + x)^{\frac{1}{2}}(1+x)} \qquad \text{[When } (1 + x)^{-\frac{1}{2}} \text{ is taken to the denominator it becomes } (1 + x)^{\frac{1}{2}}]$$

$$\frac{dy}{dx} = \frac{1 - 2x - 3x^2}{2(1 + x)^{\frac{3}{2}}} \qquad \text{(The exponents of same terms have been added together, i.e. } \frac{1}{2} + 1 = \frac{3}{2})$$

## Exercise 20

1. If $y = \dfrac{x^2 - 5x + 1}{x - 1}$ find $\dfrac{dy}{dx}$

2. Differentiate with respect to $x$, the function: $\dfrac{4x^2 - x}{2x + 3}$

3. Find the differential coefficient of $y = \dfrac{-7}{3x^2 + 1}$

4. Differentiate: $y = \dfrac{\sqrt{1 + x}}{\sqrt{1 - x}}$

5. Differentiate with respect to $x$: $y = \dfrac{(x^3 - 2)^2}{x^2}$

6. Find the derivative of $\dfrac{\sqrt{2 + 3x^2}}{x^3}$

7. If $y = \dfrac{(x^2 - x - 4)^{\frac{1}{3}}}{(2x + 1)^2}$

8. Determine $\dfrac{d}{dx}\left(\dfrac{x - 2x^3}{\sqrt{2 - x}}\right)$

9. Find the differential coefficient of $y = -\dfrac{1}{1 - 3x^2}$

10. Differentiate: $y = \dfrac{2 + x}{2 - x}$

# CHAPTER 21
## DERIVATIVE OF PARAMETRIC EQUATIONS

If y = f(t) and $x$ = g(t) are two different functions of a common variable, t, then the two equations are called parametric equations. The variable t, is the parameter.

The derivative of a parametric equation such as the one stated above is obtained as follows:

$$\frac{dy}{dx} = \frac{\frac{dy}{dt}}{\frac{dx}{dt}}$$

**Examples**

1. If y = 5 + t$^2$ and $x$ = 3 − 2t$^2$, find $\frac{dy}{dx}$

Solution

$$y = 5 + t^2$$

Hence, $\frac{dy}{dt} = 2t$

$$x = 3 - 2t^2$$

Hence, $\frac{dx}{dt} = -4t$

Therefore, $\frac{dy}{dx} = \frac{\frac{dy}{dt}}{\frac{dx}{dt}}$

$$= \frac{2t}{-4t}$$

$$= -\frac{1}{2} \qquad \text{(t cancels out)}$$

2. Find $\frac{dy}{dx}$ of the functions below which are expressed in the parametric form.

$$x = \frac{5}{t^2} \quad \text{and} \quad y = 2t^5 - 3$$

Solution

$$x = \frac{5}{t^2}$$

$$x = 5t^{-2}$$

$$\frac{dx}{dt} = -10t^{-3}$$

$$y = 2t^5 - 3$$

$$\frac{dy}{dt} = 10t^4$$

Therefore, $\frac{dy}{dx} = \frac{\frac{dy}{dt}}{\frac{dx}{dt}}$

$$= \frac{10t^4}{-10t^{-3}}$$

$$= -t^{4-(-3)} \qquad \text{(10 cancels out)}$$

$$= -t^{4+3}$$

$$= -t^7$$

3. If $V = \frac{4}{3}\pi r^3$ and $A = \pi r^2$, find $\frac{dA}{dV}$

Solution

$$V = \frac{4}{3}\pi r^3$$

$$\frac{dV}{dr} = 3\left(\frac{4}{3}\right)\pi r^2$$

$$= 4\pi r^2$$

$$A = \pi r^2$$

$$\frac{dA}{dr} = 2\pi r$$

$$\frac{dA}{dV} = \frac{\frac{dA}{dr}}{\frac{dV}{dr}}$$

$$= \frac{2\pi r}{4\pi r^2}$$

$$\frac{dA}{dV} = \frac{1}{2r}$$

4. Determine the derivative of the curve defined by the equations: $x = t^2 - 4t$ and $y = 2t^3 - 7t$.

Solution

$$y = 2t^3 - 7t$$

Hence, $\frac{dy}{dt} = 6t^2 - 7$

$$x = t^2 - 4t$$

Hence, $\frac{dx}{dt} = 2t - 4$

Therefore, $\frac{dy}{dx} = \frac{\frac{dy}{dt}}{\frac{dx}{dt}}$

$$= \frac{6t^2 - 7}{2t - 4}$$

5. Given that $v = u + at$ and $s = ut + \frac{1}{2}at^2$. Find $\frac{dv}{ds}$ if u and a are constants.

Solution

$$v = u + at$$

$$\frac{dv}{dt} = a$$

$$s = ut + \frac{1}{2}at^2$$

$$\frac{ds}{dt} = u + (2 \times \frac{1}{2} \times at)$$

$$= u + at$$

Hence, $\dfrac{dv}{ds} = \dfrac{\frac{dv}{dt}}{\frac{ds}{dt}}$

$$= \frac{a}{u+at}$$

6. If $y = \dfrac{t}{t-2}$ and $x = \dfrac{1}{t+1}$, find $\dfrac{dy}{dx}$

<u>Solution</u>

$$y = \frac{t}{t-2}$$

$$\frac{dy}{dt} = \frac{(t-2)(1) - t(1)}{(t-2)^2} \qquad \text{(Use of quotient rule where u = t and v = t – 2)}$$

$$= \frac{t - 2 - t}{(t-2)^2}$$

$$\frac{dy}{dt} = \frac{-2}{(t-2)^2}$$

$$x = \frac{1}{t+1}$$

$$\frac{dx}{dt} = \frac{(t+1)(0) - 1(1)}{(t+1)^2} \qquad \text{(Use of quotient rule where u = 1 and v = t + 1)}$$

$$= \frac{-1}{(t+1)^2}$$

Hence, $\dfrac{dy}{dx} = \dfrac{\frac{dy}{dt}}{\frac{dx}{dt}}$

$$= \frac{\frac{-2}{(t-2)^2}}{\frac{-1}{(t+1)^2}}$$

$$= \frac{-2}{(t-2)^2} \times \frac{(t+1)^2}{-1}$$

$$\frac{dy}{dx} = \frac{2(t+1)^2}{(t-2)^2}$$

7. The parametric equations of the motion of a stone are: $y = 12 + 3t - 2t^2$ and $x = 5t$. Find $\dfrac{dy}{dx}$.

<u>Solution</u>

$$y = 12 + 3t - 2t^2$$

$$\frac{dy}{dt} = 3 - 4t$$

$$x = 5t$$

$$\frac{dx}{dt} = 5$$

$$\frac{dy}{dx} = \frac{\frac{dy}{dt}}{\frac{dx}{dt}}$$

$$= \frac{3-4t}{5}$$

8. If the parametric equations of a parabola are $y = \dfrac{2mt^2}{1+t^2}$ and $x = \dfrac{2m}{1+t^2}$ where m is a constant,

find $\dfrac{dy}{dx}$.

<u>Solution</u>

$$y = \frac{2mt^2}{1+t^2}$$

$$\frac{dy}{dt} = \frac{(1+t^2)(4mt) - 2mt^2(2t)}{(1+t^2)^2}$$    (Use of quotient rule where $u = 2mt^2$ and $v = 1 + t^2$)

$$= \frac{4mt + 4mt^3 - 4mt^3}{(1+t^2)^2}$$

$$\frac{dy}{dt} = \frac{4mt}{(1+t^2)^2}$$

$$x = \frac{2m}{1+t^2}$$

$$\frac{dx}{dt} = \frac{(1+t^2)(0) - 2m(2t)}{(1+t^2)^2}$$    (Use of quotient rule where $u = 1$ and $v = t + 1$)

$$= \frac{-4mt}{(1+t^2)^2}$$

Hence, $\dfrac{dy}{dx} = \dfrac{\frac{dy}{dt}}{\frac{dx}{dt}}$

$$= \frac{\frac{4mt}{(1+t^2)^2}}{\frac{-4mt}{(1+t^2)^2}}$$

$$= \frac{4mt}{(1+t^2)^2} \quad \times \quad \frac{(1+t^2)^2}{-4mt}$$

$$\frac{dy}{dx} = -1 \quad \text{(Same terms cancels out)}$$

## Exercise 21

1. If $y = t^3 - 5$ and $x = t^2 + 1$ find $\dfrac{dy}{dx}$

2. Find $\dfrac{dy}{dx}$ of the functions below which are expressed in the parametric form:

   $x = \dfrac{1}{4 - t^3}$ and $y = 3 + t^2$

3. If $V = \dfrac{1}{3}\pi r^2$ and $A = \pi r l + \pi r^2$, find $\dfrac{dV}{dA}$

4. Find the derivative of the curve defined by the equations: $x = 5t^2 - t + 3$ and $y = t^2 - t - 1$.

5. Given that $F = \dfrac{m(v - u)}{t}$ and $s = \dfrac{(u+v)t}{2}$. Find $\dfrac{dF}{ds}$ if u, v and m are constants.

6. If $y = \dfrac{5}{t^2 - 1}$ and $x = \dfrac{3}{t^4 + 1}$, find $\dfrac{dy}{dx}$

7. The parametric equations of the motion of a stone are: $y = t - 3t^4$ and $x = t^2 + 3$. Find $\dfrac{dy}{dx}$.

8. If the parametric equations of a parabola are $y = \dfrac{at^3}{2}$ and $x = \dfrac{at}{5}$ where a is a constant, find $\dfrac{dy}{dx}$.

9. If $V = \dfrac{1}{3}\pi r^3$ and $A = 4\pi r^2$, find $\dfrac{dA}{dV}$

10. Given that $G = 2m + s^2$ and $H = m^2 - \dfrac{3}{s}$. Find $\dfrac{dG}{dH}$ if m is a constant.

# CHAPTER 22
## DERIVATIVE OF IMPLICIT FUNCTIONS

In the function y = f(x), y is said to be expressed explicitly in terms of $x$. However, in expressions such as $2xy - x^2y = 5$, the relationship between y and $x$ is said to be implicit.

In order to differentiate implicit functions, y is differentiated just like $x$ but with the addition of $\frac{dy}{dx}$ along with the value obtained.

**Examples**

1. Differentiate implicitly, the expression: $2x^2 + y^2 = 9$.

Solution

$$2x^2 + y^2 = 9$$

Follow the rule of differentiation and add $\frac{dy}{dx}$ to the value obtained whenever you differentiate y. Hence we differentiate each term in the expression above as follows:

$$\frac{d(2x^2)}{dx} + \frac{d(2y^2)}{dx} = \frac{d(9)}{dx}$$

$$4x + 2y\frac{dy}{dx} = 0 \quad \text{(The derivative of } y^2 \text{ is 2y and the addition of } \frac{dy}{dx} \text{ to 2y gives } 2y\frac{dy}{dx})$$

We now make $\frac{dy}{dx}$ the subject of the formula as follows:

$$2y\frac{dy}{dx} = -4x$$

$$\frac{dy}{dx} = \frac{-4x}{2y} \quad \text{(When both sides are divided by 2y)}$$

$$\frac{dy}{dx} = \frac{-2x}{y}$$

2. If $x^3 + y^3 = 18xy$, find $\frac{dy}{dx}$

Solution

$$x^3 + y^3 = 18xy$$

$$3x^2 + 3y^2\frac{dy}{dx} = (18x \times 1\frac{dy}{dx}) + (y \times 18)$$

Note that 18xy is differentiated by using product rule where 18x is taken as u while y is take as v. Also, the derivative of y is what gave us $1\frac{dy}{dx}$. The above differentiation now simplifies to:

$$3x^2 + 3y^2\frac{dy}{dx} = 18x\frac{dy}{dx} + 18y$$

Collect terms in $\frac{dy}{dx}$ on one side in order to make $\frac{dy}{dx}$ the subject of the formula as follows:

$$3y^2\frac{dy}{dx} - 18x\frac{dy}{dx} = 18y - 3x^2$$

Factorizing the left hand side gives:

$$\frac{dy}{dx}(3y^2 - 18x) = 18y - 3x^2$$

Divide both sides by $3y^2 - 18x$. This gives:

$$\frac{dy}{dx} = \frac{18y - 3x^2}{3y^2 - 18x}$$

$$= \frac{3(6y - x^2)}{3(y^2 - 6x)}$$

$$\frac{dy}{dx} = \frac{6y - x^2}{y^2 - 6x} \qquad \text{(3 cancels out)}$$

3. Find $\frac{dy}{dx}$ given that $x^2y^2 - 3xy + 4xy^3 = 4$

<u>Solution</u>

$$x^2y^2 - 3xy + 4xy^3 = 4$$

Apply product rule to $x^2y^2$, $3xy$ and $4xy^3$ and differentiate appropriately as follows:

$$(x^2 \times 2y\frac{dy}{dx}) + (y^2 \times 2x) - [(3x \times 1\frac{dy}{dx}) + (y \times 3)] + (4x \times 3y^2\frac{dy}{dx}) + (y^3 \times 4) = 0$$

$$2x^2y\frac{dy}{dx} + 2xy^2 - (3x\frac{dy}{dx} + 3y) + 12xy^2\frac{dy}{dx} + 4y^3 = 0$$

$$2x^2y\frac{dy}{dx} + 2xy^2 - 3x\frac{dy}{dx} - 3y + 12xy^2\frac{dy}{dx} + 4y^3 = 0$$

$$2x^2y\frac{dy}{dx} - 3x\frac{dy}{dx} + 12xy^2\frac{dy}{dx} + 2xy^2 - 3y + 4y^3 = 0$$

$$2x^2y\frac{dy}{dx} - 3x\frac{dy}{dx} + 12xy^2\frac{dy}{dx} = 3y - 2xy^2 - 4y^3$$

Factorizing the left hand side gives:

$$\frac{dy}{dx}(2x^2y - 3x + 12xy^2) = 3y - 2xy^2 - 4y^3$$

$$\frac{dy}{dx} = \frac{3y - 2xy^2 - 4y^3}{2x^2y - 3x + 12xy^2}$$

4. Differentiate $x^4 + 6x^2y^2 - 5 = 0$ implicitly with respect to $x$.

<u>Solution</u>

$$x^4 + 6x^2y^2 - 5 = 0$$

We differentiate accordingly and apply product rule to $6x^2y^2$ as follows:

$$4x^3 + (6x^2 \times 2y\frac{dy}{dx}) + (y^2 \times 12x) - 0 = 0$$

$$4x^3 + 12x^2y\frac{dy}{dx} + 12xy^2 = 0$$

$$12x^2y\frac{dy}{dx} = -4x^3 - 12xy^2$$

$$\frac{dy}{dx} = \frac{-4x^3 - 12xy^2}{12x^2y}$$

$$= \frac{-4x(x^2 + 3y^2)}{12x^2y}$$

$$\frac{dy}{dx} = \frac{-(x^2 + 3y^2)}{3xy} \qquad \text{(4 and } x \text{ cancels out)}$$

5. Find $\frac{dy}{dx}$ if $\frac{x^2}{16} + \frac{y^2}{25} = 1$

Solution

$$\frac{x^2}{16} + \frac{y^2}{25} = 1$$

$$\frac{2x}{16} + \frac{2y}{25}\frac{dy}{dx} = 0$$

$$\frac{2y}{25}\frac{dy}{dx} = -\frac{2x}{16}$$

$$\frac{2y}{25}\frac{dy}{dx} = -\frac{x}{8}$$

$$\frac{dy}{dx} = \frac{-\frac{x}{8}}{\frac{2y}{25}}$$

$$= -\frac{x}{8} \times \frac{25}{2y}$$

$$\frac{dy}{dx} = -\frac{25x}{16y}$$

## Exercise 22

1. Differentiate implicitly, the expression: $5x^3 + 3y = y^2$.

2. If $2x^2 + 3y^3 = 10$, find $\frac{dy}{dx}$

3. Find $\frac{dy}{dx}$ given that $xy - 3xy^2 + 4x^3 = 4$

4. Differentiate $2x^2 + xy^2 - 5 = x^3$ implicitly with respect to $x$.

5. Find $\frac{dy}{dx}$ if $\frac{x^3}{3} - \frac{y^2}{2} = 0$

6. Differentiate implicitly, the expression: $2x^2y + 5y = 1$.

7. If $2x^2y + 4y^3 = 2y$, find $\frac{dy}{dx}$

8. Find $\frac{dy}{dx}$ given that $5y^2 - 4xy = y$

9. Differentiate $x + y^2 - x^2y^2 = 7$ implicitly with respect to $x$.

10. Find $\dfrac{dy}{dx}$ if $x^5 - \dfrac{3}{y} = 2$

# CHAPTER 23
## DERIVATIVE OF TRIGONOMETRIC FUNCTIONS

The derivatives of trigonometric functions are as given below:

If $y = \sin x$, then $\dfrac{dy}{dx} = \cos x$

If $y = \cos x$, then $\dfrac{dy}{dx} = -\sin x$

If $y = \tan x$, then $\dfrac{dy}{dx} = \sec^2 x$

If $y = \cot x$, then $\dfrac{dy}{dx} = -\csc^2 x$

If $y = \sec x$, then $\dfrac{dy}{dx} = \sec x \tan x$

If $y = \csc x$, then $\dfrac{dy}{dx} = -\cot x \csc x$

**Example**

1. Find the derivative of $\cos 5x$

Solution

Let $y = \cos 5x$

We have to use chain rule from composite function due to $5x$ which is a function of $x$.

Hence, let $u = 5x$

Therefore, $y = \cos u$     (By replacing $5x$ with $u$)

$\dfrac{du}{dx} = 5$

$\dfrac{dy}{du} = -\sin u$     (Recall that the derivative of $\cos x$ is $-\sin x$)

Therefore, $\dfrac{dy}{dx} = \dfrac{dy}{du} \times \dfrac{du}{dx}$     (Chain rule)

$= -\sin u \times 5$

$= -5\sin u$

$\dfrac{dy}{dx} = -5\sin 5x$     (Since $u = 5x$)


2. If $y = \sin \dfrac{1}{2} x$, find $\dfrac{dy}{dx}$

Solution

$y = \sin \dfrac{1}{2} x$

Let $u = \dfrac{1}{2} x$

Therefore, $y = \sin u$     (By replacing $\dfrac{1}{2} x$ with $u$)

$$\frac{du}{dx} = \frac{1}{2}$$

$$\frac{dy}{du} = \cos u \qquad \text{(Recall that the derivative of sin}x \text{ is cos}x)$$

Therefore, $\frac{dy}{dx} = \frac{dy}{du} \times \frac{du}{dx}$      (Chain rule)

$$= \cos u \times \frac{1}{2}$$

$$= \frac{1}{2}\cos u$$

$$\frac{dy}{dx} = \frac{1}{2}\cos\frac{1}{2}x \qquad \text{(Since } u = \frac{1}{2}x\text{)}$$

3. Find the derivative of $5\cos 3x$

<u>Solution</u>

$$y = 5\cos 3x$$

Let $u = 3x$

Therefore, $y = 5\cos u$

$$\frac{du}{dx} = 3$$

$$\frac{dy}{du} = 5(-\sin u) \qquad \text{(Note that the constant term i.e. 5 should be used to multiply the derivative)}$$

$$= -5\sin u$$

Therefore, $\frac{dy}{dx} = \frac{dy}{du} \times \frac{du}{dx}$

$$= -5\sin u \times 3$$

$$= -15\sin u$$

$$\frac{dy}{dx} = -15\sin 3x \qquad \text{(Since } u = 3x\text{)}$$

4. Find the derivative of $\sin^2 x$

<u>Solution</u>

$$y = \sin^2 x$$

Note that $\sin^2 x = \sin x \times \sin x$

Hence, let $u = \sin x$

Therefore, $y = u^2$    (i.e. u x u from $\sin x \times \sin x$)

$$\frac{du}{dx} = \cos x$$

$$\frac{dy}{du} = 2u$$

Therefore, $\frac{dy}{dx} = \frac{dy}{du} \times \frac{du}{dx}$

$$= 2u \times \cos x$$

$$= 2u\cos x$$

$$\frac{dy}{dx} = 2\sin x \cos x \qquad (\text{Since } u = \sin x)$$

5. Differentiate with respect to $x$: $y = \tan 2x$

Solution

$$y = \tan 2x$$

$$\frac{dy}{dx} = \sec^2 2x \times \frac{d(2x)}{dx} \qquad (\text{Note that the derivative of } \tan x \text{ is } \sec^2 x)$$

$$= \sec^2 2x \times 2$$

$$= 2\sec^2 2x$$

6. If $y = \cos^3 6x^2$, differentiate y with respect to $x$.

Solution

$$y = \cos^3 6x^2$$

Let us solve this problem without the use of v as follows:

Let $u = \cos 6x^2$

Hence, $y = u^3$

$$\frac{du}{dx} = 12x(-\sin 6x^2) \qquad (\text{Note that } 12x \text{ is from the derivative of } 6x^2)$$

$$= -12x\sin 6x^2$$

$$\frac{dy}{du} = 3u^2$$

Therefore, $\dfrac{dy}{dx} = \dfrac{dy}{du} \times \dfrac{du}{dx}$

$$= 3u^2 \times (-12x\sin 6x^2)$$

$$= -36xu^2\sin 6x^2$$

$$= -36x(\cos 6x^2)^2\sin 6x^2 \qquad (\text{u has been replaced with } \cos 6x^2)$$

$$\frac{dy}{dx} = -36x\cos^2 6x^2\sin 6x^2$$

7. Find the derivative of $y = \sec 3x$

Solution

$$y = \sec 3x$$

$$\frac{dy}{dx} = \sec 3x\tan 3x \times \frac{d(3x)}{dx} \qquad (\text{Note that the derivative of } \sec x \text{ is } \sec x\tan x)$$

$$= \sec 3x\tan 3x \times 3$$

$$\frac{dy}{dx} = 3\sec 3x\tan 3x$$

8. Find the derivative of $\operatorname{cosec}4x^3$

<u>Solution</u>

$$y = \csc 4x^3$$

$$\frac{dy}{dx} = -\csc 4x^3 \cot 4x^3 \times \frac{d(4x^3)}{dx} \qquad \text{(Note that the derivative of cosec is } -\text{cosec cot)}$$

$$= -\csc 4x^3 \cot 4x^3 \times 12x^2$$

$$= -12x^2 \csc 4x^3 \cot 4x^3$$

9. Find the derivative of $\cot^2 2x^4$

<u>Solution</u>

$$y = \cot^2 2x^4$$

This can also be written as: $y = \cot 2x^4 \times \cot 2x^4$

Let $u = 2x^4$ (Take the function of $x$)

Also, let $v = \cot u$ (Take a function of $u$ without taking the exponent)

Hence, $y = v^2$ (Since $v^2 = (\cot u)^2 = (\cot 2x^4)^2 = \cot^2 2x^4$. Hence, $y = v^2$)

$$\frac{du}{dx} = 8x^3$$

$$\frac{dv}{du} = -\csc^2 u$$

$$\frac{dy}{dv} = 2v$$

Therefore, $\dfrac{dy}{dx} = \dfrac{dy}{dv} \times \dfrac{dv}{du} \times \dfrac{du}{dx}$

$$= 2v \times -\csc^2 u \times 8x^3$$

$$= -16x^3 v \csc^2 u$$

$$= -16x^3 \cot u \csc^2 u \qquad \text{(Since } v = \cot u)$$

Substituting in the original value of $u$ gives:

$$\frac{dy}{dx} = -16x^3 \cot 2x^4 \csc^2 2x^4$$

10. Find the derivative of $5x \sin 2x$

<u>Solution</u>

$$y = 5x \sin 2x$$

We are going to apply the product rule of differentiation.

Let $u = 5x$

and $v = \sin 2x$

$$\frac{du}{dx} = 5$$

$$\frac{dv}{dx} = 2\cos 2x \qquad \text{(Note that the differentiation of } 2x \text{ gives 2)}$$

Hence, $\dfrac{dy}{dx} = u\dfrac{dv}{dx} + v\dfrac{du}{dx}$

$$= (5x \times 2\cos 2x) + (\sin 2x \times 5)$$

$$= 10x\cos 2x + 5\sin 2x$$

$$\frac{dy}{dx} = 5(2x\cos 2x + \sin 2x)$$

11. Find $\frac{dy}{dx}$ if $y = \frac{1}{x}\sec x$

Solution

$$y = \frac{1}{x}\sec x$$

We apply product rule as follows:

$$u = \frac{1}{x}$$

$$= x^{-1}$$

$$v = \sec 3x$$

$$\frac{du}{dx} = -1x^{-2}$$

$$= \frac{-1}{x^2}$$

$$\frac{dv}{dx} = \sec 3x\tan 3x \times 3$$

$$= 3\sec 3x\tan 3x$$

Note that the derivative of $\sec x$ is $\sec x\tan x$ and 3 is from the derivative of $3x$

Hence, $\frac{dy}{dx} = u\frac{dv}{dx} + v\frac{du}{dx}$

$$= \frac{1}{x}(3\sec 3x\tan 3x) + \sec 3x\left(\frac{-1}{x^2}\right)$$

$$\frac{dy}{dx} = \frac{3}{x}\sec 3x\tan 3x - \frac{1}{x^2}\sec 3x$$

12. Find the derivative of $3\operatorname{cosec} x^6$

Solution

$$y = 3\operatorname{cosec} x^6$$

Let $u = x^6$

Hence, $y = 3\operatorname{cosec} u$

$$\frac{du}{dx} = 6x^5$$

$$\frac{dy}{du} = -3\cot u \operatorname{cosec} u \quad \text{(The constant term i.e. 5 should be used to multiply the derivative)}$$

Therefore, $\frac{dy}{dx} = \frac{dy}{du} \times \frac{du}{dx}$

$$= -3\cot u \operatorname{cosec} u \times 6x^5$$

$$= -18x^5 \cot u \operatorname{cosec} u$$

$$\frac{dy}{dx} = -18x^5 \cot x^6 \operatorname{cosec} x^6 \quad \text{(Since } u = x^6\text{)}$$

13. If $y = \dfrac{1 + \cos 2x}{\sin 2x}$ find $\dfrac{dy}{dx}$.

<u>Solution</u>

$$y = \frac{1 + \cos 2x}{\sin 2x}$$

We have to apply quotient rule on this as follows:

$u = 1 + \cos 2x$

$v = \sin 2x$

$\dfrac{du}{dx} = -2\sin 2x$

$\dfrac{dv}{dx} = 2\cos 2x$

Hence, $\dfrac{dy}{dx} = \dfrac{v\frac{du}{dx} - u\frac{dv}{dx}}{v^2}$  (Quotient rule)

$$= \frac{\sin 2x(-2\sin 2x) - (1 + \cos 2x)(2\cos 2x)}{(\sin 2x)^2}$$

$$= \frac{-2\sin^2 2x - (2\cos 2x + 2\cos^2 2x)}{\sin^2 2x} \qquad \text{(Note that } (\sin 2x)^2 = \sin^2 2x)$$

$$= \frac{-2\sin^2 2x - 2\cos 2x - 2\cos^2 2x}{\sin^2 2x}$$

$$= \frac{-2\sin^2 2x - 2\cos^2 2x - 2\cos 2x}{\sin^2 2x}$$

$$= \frac{-2(\sin^2 2x + \cos^2 2x) - 2\cos 2x}{\sin^2 2x}$$

$$= \frac{-2(1) - 2\cos 2x}{\sin^2 2x} \qquad \text{(Note that } \sin^2 x + \cos^2 x = 1, \text{ hence } \sin^2 2x + \cos^2 2x = 1)$$

$$\frac{dy}{dx} = \frac{-2(1 + \cos 2x)}{\sin^2 2x}$$

14. Differentiate $y = \dfrac{\sin^2 x}{x}$

<u>Solution</u>

$$y = \frac{\sin^2 x}{x}$$

This is also quotient rule.

Therefore, $u = \sin^2 x$

$v = x$

Let us follow a direct and systematic way of differentiating trigonometric functions.

Hence, in order to differentiate $\sin 2x$:

First, differentiate the exponent of $\sin^2$ without changing the trigonometric term. This gives:

$2\sin^{2-1}$

$= 2\sin$

Then add the term in $x$. This gives:

$2\sin x$

The next step is to differentiate sin which gives cos. Also add the term in $x$ to obtain $\cos x$.

Finally, differentiate the term in $x$. Hence, differentiating $x$ gives 1.

Now multiply the three terms obtained in the three steps above. This gives:

$2\sin x \ \times \ \cos x \ \times \ 1$

$= 2\sin x \cos x$

Hence, $\dfrac{du}{dx} = 2\sin x \cos x$     (As obtained above)

Since, $v = x$

Then, $\dfrac{dv}{dx} = 1$

Hence, $\dfrac{dy}{dx} = \dfrac{v\frac{du}{dx} - u\frac{dv}{dx}}{v^2}$

$= \dfrac{x(2\sin x\cos x) - \sin^2 x(1)}{x^2}$

$= \dfrac{2x\sin x\cos x - \sin^2 x}{x^2}$

$= \dfrac{\sin x(2x\cos x - \sin x)}{x^2}$

15. Differentiate with respect to $x$: $\dfrac{\sec 2x}{x^3 + 1}$

<u>Solution</u>

$y = \dfrac{\sec 2x}{x^3 + 1}$

$u = \sec 2x$

$v = x^3 + 1$

Hence, $\dfrac{du}{dx} = 2\sec 2x\tan 2x$

Then, $\dfrac{dv}{dx} = 3x^2$

Hence, $\dfrac{dy}{dx} = \dfrac{v\frac{du}{dx} - u\frac{dv}{dx}}{v^2}$

$= \dfrac{(x^3 + 1)(2\sec 2x\tan 2x) - \sec 2x(3x^2)}{(x^3 + 1)^2}$

$= \dfrac{(x^3 + 1)(2\sec 2x\tan 2x) - 3x^2\sec 2x}{(x^3 + 1)^2}$

16. Find the derivative of $\dfrac{\cos \sqrt{x}}{1 + x}$

<u>Solution</u>

$$y = \frac{\cos\sqrt{x}}{1+x}$$

$$u = \cos\sqrt{x}$$

$$= \cos x^{\frac{1}{2}}$$

$$v = 1 + x$$

Hence, $\dfrac{du}{dx} = \dfrac{d(x^{\frac{1}{2}})}{dx} \times \dfrac{d(\cos)}{dx}$

Note that $\dfrac{d(\cos)}{dx}$ means the derivative of cos which gives $-\sin$, and then $x^{\frac{1}{2}}$ is added to it to give $-\sin x^{\frac{1}{2}}$. Hence we continue as follows:

$$\frac{du}{dx} = \frac{1}{2} x^{-\frac{1}{2}} \times -\sin x^{\frac{1}{2}}$$

$$= \frac{1}{2x^{\frac{1}{2}}} \times -\sin x^{\frac{1}{2}}$$

$$\frac{du}{dx} = \frac{-\sin x^{\frac{1}{2}}}{2x^{\frac{1}{2}}}$$

Also, $\dfrac{dv}{dx} = 1$

Hence, $\dfrac{dy}{dx} = \dfrac{v\dfrac{du}{dx} - u\dfrac{dv}{dx}}{v^2}$

$$= \frac{(1+x)\left(\dfrac{-\sin x^{\frac{1}{2}}}{2x^{\frac{1}{2}}}\right) - \cos x^{\frac{1}{2}}(1)}{(1+x)^2}$$

$$= \frac{-\sin\sqrt{x}}{2\sqrt{x}(1+x)} - \frac{\cos\sqrt{x}}{(1+x)^2} \qquad \text{(When the fractions are separated)}$$

$$\frac{dy}{dx} = \frac{-(1+x)\sin\sqrt{x} - 2\sqrt{x}\cos\sqrt{x}}{2\sqrt{x}(1+x)^2} \qquad \text{(When the fractions are combined)}$$

17. Find $\dfrac{dy}{dx}$ if $y = \dfrac{1-x^2}{1+\cos x}$

<u>Solution</u>

$$y = \frac{1-x^2}{1+\cos x}$$

$$u = 1 - x^2$$

$$v = 1 + \cos x$$

$$\frac{du}{dx} = -2x$$

$$\frac{dv}{dx} = -\sin x$$

Hence, $\dfrac{dy}{dx} = \dfrac{v\frac{du}{dx} - u\frac{dv}{dx}}{v^2}$

$= \dfrac{(1+\cos x)(-2x) - (1-x^2)(-\sin x)}{(1+\cos x)^2}$

$\dfrac{dy}{dx} = \dfrac{-2x(1+\cos x) + (1-x^2)\sin x}{(1+\cos x)^2}$

Note that the negative sign from $-\sin x$ changed the negative sign at the middle to a positive sign since negative sign multiplied by negative sign gives a positive sign.

18. Find the derivative of $8x\sin x^2$

Solution

$y = 8x\sin x^2$

We apply product rule as follows:

$u = 8x$

$\dfrac{du}{dx} = 8$

$v = \sin x^2$

$\dfrac{dv}{dx} = \dfrac{d(\sin)}{dx} \times \dfrac{d(x^2)}{dx}$

$= \cos x^2 \times 2x$

$= 2x\cos x^2$

Note that $\dfrac{d(\sin)}{dx}$ gives cos which result to $\cos x^2$ when $x^2$ from the question is added

Hence, $\dfrac{dy}{dx} = u\dfrac{dv}{dx} + v\dfrac{du}{dx}$

$= 8x(2x\cos x^2) + \sin x^2(8)$

$= 16x^2\cos x^2 + 8\sin x^2$

$\dfrac{dy}{dx} = 8(2x^2\cos x^2 + \sin x^2)$

19. Differentiate with respect to $x$: $\sec 2x\sin^3 2x$

Solution

$y = \sec 2x\sin^3 2x$

We apply product rule as follows:

$u = \sec 2x$

$\dfrac{du}{dx} = \dfrac{d(\sec)}{dx} \times \dfrac{d(2x)}{dx}$

$= \sec 2x\tan 2x \times 2$

$\dfrac{du}{dx} = 2\sec 2x\tan 2x$

Note that $\dfrac{d(\sec)}{dx}$ gives sectan, but remember to add $2x$ after sec and tan respectively to obtain

sec2xtan2x

$v = \sin^3 2x$

In order to directly differentiate a trigonometric term with exponent like this (i.e. $\sin^3$) we have three solutions to multiply as follows:

**First solution**: consider only the exponent and differentiate $\sin^3$. This gives:

$$\dfrac{d(\sin^3)}{dx} = 3\sin^2$$

We now add $2x$ from the question to obtain $3\sin^2 2x$

**Second solution**: $\dfrac{d(\sin)}{dx} = \cos$ which gives $\cos 2x$

**Third solution**: $\dfrac{d(2x)}{dx} = 2$

Multiply theses three solutions to give the derivative of $v = \sin^3 2x$ as follows:

$$\dfrac{dv}{dx} = 3\sin^2 2x \times \cos 2x \times 2$$

$$= 6\sin^2 2x\cos 2x$$

Now, $\dfrac{dy}{dx} = u\dfrac{dv}{dx} + v\dfrac{du}{dx}$

$= \sec 2x(6\sin^2 2x\cos 2x) + \sin^3 2x(2\sec 2x\tan 2x)$

$= \dfrac{1}{\cos 2x}(6\sin^2 2x\cos 2x) + \sin^3 2x(2\dfrac{1}{\cos 2x}\dfrac{\sin 2x}{\cos 2x})$   (Note that $\sec 2x = \dfrac{1}{\cos 2x}$ and $\tan 2x = \dfrac{\sin 2x}{\cos 2x}$)

$= 6\sin^2 2x + 2\sin^2 2x(\dfrac{\sin 2x}{\cos 2x}\dfrac{\sin 2x}{\cos 2x})$

Note that one $\sin 2x$ has been taken out of $\sin^3 2x$ and placed inside the bracket.

$= 6\sin^2 2x + 2\sin^2 2x(\tan 2x\tan x)$     (Since $\dfrac{\sin 2x}{\cos 2x} = \tan 2x$)

$= 6\sin^2 2x + 2\sin^2 2x(\tan^2 2x)$

$\dfrac{dy}{dx} = 2\sin^2 2x(3 + \tan^2 2x)$     (After factorization)

20. Differentiate with respect to $x$: $\dfrac{\sec x - \tan x}{\sec x + \tan x}$

Solution

$y = \dfrac{\sec x - \tan x}{\sec x + \tan x}$

$u = \sec x - \tan x$

$\dfrac{du}{dx} = \sec x\tan x - \sec^2 x$

$v = \sec x + \tan x$

$\dfrac{dv}{dx} = \sec x\tan x + \sec^2 x$

$$\frac{dy}{dx} = \frac{v\frac{du}{dx} - u\frac{dv}{dx}}{v^2}$$

$$= \frac{(\sec x + \tan x)(\sec x\tan x - \sec 2x) - (\sec x - \tan x)(\sec x\tan x + \sec 2x)}{(\sec x + \tan x)^2}$$

Expanding bracket in the numerator gives:

$$\frac{dy}{dx} = \frac{\sec^2 x\tan x - \sec^3 x + \sec x\tan^2 x - \sec^2 x\tan x - [\sec^2 x\tan x + \sec^3 x - \sec x\tan^2 x - \sec^2 x\tan x]}{(\sec x + \tan x)^2}$$

$$= \frac{\sec^2 x\tan x - \sec^3 x + \sec x\tan^2 x - \sec^2 x\tan x - \sec^2 x\tan x - \sec^3 x + \sec x\tan^2 x + \sec^2 x\tan x]}{(\sec x + \tan x)^2}$$

$$= \frac{2\sec x\tan^2 x - 2\sec^3 x}{(\sec x + \tan x)^2} \qquad \text{(Note that all the } \sec^2 x\tan x \text{ have cancelled out)}$$

$$= \frac{2\sec x(\tan^2 x - \sec^2 x)}{(\sec x + \tan x)^2}$$

$$= \frac{2\sec x(-1)}{(\sec x + \tan x)^2} \qquad \text{(Note that } \tan^2 x - \sec^2 x = -1)$$

$$= \frac{-2\sec x}{(\sec x + \tan x)^2}$$

$$\frac{dy}{dx} = -\frac{2\sec x}{(\sec x + \tan x)^2}$$

21. Find the derivative of $\sqrt{\dfrac{\cos 2x}{1 + \sin 2x}}$

Solution

$$y = \sqrt{\frac{\cos 2x}{1 + \sin 2x}}$$

This can also be written as:

$$y = \frac{(\cos 2x)^{\frac{1}{2}}}{(1 + \sin 2x)^{\frac{1}{2}}}$$

$$u = (\cos 2x)^{\frac{1}{2}}$$

Using the chain rule, we obtain $\dfrac{du}{dx}$ as follows:

$$\frac{du}{dx} = \frac{1}{2}(\cos 2x)^{\frac{1}{2} - 1} \times \frac{d(\cos 2x)}{dx}$$

$$= \frac{1}{2}(\cos 2x)^{-\frac{1}{2}} \times -2\sin 2x$$

$$= \frac{-2\sin 2x}{2(\cos 2x)^{\frac{1}{2}}}$$

$$= \frac{-\sin 2x}{(\cos 2x)^{\frac{1}{2}}}$$

$$v = (1 + \sin 2x)^{\frac{1}{2}}$$

254

$$\frac{dv}{dx} = \frac{1}{2}(1 + \sin 2x)^{\frac{1}{2}-1} \times \frac{d(1+\sin 2x)}{dx}$$

$$= \frac{1}{2}(1 + \sin 2x)^{-\frac{1}{2}} \times 2\cos 2x$$

$$= \frac{2\cos 2x}{2(1 + \sin 2x)^{\frac{1}{2}}}$$

$$= \frac{\cos 2x}{(1 + \sin 2x)^{\frac{1}{2}}}$$

$$\frac{dy}{dx} = \frac{v\frac{du}{dx} - u\frac{dv}{dx}}{v^2}$$

$$= \frac{(1 + \sin 2x)^{\frac{1}{2}}\left(\dfrac{-\sin 2x}{(\cos 2x)^{\frac{1}{2}}}\right) - (\cos 2x)^{\frac{1}{2}}\left(\dfrac{\cos 2x}{(1 + \sin 2x)^{\frac{1}{2}}}\right)}{[(1 + \sin 2x)^{\frac{1}{2}}]^2}$$

$$= \frac{\left(\dfrac{-\sin 2x(1 + \sin 2x)^{\frac{1}{2}}}{(\cos 2x)^{\frac{1}{2}}}\right) - \left(\dfrac{\cos 2x(\cos 2x)^{\frac{1}{2}}}{(1 + \sin 2x)^{\frac{1}{2}}}\right)}{1 + \sin 2x}$$

$$= \frac{\left(\dfrac{-\sin 2x(1 + \sin 2x) - \cos 2x(\cos 2x)}{(\cos 2x)^{\frac{1}{2}}(1 + \sin 2x)^{\frac{1}{2}}}\right)}{1 + \sin 2x}$$

Note that $(\cos 2x)^{\frac{1}{2}} \times (\cos 2x)^{\frac{1}{2}} = \cos 2x$ (Since the exponents are added). Similarly:

$(1 + \sin 2x)^{\frac{1}{2}} \times (1 + \sin 2x)^{\frac{1}{2}} = 1 + \sin 2x$. Simplifying the above expression further, gives:

$$\frac{dy}{dx} = \frac{-\sin 2x(1 + \sin 2x) - \cos^2 2x}{(\cos 2x)^{\frac{1}{2}}(1 + \sin 2x)^{\frac{1}{2}}(1 + \sin 2x)}$$

$$= \frac{-\sin 2x - \sin^2 2x - \cos^2 2x}{(\cos 2x)^{\frac{1}{2}}(1 + \sin 2x)^{\frac{1}{2}}(1 + \sin 2x)}$$

$$= \frac{-\sin 2x - (\sin^2 2x + \cos^2 2x)}{(\cos 2x)^{\frac{1}{2}}(1 + \sin 2x)^{\frac{1}{2}}(1 + \sin 2x)}$$

$$= \frac{-\sin 2x - 1}{(\cos 2x)^{\frac{1}{2}}(1 + \sin 2x)^{\frac{1}{2}}(1 + \sin 2x)} \qquad \text{(Note that } \sin^2 2x + \cos^2 2x = 1)$$

$$= \frac{-(1 + \sin 2x)}{(\cos 2x)^{\frac{1}{2}}(1 + \sin 2x)^{\frac{1}{2}}(1 + \sin 2x)}$$

$$= \frac{-1}{(\cos 2x)^{\frac{1}{2}}(1 + \sin 2x)^{\frac{1}{2}}} \qquad \text{(Note that } 1 + \sin 2x \text{ cancels out)}$$

22. Differentiate with respect to $x$: $\sin x - 2x\cos x$

Solution

$$y = \sin x - 2x\cos x$$

Treat $2x\cos x$ using product rule.

$$\frac{dy}{dx} = \frac{d(\sin x)}{dx} - \left(2x\frac{d(\cos x)}{dx} + \cos x\frac{d(2x)}{dx}\right)$$
$$= \cos x - [2x(-\sin x) + \cos x(2)]$$
$$= \cos x + 2x\sin x - 2\cos x$$
$$= 2x\sin x - \cos x$$

23. Find the derivative of $\cos^5 3x^4$

Solution

$$y = \cos^5 3x^4$$

A direct way of differentiating this problem is applied as follows:

$$\frac{dy}{dx} = \frac{d(\cos^5)}{dx} \times \frac{d(\cos)}{dx} \times \frac{d(3x^4)}{dx}$$
$$= 5\cos^4 3x^4 \times (-\sin 3x^4) \times 12x^3$$

Note that the derivative of $\cos^5$ gives $5\cos^4$ (do not change the trigonometric term, i.e. cos) and then the addition of $3x^4$ from the question gives $5\cos^4 3x^4$.

Similarly the derivative of cos gives $-\sin$ and the addition of $3x^4$ from the question gives $-\sin 3x^4$.

Hence, multiplying the terms above gives:

$$\frac{dy}{dx} = -60x^3\cos^4 3x^4\sin 3x^4$$

24. Differentiate with respect to $x$: $\sin^8 15x^6$

Solution

$$y = \sin^8 15x^6$$

We can also differentiate this problem directly as follows:

$$\frac{dy}{dx} = \frac{d(\sin^8)}{dx} \times \frac{d(\sin)}{dx} \times \frac{d(15x^6)}{dx}$$
$$= 8\sin^7 15x^6 \times \cos 15x^6 \times 90x^5$$

Note that the derivative of $\sin^8$ gives $8\sin^7$ (in this case do not change the trigonometric term, i.e. sin) and then the addition of $15x^6$ from the question gives $8\sin^7 15x^6$.

Similarly the derivative of sin gives cos and the addition of $15x^6$ from the question gives $\cos 15x^6$. Hence, we continue as follows:

$$\frac{dy}{dx} = 8\sin^7 15x^6 \times \cos 15x^6 \times 90x^5$$
$$= 720x^5\sin^7 15x^6\cos 15x^6$$

## Exercise 23

1. Find the derivative of $\tan 2x$

2. If $y = \cos\frac{1}{5}x$, find $\frac{dy}{dx}$

3. Find the derivative of $10\cos 5x$

4. Find the derivative of $\sin^3 x$

5. Differentiate with respect to $x$: $y = \tan^2 x$

6. If $y = \sin^4 3x^5$, differentiate y with respect to $x$.

7. Find the derivative of $y = \csc 6x^2$

8. Find the derivative of $\sec 2x^5$

9. Find the derivative of $\tan 3x^2$

10. Find the derivative of $x^2\cos 3x$

11. Find $\frac{dy}{dx}$ if $y = \frac{3}{x^3}\sin 2x$

12. Find the derivative of $12\sec x^4$

13. If $y = \frac{2 - \sin 5x}{\tan x}$ find $\frac{dy}{dx}$ .

14. Differentiate $y = \frac{\sin^3 x}{2x}$

15. Differentiate with respect to $x$: $\frac{\cot 2x}{x + 3}$

16. Find the derivative of $\frac{\sin\sqrt[3]{3}}{5x}$

17. Find $\frac{dy}{dx}$ if $y = \frac{x^2 - 3}{\sec 2x}$

18. Find the derivative of $3x^2\cos 3x^2$

19. Differentiate with respect to $x$: $\cot x\cos^2 x$

20. Differentiate with respect to $x$: $\frac{\sin x - \cos x}{\sin x + \cos x}$

21. Find the derivative of $\frac{\sec 4x}{\cos 4x - 2}$

22. Differentiate with respect to $x$: $\cos 3x - x^2\sec x$

23. Find the derivative of $\sin^9 x^3$

24. Differentiate with respect to $x$: $\sin 6x^{\frac{1}{2}}$

25. Find $\frac{dy}{dx}$ if $y = \frac{1}{x^2}\sin x^3$

26. Find the derivative of $\cos^3 2x^5$

27. If $y = \frac{\cos 10x}{\sin 2x}$ find $\frac{dy}{dx}$ .

28. Differentiate $y = \frac{3\sin x^4}{2x}$

29. Differentiate with respect to $x$: $\dfrac{\tan 5x}{2x-1}$

30. Find the derivative of $\dfrac{\tan^2 x}{2x}$

# CHAPTER 24
## DERIVATIVE OF INVERSE FUNCTIONS

If the derivative of a function is given by $\dfrac{dy}{dx}$, then the derivative of the inverse function is given

by: $\dfrac{1}{\frac{dy}{dx}} = \dfrac{dx}{dy}$

Or, $\dfrac{dy}{dx} = \dfrac{1}{\frac{dx}{dy}}$

**Examples**

1. Find $\dfrac{dx}{dy}$ if $y = \sqrt[3]{x}$

<u>Solution</u>

**Method 1**

$y = \sqrt[3]{x}$

Or $y = x^{\frac{1}{3}}$

Let us make $x$ the subject of the formula. The inverse of $\dfrac{1}{3}$ is 3. Hence raise both sides to the

exponent 3 as follows:

$y^3 = (x^{\frac{1}{3}})^3$

$y^3 = x^1$     (Note that 1 was obtained from $\dfrac{1}{3}$ x 3)

Hence, $x = y^3$

Therefore, $\dfrac{dx}{dy} = 3y^2$

**Method 2**

$y = \sqrt[3]{x}$

Or $y = x^{\frac{1}{3}}$

$\dfrac{dy}{dx} = \dfrac{1}{3}x^{\frac{1}{3}-1}$

$= \dfrac{1}{3}x^{-\frac{2}{3}}$

$\dfrac{dy}{dx} = \dfrac{1}{3x^{\frac{2}{3}}}$

Hence, $\dfrac{dx}{dy} = \dfrac{1}{\frac{dy}{dx}}$

$= \dfrac{3x^{\frac{2}{3}}}{1}$     (This means the inverse of $\dfrac{1}{3x^{\frac{2}{3}}}$)

$$= 3x^{\frac{2}{3}}$$

$$= 3(\sqrt[3]{x})^2 \qquad \text{[Recall from indices that } x^{\frac{a}{b}} = (\sqrt[b]{x})^a]$$

$$\frac{dx}{dy} = 3y^2 \qquad \text{(Since } y = \sqrt[3]{x})$$

2. If $y = \sqrt[5]{2x - 3}$ find $\dfrac{dx}{dy}$

Solution

$$y = \sqrt[5]{2x - 3}$$

$$\text{Or } y = (2x - 3)^{\frac{1}{5}}$$

Let us make $x$ the subject of the formula. The inverse of $\dfrac{1}{5}$ is 5. Hence raise both sides to the exponent 5 as follows:

$$y^5 = [(2x - 3)^{\frac{1}{5}}]^5$$

$$y^5 = 2x - 3 \qquad \text{(Note that } \frac{1}{5} \times 5 = 1, \text{ which cancels the fractional exponent)}$$

Hence, $2x = y^5 + 3$

$$x = \frac{y^5 + 3}{2}$$

$$x = \frac{y^5}{2} + \frac{3}{2} \qquad \text{(When we separate into fractions by dividing each part by the denominator)}$$

Therefore, $\dfrac{dx}{dy} = \dfrac{5y^4}{2}$

3. If $y = x^3 - 5$. Find $\dfrac{dx}{dy}$

Solution

$$y = x^3 - 5$$

$$y + 5 = x^3$$

$$x^3 = y + 5$$

$$x = (y + 5)^{\frac{1}{3}} \qquad \text{(By raising both sides to an exponent of the inverse of 3 which is } \frac{1}{3})$$

$$\frac{dx}{dy} = \frac{1}{3}(y + 5)^{\frac{1}{3} - 1} \times 1$$

Note that the 1 was obtained from the derivative of $y + 5$ since chain rule was used.

$$\frac{dx}{dy} = \frac{1}{3}(y + 5)^{-\frac{2}{3}}$$

$$\frac{dx}{dy} = \frac{1}{3(y+5)^{\frac{2}{3}}}$$

Recall from indices that $x^{\frac{a}{b}} = (\sqrt[b]{x})^a$. Applying this rule gives:

$$\frac{dx}{dy} = \frac{1}{3[\sqrt[3]{(y+5)^2}]}$$

4. If $y = \frac{1}{2}x^4 + 3$, find $\frac{dx}{dy}$

Solution

$$y = \frac{1}{2}x^4 + 3$$

$$y - 3 = \frac{1}{2}x^4$$

$$2(y - 3) = x^4$$

$$x^4 = 2y - 6$$

$x = (2y - 6)^{\frac{1}{4}}$   (This is obtained by raising both sided to an exponent of the inverse of 4, i.e. $\frac{1}{4}$)

Hence we use chain rule to determine $\frac{dx}{dy}$ as follows:

$$\frac{dx}{dy} = \frac{1}{4}(2y - 6)^{\frac{1}{4} - 1} \times 2 \qquad \text{(Note that 2 is from the derivative of } 2y - 6)$$

$$\frac{dx}{dy} = \frac{1}{2}(2y - 6)^{-\frac{3}{4}}$$

$$\frac{dx}{dy} = \frac{1}{2(2y - 6)^{\frac{3}{4}}}$$

Hence, $\frac{dx}{dy} = \frac{1}{2[\sqrt[4]{(2y-6)^3}]}$   [This is obtained from the law of indices given by: $x^{\frac{a}{b}} = (\sqrt[b]{x})^a$]

5. If $y = \frac{x + 2}{x}$, find $\frac{dx}{dy}$.

Solution

$$y = \frac{x + 2}{x}$$

$xy = x + 2$   (When we cross multiply)

$xy - x = 2$

Factorizing the left hand side gives:

$x(y - 1) = 2$

$$x = \frac{2}{y - 1}$$

$x = 2(y - 1)^{-1}$   (Take note of the use of negative exponent when the denominator goes up)

Hence we use chain rule to determine $\frac{dx}{dy}$ as follows:

$$\frac{dx}{dy} = -1 \times 2(y - 1)^{-1-1} \times 1 \qquad \text{(Note that 1 is from the derivative of } y - 1)$$

$$\frac{dx}{dy} = -2(y-1)^{-2}$$

$$\frac{dx}{dy} = \frac{-2}{(y-1)^2}$$

6. Find $\frac{dx}{dy}$ if $y = \frac{1}{x+2}$

<u>Solution</u>

$$y = \frac{1}{x+2}$$

$$x + 2 = \frac{1}{y}$$

$$x = \frac{1}{y} - 2$$

$$x = y^{-1} - 2$$

$$\frac{dx}{dy} = -1 \times y^{-1-1}$$

$$= -y^{-2}$$

$$\frac{dx}{dy} = \frac{-1}{y^2}$$

7. Find $\frac{dx}{dy}$ if $y = x^{\frac{2}{3}}$

<u>Solution</u>

$$y = x^{\frac{2}{3}}$$

Raise both sides to the exponent $\frac{3}{2}$ i.e. the inverse of $\frac{2}{3}$. This gives:

$$y^{\frac{3}{2}} = (x^{\frac{2}{3}})^{\frac{3}{2}}$$

$$y^{\frac{3}{2}} = x \quad \text{(Note that } \frac{2}{3} \times \frac{3}{2} = 1, \text{ and } x^1 = x)$$

$$x = y^{\frac{3}{2}}$$

$$\frac{dx}{dy} = \frac{3}{2} \times y^{\frac{3}{2}-1}$$

$$= \frac{3}{2} y^{\frac{1}{2}}$$

$$\frac{dx}{dy} = \frac{3\sqrt{y}}{2}$$

8. If $y = (5x + 7)^{\frac{3}{10}}$ find $\frac{dx}{dy}$.

<u>Solution</u>

$$y = (5x + 7)^{\frac{3}{10}}$$

Raise both sides to the exponent $\dfrac{10}{3}$ i.e. the inverse of $\dfrac{3}{10}$. This gives:

$$y^{\frac{10}{3}} = [(5x + 7)^{\frac{3}{10}}]^{\frac{10}{3}}$$

$$y^{\frac{10}{3}} = 5x + 7 \quad \text{(Note that } \dfrac{3}{10} \times \dfrac{10}{3} = 1)$$

$$y^{\frac{10}{3}} - 7 = 5x$$

$$x = \frac{y^{\frac{10}{3}} - 7}{5}$$

$$= \frac{y^{\frac{10}{3}}}{5} - \frac{7}{5}$$

$$x = \frac{1}{5}y^{\frac{10}{3}} - \frac{7}{5}$$

$$\frac{dx}{dy} = \frac{10}{3} \times \frac{1}{5}y^{\frac{10}{3} - 1}$$

$$= \frac{2}{3}y^{\frac{7}{3}}$$

$$\frac{dx}{dy} = \frac{2\sqrt[3]{y^7}}{3}$$

## Exercise 24

1. Find $\dfrac{dx}{dy}$ if $y = 7x^3$

2. If $y = \sqrt[3]{1 + x^2}$ find $\dfrac{dx}{dy}$

3. If $y = 2x^5 - 3$. Find $\dfrac{dx}{dy}$

4. If $y = \dfrac{2}{3}x^3 - 9$, find $\dfrac{dx}{dy}$

5. If $y = \dfrac{2x + 1}{x}$, find $\dfrac{dx}{dy}$.

6. Find $\dfrac{dx}{dy}$ if $y = \dfrac{3}{x^2 - 5}$

7. Find $\dfrac{dx}{dy}$ if $y = 2x^{\frac{1}{4}}$

8. If $y = (x - 3)^{\frac{1}{5}}$ find $\dfrac{dx}{dy}$.

9. If $y = \dfrac{4x + 5}{x}$, find $\dfrac{dx}{dy}$.

10. Find $\dfrac{dx}{dy}$ if $y = \dfrac{1}{x^3 + 8}$

# CHAPTER 25
## DERIVATIVES OF INVERSE TRIGONOMETRIC FUNCTIONS

Recall that if $\sin x = y$, then $x = \sin^{-1}y$. This is referred to as inverse trigonometric function. The derivatives of inverse trigonometric functions are given below.

If $y = \sin^{-1}x$, then $\dfrac{dy}{dx} = \dfrac{1}{\sqrt{1-x^2}}$

If $y = \cos^{-1}x$, then $\dfrac{dy}{dx} = \dfrac{-1}{\sqrt{1-x^2}}$

If $y = \tan^{-1}x$, then $\dfrac{dy}{dx} = \dfrac{1}{1+x^2}$

If $y = \cot^{-1}x$, then $\dfrac{dy}{dx} = \dfrac{-1}{1+x^2}$

If $y = \sec^{-1}x$, then $\dfrac{dy}{dx} = \dfrac{1}{x\sqrt{x^2-1}}$

If $y = \csc^{-1}x$, then $\dfrac{dy}{dx} = \dfrac{-1}{x\sqrt{x^2-1}}$

Note that $\sin^{-1}x$ can also be written as arcsin$x$. Other inverse function can be written in a similar way.

**Examples**

1. If $y = \cos^{-1}x$ find the $\dfrac{dy}{dx}$.

Solution

$\quad y = \cos^{-1}x$

$\dfrac{dy}{dx} = \dfrac{-1}{\sqrt{1-x^2}}$

2. Find the derivative of $\sin^{-1}3x$

Solution

$\quad y = \sin^{-1}3x$

Let us use the chain rule to solve this problem

Let $u = 3x$

Hence, $y = \sin^{-1}u$

$\quad \dfrac{du}{dx} = 3$

$\quad \dfrac{dy}{du} = \dfrac{1}{\sqrt{1-u^2}}$

Therefore, $\dfrac{dy}{dx} = \dfrac{dy}{du} \times \dfrac{du}{dx}$ $\qquad$ (Chain rule)

$$= \frac{1}{\sqrt{1-u^2}} \times 3$$

$$= \frac{3}{\sqrt{1-u^2}}$$

$$= \frac{3}{\sqrt{1-(3x)^2}} \qquad \text{(Since } u = 3x\text{)}$$

$$\frac{dy}{dx} = \frac{3}{\sqrt{1-9x^2}}$$

3. Find the derivative of $\cot^{-1}x^2$

Solution

$$y = \cot^{-1}x^2$$

Let $u = x^2$

Hence, $y = \cot^{-1}u$

$$\frac{du}{dx} = 2x$$

$$\frac{dy}{du} = \frac{-1}{1+u^2}$$

Therefore, $\dfrac{dy}{dx} = \dfrac{dy}{du} \times \dfrac{du}{dx}$

$$= \frac{-1}{1+u^2} \times 2x$$

$$= \frac{-2x}{1+u^2}$$

$$= \frac{-2x}{1+(x^2)^2} \qquad \text{(since } u = x^2\text{)}$$

$$\frac{dy}{dx} = \frac{-2x}{1+x^4}$$

4. If $y = \sec^{-1}2x^3$, find $y = \dfrac{dy}{dx}$

Solution

$$y = \sec^{-1}2x^3$$

Let $u = 2x^3$

Hence, $y = \sec^{-1}u$

$$\frac{du}{dx} = 6x^2$$

$$\frac{dy}{du} = \frac{1}{u\sqrt{u^2-1}}$$

Therefore, $\dfrac{dy}{dx} = \dfrac{dy}{du} \times \dfrac{du}{dx}$

$$= \frac{1}{u\sqrt{u^2 - 1}} \times 6x^2$$

$$= \frac{6x^2}{u\sqrt{u^2 - 1}}$$

$$= \frac{6x^2}{2x^3\sqrt{(2x^3)^2 - 1}} \qquad \text{(since } u = 2x^3\text{)}$$

$$\frac{dy}{dx} = \frac{3}{x\sqrt{4x^6 - 1}} \qquad \text{(Note that } \frac{6x^2}{2x^3} = \frac{3}{x}\text{)}$$

5. Find the derivative $x^2\tan^{-1}x$

<u>Solution</u>

$$y = x^2\tan^{-1}x$$

we apply the product rule as follows:

$$u = x^2$$

$$v = \tan^{-1}x$$

$$\frac{du}{dx} = 2x$$

$$\frac{dv}{dx} = \frac{1}{1+x^2}$$

$$\frac{dy}{dx} = u\frac{dv}{dx} + v\frac{du}{dx} \qquad \text{(product rule)}$$

$$= x^2\left(\frac{1}{1+x^2}\right) + \tan^{-1}x(2x)$$

$$\frac{dy}{dx} = \frac{x^2}{1+x^2} + 2x\tan^{-1}x$$

6. If $y = 3x - 1 \, \text{cosec}^{-1}x^3$, find $\frac{dy}{dx}$.

<u>Solution</u>

$$y = 3x - 1 \, \text{cosec}^{-1}x^3$$

$$u = 3x - 1$$

$$\frac{du}{dx} = 3$$

$$v = \text{cosec}^{-1}x^3$$

$$\frac{dv}{dx} = \frac{-1}{x^3\sqrt{(x^3)^2 - 1}} \times 3x^2 \qquad \text{(From use of chain rule. Note that } 3x^2 \text{ is from the derivative of } x^3\text{)}$$

$$= \frac{-3x^2}{x^3\sqrt{x^6 - 1}}$$

$$\frac{dv}{dx} = \frac{-3}{x\sqrt{x^6 - 1}}$$

$$\frac{dy}{dx} = u\frac{dv}{dx} + v\frac{du}{dx} \qquad \text{(product rule)}$$

$$= 3x - 1\left(\frac{-3}{x\sqrt{x^6 - 1}}\right) + \text{cosec}^{-1}x^3(3)$$

$$\frac{dy}{dx} = \frac{-3(3x - 1)}{x\sqrt{x^6 - 1}} + 3\text{cosec}^{-1}x^3$$

**7. Find the derivative of $\cos^{-1}(5x - 3)$**

<u>Solution</u>

$$y = \cos^{-1}(5x - 3)$$

Let $u = 5x - 3$

Hence, $y = \cos^{-1}u$

$$\frac{du}{dx} = 5$$

$$\frac{dy}{du} = \frac{-1}{\sqrt{1 - u^2}}$$

Therefore, $\dfrac{dy}{dx} = \dfrac{dy}{du} \times \dfrac{du}{dx}$

$$= \frac{-1}{\sqrt{1 - u^2}} \times 5$$

$$= \frac{-5}{\sqrt{1 - u^2}}$$

$$= \frac{-5}{\sqrt{1 - (5x - 3)^2}} \qquad \text{(Since } u = 5x - 3\text{)}$$

Expanding the bracket gives:

$$\frac{dy}{dx} = \frac{-5}{\sqrt{1 - (25x^2 - 15x - 15x + 9)}}$$

$$= \frac{-5}{\sqrt{1 - 25x^2 + 15x + 15x - 9}}$$

$$\frac{dy}{dx} = \frac{-5}{\sqrt{-25x^2 + 30x - 8}}$$

**8. If $y = \tan^{-1}\left(\dfrac{1}{m}\right)$, find $\dfrac{dy}{dm}$**

<u>Solution</u>

$$y = \tan^{-1}\left(\frac{1}{m}\right)$$

Let $u = \dfrac{1}{m} = m^{-1}$

Hence, $y = \tan^{-1}u$

$$\frac{du}{dm} = -1m^{-2}$$

$$\frac{du}{dm} = \frac{-1}{m^2}$$

$$\frac{dy}{du} = \frac{1}{1 + u^2}$$

Therefore, $\dfrac{dy}{dm} = \dfrac{dy}{du} \times \dfrac{du}{dm}$

$= \dfrac{1}{1+u^2} \times \dfrac{-1}{m^2}$

$= \dfrac{1}{1+\left(\frac{1}{m}\right)^2} \times \dfrac{-1}{m^2}$

$= \dfrac{-1}{m^2\left[1+\left(\frac{1}{m}\right)^2\right]}$

$= \dfrac{-1}{m^2\left(1+\frac{1}{m^2}\right)}$

$\dfrac{dy}{dm} = \dfrac{-1}{m^2+1}$

9. If $y = \sin 3x$, find $\dfrac{dx}{dy}$.

<u>Solution</u>

$y = \sin 3x$

$\sin^{-1} y = 3x$  (Recall that if $\sin a = b$, then $a = \sin^{-1} b$)

$x = \dfrac{\sin^{-1} y}{3}$

$x = \dfrac{1}{3}\sin^{-1} y$

Hence, $\dfrac{dx}{dy} = \dfrac{1}{3}\,\dfrac{1}{\sqrt{1-y^2}}$

$\dfrac{dx}{dy} = \dfrac{1}{3\sqrt{1-y^2}}$

Note that this example asked us to find $\dfrac{dx}{dy}$ and not $\dfrac{dy}{dx}$

10. If $y = (\cos 5x)^2$ find $\dfrac{dx}{dy}$.

<u>Solution</u>

$y = (\cos 5x)^2$

This can also be written as $\cos^2 5x$. Hence:

$y = \cos^2 5x$.

Or, $\cos^2 5x = y$

taking the square root of both sides gives:

$\cos 5x = \sqrt{y}$

$5x = \cos^{-1}\sqrt{y}$

$$x = \frac{\cos^{-1}\sqrt{y}}{5}$$

$$= \frac{1}{5}\cos^{-1}\sqrt{y}$$

$$x = \frac{1}{5}\cos^{-1}y^{\frac{1}{2}}$$

Hence, $\dfrac{dx}{dy} = \dfrac{1}{5} \dfrac{-1}{\sqrt{1-(y^{\frac{1}{2}})^2}} \times \dfrac{1}{2}y^{-\frac{1}{2}}$ \quad (Note that $\dfrac{1}{2}y^{-\frac{1}{2}}$ is from the derivative of $y^{\frac{1}{2}}$)

$$= \frac{-1}{5\sqrt{1-y}} \times \frac{1}{2y^{\frac{1}{2}}}$$

$$= \frac{-1}{10y^{\frac{1}{2}}\sqrt{1-y}}$$

$$= \frac{-1}{10\sqrt{y}\sqrt{1-y}}$$

$$\frac{dx}{dy} = \frac{-1}{10\sqrt{y(1-y)}}$$

## Exercise 25

1. If $y = \sin^{-1}2x$ find the $\dfrac{dy}{dx}$.

2. Find the derivative of $\cos^{-1}x$

3. Find the derivative of $\cot^{-1}3x^2$

4. If $y = \cot^{-1}x^4$, find $y = \dfrac{dy}{dx}$

5. Find the derivative $3x\sec^{-1}x$

6. If $y = x^2 + 3\tan^{-1}x$, find $\dfrac{dy}{dx}$.

7. Find the derivative of $\tan^{-1}(x + 5)$

8. If $y = \cos^{-1}\left(\dfrac{1}{x}\right)$, find $\dfrac{dy}{dx}$

9. If $y = \cos5x$, find $\dfrac{dx}{dy}$.

10. If $y = (\sec x)^2$ find $\dfrac{dx}{dy}$.

11. Find the derivative $5x\tan^{-1}3x$

12. If $y = x^2 + 4\sin^{-1}x^5$, find $\dfrac{dy}{dx}$.

13. Find the derivative of $\csc^{-1}x^3$

14. If $y = \tan 2x$, find $\dfrac{\mathrm{d}x}{\mathrm{d}y}$.

15. If $y = \sin^3 5x$, find $\dfrac{\mathrm{d}x}{\mathrm{d}y}$.

# CHAPTER 26
## DERIVATIVES OF HYPERBOLIC FUNCTIONS

Hyperbolic functions are functions in calculus which are expressed by the combination of the exponential functions $e^x$ and $e^{-x}$. The derivatives of the six main hyperbolic functions are as given below.

1. If $y = \sinh x$, then $\dfrac{dy}{dx} = \cosh x$

2. If $y = \cosh x$, then $\dfrac{dy}{dx} = \sinh x$

3. If $y = \tanh x$, then $\dfrac{dy}{dx} = \operatorname{sech}^2 x$

4. If $y = \coth x$, then $\dfrac{dy}{dx} = -\operatorname{cosech}^2 x$

5. If $y = \operatorname{sech} x$, then $\dfrac{dy}{dx} = -\operatorname{sech} x \tanh x$

6. If $y = \operatorname{cosech} x$, then $\dfrac{dy}{dx} = -\operatorname{cosech} x \coth x$

Note that $\sinh x = \dfrac{e^x - e^{-x}}{2}$ and $\cosh x = \dfrac{e^x + e^{-x}}{2}$

**Examples**

1. If $y = \cosh x - 5\sinh x$, find $\dfrac{dy}{dx}$.

Solution

$\quad y = \cosh x - 5\sinh x$

$\dfrac{dy}{dx} = \sinh x - 5\cosh x$


2. Find the derivative of $2x^3 \coth x$

Solution

$\quad y = 2x^3 \coth x$

Using product rule gives $\dfrac{dy}{dx}$ as follows:

$\dfrac{dy}{dx} = 2x^3\left[\dfrac{d(\coth x)}{dx}\right] + \coth x\left[\dfrac{d(2x^3)}{dx}\right]$

$\qquad = 2x^3(-\operatorname{cosech}^2 x) + \coth x(6x^2)$

$\qquad = -2x^3\operatorname{cosech}^2 x + 6x^2\coth x$

$\dfrac{dy}{dx} = -2x^2(x\operatorname{cosech}^2 x - 3\coth x)$

Take note of the change in sign of the term in the bracket. This is due to the negative sign outside the bracket. Expanding the bracket gives the original expression that was factorized.

3. If $y = \dfrac{\cosh x}{x^2 + 1}$ find $\dfrac{dy}{dx}$.

Solution

Let us use product rule to obtained $\dfrac{dy}{dx}$ as follows:

$$\frac{dy}{dx} = \frac{(x^2 + 1)\left[\frac{d(\cosh x)}{dx}\right] - \cosh x\left[\frac{d(x^2 + 1)}{dx}\right]}{(x^2 + 1)^2}$$

$$= \frac{(x^2 + 1)(\sinh x) - \cosh x(2x)}{(x^2 + 1)^2}$$

$$\frac{dy}{dx} = \frac{\sinh x(x^2 + 1) - 2x\cosh x}{(x^2 + 1)^2}$$

4. Find the derivative of $(\sinh 3x)^2$

Solution

$\quad y = (\sinh 3x)^2 \qquad$ (This can also be written as $\sinh^2 3x$)

Let $u = \sinh 3x$

Hence, $y = u^2$

$\quad \dfrac{du}{dx} = 3\cosh 3x \qquad$ (Note that 3 is from the derivative of $3x$)

$\quad \dfrac{dy}{du} = 2u$

$\quad \dfrac{dy}{dx} = \dfrac{dy}{du} \times \dfrac{du}{dx}$

$\qquad = 2u \times 3\cosh 3x$

$\qquad = 6u\cosh 3x$

$\qquad = 6\sinh 3x \cosh 3x \qquad$ (u has been replaced with $\sinh 3x$)

5. Find the derivative of $\sinh^4 2x^3$

Solution

$\quad y = \sinh^4 2x^3$

Let $u = \sinh 2x^3$

Hence, $y = u^4$

$\quad \dfrac{du}{dx} = 6x^2\cosh 2x^3 \qquad$ (Note that $6x^2$ is from the derivative of $2x^3$)

$\quad \dfrac{dy}{du} = 4u^3$

$\quad \dfrac{dy}{dx} = \dfrac{dy}{du} \times \dfrac{du}{dx}$

$\qquad = 4u^3 \times 6x^2\cosh 2x^3$

$\qquad = 24x^2 u^3\cosh 2x^3$

$\quad \dfrac{dy}{dx} = 24x^2\sinh^3 2x^3\cosh 2x^3 \qquad$ (u has been replaced with $\sinh 2x^3$)

273

6. Find the derivative of $\text{cosech}4x^3$

Solution

$$y = \text{cosech}4x^3$$

$$\frac{dy}{dx} = -\text{cosech}4x^3\text{coth}4x^3 \times \frac{d(4x^3)}{dx}$$  (Note that the derivative of cosech is $-$cosechcoth)

$$= -\text{cosech}4x^3\text{coth}4x^3 \times 12x^2$$

$$= -12x^2\text{cosech}4x^3\text{coth}4x^3$$

7. Find the derivative of $\text{coth}^2 2x^4$

Solution

$$y = \text{coth}^2 2x^4$$

Let $u = \text{coth}2x^4$

Hence, $y = u^2$

$$\frac{du}{dx} = -\text{cosech}^2 2x^4 \times \frac{d(2x^4)}{dx}$$

$$= -\text{cosech}^2 2x^4 \times 8x^3$$

$$\frac{du}{dx} = -8x^3\text{cosech}^2 2x^4$$

$$\frac{dy}{du} = 2u$$

Therefore, $\dfrac{dy}{dx} = \dfrac{dy}{du} \times \dfrac{du}{dx}$

$$= 2u \times (-8x^3\text{cosech}^2 2x^4)$$

$$= -16x^3 u\ \text{cosech}^2 2x^4$$

$$\frac{dy}{dx} = -16x^3\text{coth}2x^4\text{cosech}^2 2x^4$$  (Since $u = \text{coth}2x^4$)

8. Find the derivative of $5x\text{sinh}2x$

Solution

$$y = 5x\text{sinh}2x$$

We apply the product rule of differentiation as follows:

Let $u = 5x$

and $v = \text{sinh}2x$

$$\frac{du}{dx} = 5$$

$$\frac{dv}{dx} = 2\text{cosh}2x$$  (Note that the differentiation of $2x$ gives 2)

Hence, $\dfrac{dy}{dx} = u\dfrac{dv}{dx} + v\dfrac{du}{dx}$

$$= (5x \times 2\text{cosh}2x) + (\text{sinh}2x \times 5)$$

$$= 10x\text{cosh}2x + 5\text{sinh}2x$$

$$\frac{dy}{dx} = 5(2x\cosh 2x + \sinh 2x)$$

9. If $y = \dfrac{1 + \cosh 2x}{\sinh 2x}$ find $\dfrac{dy}{dx}$.

<u>Solution</u>

$$y = \frac{1 + \cosh 2x}{\sinh 2x}$$

We apply quotient rule as follows:

$$u = 1 + \cosh 2x$$

$$v = \sinh 2x$$

$$\frac{du}{dx} = 2\sinh 2x$$

$$\frac{dv}{dx} = 2\cosh 2x$$

Hence, $\dfrac{dy}{dx} = \dfrac{v\frac{du}{dx} - u\frac{dv}{dx}}{v^2}$  (Quotient rule)

$$= \frac{\sinh 2x(2\sinh 2x) - (1 + \cosh 2x)(2\cosh 2x)}{(\sinh 2x)^2}$$

$$= \frac{2\sin^2 2x - (2\cosh 2x + 2\cosh^2 2x)}{\sinh^2 2x} \qquad \text{(Note that } (\sinh 2x)^2 = \sinh^2 2x)$$

$$= \frac{2\sinh^2 2x - 2\cosh 2x - 2\cosh^2 2x}{\sinh^2 2x}$$

$$= \frac{2\sinh^2 2x - 2\cosh^2 2x - 2\cosh 2x}{\sinh^2 2x}$$

$$= \frac{-2(\cosh^2 2x - \sinh^2 2x) - 2\cosh 2x}{\sinh^2 2x}$$

$$= \frac{-2(1) - 2\cosh 2x}{\sinh^2 2x} \qquad \text{(Note that } \cosh^2 2x - \sinh^2 2x = 1)$$

$$\frac{dy}{dx} = \frac{-2(1 + \cosh 2x)}{\sinh^2 2x}$$

Or, $\dfrac{dy}{dx} = \dfrac{-2(1 + \cosh 2x)}{\cosh^2 2x - 1}$  (Note that since $\cosh^2 2x - \sinh^2 2x = 1$, then $\cosh^2 2x - 1 = \sinh^2 2x$)

$$= \frac{-2(1 + \cosh 2x)}{(\cosh 2x + 1)(\cosh 2x - 1)}$$

Note that from difference of two squares we have: $a^2 - b^2 = (a + b)(a - b)$.

Hence, $\cosh^2 2x - 1$ is also a difference of two squares since 1 is also $1^2$. Therefore, $\cosh^2 2x - 1 =$ $(\cosh 2x + 1)(\cosh 2x - 1)$ as represented above. Hence the expression above becomes:

$$\frac{dy}{dx} = \frac{-2}{(\cosh 2x - 1)}$$

10. Differentiate $y = \dfrac{\sinh^2 x}{x}$

275

## Solution

$$y = \frac{\sinh^2 x}{x}$$

We use quotient rule as follows:

Therefore, u = $\sinh^2 x$

    v = $x$

Hence, $\dfrac{du}{dx}$ = 2sinh$x$cosh$x$

Since, v = $x$

Then, $\dfrac{dv}{dx}$ = 1

Hence, $\dfrac{dy}{dx} = \dfrac{v\frac{du}{dx} - u\frac{dv}{dx}}{v^2}$

$$= \frac{x(2\sinh x\cosh x) - \sinh^2 x(1)}{x^2}$$

$$= \frac{2x\sinh x\cosh x - \sin^2 x}{x^2}$$

$$= \frac{\sinh x(2x\cosh x - \sinh x)}{x^2}$$

## Exercise 26

1. If y = sinh$3x$ – 2cosh$x$, find $\dfrac{dy}{dx}$.

2. Find the derivative of $5x^2$sech$x$

3. If y = $\dfrac{\sinh 2x}{3x^2}$ find $\dfrac{dy}{dx}$.

4. Find the derivative of $\cosh^2 3x$

5. Find the derivative of $\cosh^3 4x^5$

6. Find the derivative of coth$2x^5$

7. Find the derivative of $(\text{sech}2x^2)^3$

8. Find the derivative of $x^2$cosh$5x$

9. If y = $\dfrac{\tanh 5x}{\coth 3x}$ find $\dfrac{dy}{dx}$.

10. Differentiate y = $\dfrac{2\cosh^3 x}{3x}$

# CHAPTER 27
## DERIVATIVE OF LOGARITHMIC FUNCTIONS

If $y = \log_a x$, then $\dfrac{dy}{dx} = \dfrac{1}{x}\log_a e$, where a is any base.

If $y = \log_e x$, then $\dfrac{dy}{dx} = \dfrac{1}{x}$

Note that $\log_e x$ can also be represented as $\ln x$ and the value of e is 2.718 (to 3 decimal places)

**Examples**

1. Find the derivative of $\log_a 2x$.

**Solution**

$\quad y = \log_a 2x$

We will use chain rule since we have a function that is not just $x$ but $2x$.

Let $u = 2x$

Therefore, $y = \log_a u$

$\dfrac{du}{dx} = 2$

$\dfrac{dy}{du} = \dfrac{1}{u}\log_a e$

$\dfrac{dy}{dx} = \dfrac{dy}{du} \times \dfrac{du}{dx}$

$\quad = \dfrac{1}{u}\log_a e \times 2$

$\quad = \dfrac{2}{u}\log_a e$

$\quad = \dfrac{2}{2x}\log_a e \qquad \text{(since } u = 2x)$

$\dfrac{dy}{dx} = \dfrac{1}{x}\log_a e$

2. Find the derivative of $\log_a(5x - 1)$

Solution

$\quad y = \log_a(5x - 1)$

Let $u = 5x - 1$

Therefore, $y = \log_a u$

$\dfrac{du}{dx} = 5$

$\dfrac{dy}{du} = \dfrac{1}{u}\log_a e$

$\dfrac{dy}{dx} = \dfrac{dy}{du} \times \dfrac{du}{dx}$

$$= \frac{1}{u} \log_a e \times 5$$

$$= \frac{5}{u} \log_a e$$

$$\frac{dy}{dx} = \frac{5}{5x-1} \log_a e \qquad \text{(since } u = 5x - 1\text{)}$$

3. Differentiate $\log_a(4x - 3)^2$ with respect to $x$

<u>Solution</u>

$$y = \log_a(4x - 3)^2$$

Let $u = (4x - 3)^2$

Therefore, $y = \log_a u$

$$\frac{du}{dx} = 2(4x - 3)^{2-1} \times 4 \qquad \text{(Note that 4 is from the derivative of } 4x - 3\text{)}$$

$$\frac{du}{dx} = 8(4x - 3)$$

$$\frac{dy}{du} = \frac{1}{u} \log_a e$$

$$\frac{dy}{dx} = \frac{dy}{du} \times \frac{du}{dx}$$

$$= \frac{1}{u} \log_a e \times 8(4x - 3)$$

$$= \frac{8(4x-3)}{u} \log_a e$$

$$= \frac{8(4x-3)}{(4x-3)^2} \log_a e \qquad \text{[since } u = (4x - 3)^2\text{]}$$

$$\frac{dy}{dx} = \frac{8}{4x-3} \log_a e \qquad \text{(One } 4x - 3 \text{ cancels out)}$$

4. If $y = \log_a \sqrt{1 + 2x}$, find $\frac{dy}{dx}$

<u>Solution</u>

$$y = \log_a \sqrt{1 + 2x}$$

Let $u = \sqrt{1 + 2x} = (1 + 2x)^{\frac{1}{2}}$

Therefore, $y = \log_a u$

$$\frac{du}{dx} = \frac{1}{2}(1 + 2x)^{\frac{1}{2}-1} \times 2$$

$$= (1 + 2x)^{-\frac{1}{2}} \qquad \text{(Note that } \frac{1}{2} \times 2 = 1\text{)}$$

$$\frac{du}{dx} = \frac{1}{(1+2x)^{\frac{1}{2}}}$$

$$\frac{dy}{du} = \frac{1}{u} \log_a e$$

$$\frac{dy}{dx} = \frac{dy}{du} \times \frac{du}{dx}$$

$$= \frac{1}{u} \log_a e \times \frac{1}{(1+2x)^{\frac{1}{2}}}$$

$$= \frac{1}{(1+2x)^{\frac{1}{2}}} \log_a e \times \frac{1}{(1+2x)^{\frac{1}{2}}} \qquad \text{[Note that u has been replaced with } (1+2x)^{\frac{1}{2}}]$$

$$\frac{dy}{dx} = \frac{1}{1+2x} \log_a e \qquad \text{[Note that } (1+2x)^{\frac{1}{2}} \times (1+2x)^{\frac{1}{2}} = (1+2x)^1 \text{ by adding exponents]}$$

5. Find $\frac{dy}{dx}$ given that $y = \log_a\frac{1-3x}{1+3x}$

Solution

$$y = \log_a\frac{1-3x}{1+3x}$$

Or, $y = \log_a(1 - 3x) - \log_a(1 + 3x) \qquad \left(\text{Recall that } \log_x\left(\frac{a}{b}\right) = \log_x a - \log_x b\right)$

$$\frac{dy}{dx} = \frac{-3}{1-3x} \log_a e - \frac{3}{1+3x} \log_a e$$

$$= -\log_a e \left(\frac{3}{1-3x} + \frac{3}{1+3x}\right)$$

$$= -\log_a e \left[\frac{3(1+3x) + 3(1-3x)}{(1-3x)(1+3x)}\right]$$

$$= -\log_a e \left[\frac{3 + 9x + 3 - 9x}{1 - 9x^2}\right] \qquad \text{[Note that } (1 - 3x)(1 + 3x) = 1 - 9x^2]$$

$$= -\log_a e\left(\frac{6}{1 - 9x^2}\right)$$

$$\frac{dy}{dx} = \frac{-6}{1 - 9x^2} \log_a e$$

6. If $y = \log_{10}(x^2 - 2)$, find $\frac{dy}{dx}$.

Solution

$$y = \log_{10}(x^2 - 2)$$

The value 10 represents a in other examples. So we are going to differentiate y like the examples above except that we will write 10 wherever 'a' should be.

$$y = \log_{10}(x^2 - 2)$$
$$u = x^2 - 2$$

Hence, $y = \log_{10}u$

$$\frac{du}{dx} = 2x$$

$$\frac{dy}{du} = \frac{1}{u} \log_{10}e \qquad \text{(Just like differentiating } \log_a u)$$

$$\frac{dy}{dx} = \frac{dy}{du} \times \frac{du}{dx}$$

$$= \frac{1}{u} \log_{10}e \ \times \ 2x$$

$$= \frac{2x}{u} \log_{10}e$$

$$\frac{dy}{dx} = \frac{2x}{x^2 - 2} \log_{10}e$$

7. Find $\frac{dy}{dx}$ if $y = \log_{10}\frac{1}{x}$

Solution

$$y = \log_{10}\frac{1}{x}$$

Let us solve this question directly without using $u = \frac{1}{x}$ as follows:

$$\frac{dy}{dx} = \frac{\frac{d\left(\frac{1}{x}\right)}{dx}}{\frac{1}{x}} \ \times \ \log_{10}e$$

$$= \frac{\frac{d(x^{-1})}{dx}}{x^{-1}} \ \times \ \log_{10}e$$

$$= \frac{-x^{-2}}{x^{-1}} \ \times \ \log_{10}e \qquad \text{(Note that the derivative of } x^{-1} \text{ is } -x^{-2}\text{)}$$

$$= -x^{-2} \ \times \ x^{1} \ \times \ \log_{10}e \qquad \text{(Since } \frac{1}{x^{-1}} = x^{1}\text{)}$$

$$= -x^{-1} \ \times \ \log_{10}e \qquad \text{(After adding the exponents of } x\text{)}$$

$$\frac{dy}{dx} = -\frac{1}{x} \log_{10}e$$

8. Find the derivative of $\log_{e}(2 - x^3)$

Solution

$$y = \log_{e}(2 - x^3)$$

Note that the base here is 'e' and not 'a'.

Let $u = 2 - x^3$

Hence, $y = \log_{e}u$

$$\frac{du}{dx} = -3x^2$$

$$\frac{dy}{du} = \frac{1}{u} \qquad \text{(Recall that the derivative of } \log_{e}x = \frac{1}{x}\text{)}$$

$$\frac{dy}{dx} = \frac{dy}{du} \ \times \ \frac{du}{dx}$$

$$= \frac{1}{u} \ \times \ -3x^2$$

$$= \frac{-3x^2}{u}$$

$$\frac{dy}{dx} = \frac{-3x^2}{2 - x^3}$$

9. Find the derivative of $(\log_e 5x)^2$

Solution

$$y = (\log_e 5x)^2$$

Let $u = \log_e 5x$

Hence, $y = u^2$

$$\frac{du}{dx} = \frac{\frac{d(5x)}{dx}}{5x}$$

$$= \frac{5}{5x}$$

$$\frac{du}{dx} = \frac{1}{x}$$

$$\frac{dy}{du} = 2u$$

$$\frac{dy}{dx} = \frac{dy}{du} \times \frac{du}{dx}$$

$$= 2u \times \frac{1}{x}$$

$$= \frac{2u}{x}$$

$$\frac{dy}{dx} = \frac{2}{x}\log_e 5x \quad \text{(Since } u = \log_e 5x)$$

10. If $y = \ln\sqrt{3x^2 - 4}$, find $\frac{dy}{dx}$.

Solution

$$y = \ln\sqrt{3x^2 - 4}$$

Note that $\ln\sqrt{3x^2 - 4}$ is also the same as $\log_e\sqrt{3x^2 - 4}$

Hence, $y = \log_e\sqrt{3x^2 - 4}$

Or, $y = \log_e(3x^2 - 4)^{\frac{1}{2}}$

Let $u = (3x^2 - 4)^{\frac{1}{2}}$

Hence, $y = \log_e u$

$$\frac{du}{dx} = \frac{1}{2}(3x^2 - 4)^{\frac{1}{2}-1} \times 6x \quad \text{(Note that } 6x \text{ is from the derivative of } 3x^2 - 4)$$

$$= \frac{6x}{2}(3x^2 - 4)^{-\frac{1}{2}}$$

$$\frac{du}{dx} = \frac{3x}{(3x^2 - 4)^{\frac{1}{2}}}$$

$$\frac{dy}{du} = \frac{1}{u}$$

$$\frac{dy}{dx} = \frac{dy}{du} \times \frac{du}{dx}$$

$$= \frac{1}{u} \times \frac{3x}{(3x^2-4)^{\frac{1}{2}}}$$

$$= \frac{1}{(3x^2-4)^{\frac{1}{2}}} \times \frac{3x}{(3x^2-4)^{\frac{1}{2}}}$$

$$\frac{dy}{dx} = \frac{3x}{3x^2-4}$$   (Note that $(3x^2-4)^{\frac{1}{2}} \times (3x^2-4)^{\frac{1}{2}} = 3x^2-4$, after adding their exponents)

11. Find $\frac{dy}{dx}$ if $y = \sqrt[3]{x}\ \ln 2x$.

Solution

$$y = \sqrt[3]{x}\ \ln 2x$$

We are going to apply product rule here.

$$u = \sqrt[3]{x} = x^{\frac{1}{3}}$$

$$v = \ln 2x \qquad \text{(Note that } \ln 2x \text{ is the same as } \log_e 2x)$$

$$\frac{du}{dx} = \frac{1}{3}x^{\frac{1}{3}-1}$$

$$= \frac{1}{3}x^{-\frac{2}{3}}$$

$$\frac{dv}{dx} = \frac{2}{2x}$$

$$= \frac{1}{x}$$

Hence, $\frac{dy}{dx} = u\frac{dv}{dx} + v\frac{du}{dx}$    (Product rule)

$$= x^{\frac{1}{3}}\left(\frac{1}{x}\right) + \ln 2x \left(\frac{1}{3}x^{-\frac{2}{3}}\right)$$

$$= x^{\frac{1}{3}}(x^{-1}) + \frac{1}{3}x^{-\frac{2}{3}}\ln 2x$$

$$= x^{-\frac{2}{3}}\left(1 + \frac{\ln 2x}{3}\right)$$

$$= x^{-\frac{2}{3}}\left(\frac{3 + \ln 2x}{3}\right)$$

$$\frac{dy}{dx} = \frac{1}{x^{\frac{2}{3}}}\left(\frac{3 + \ln 2x}{3}\right)$$

Or, $\frac{dy}{dx} = \frac{1}{\sqrt[3]{x^2}}\left(\frac{3 + \ln 2x}{3}\right)$

12. Find $\frac{dy}{dx}$ given that $y = \ln(1 + 2x)^2$

Solution

$$y = \ln(1 + 2x)^2 \qquad [\text{Also } y = \log_e(1 + 2x)^2]$$

282

Let $u = (1 + 2x)^2$

Hence, $y = \ln u$

$$\frac{du}{dx} = 2(1 + 2x) \times \frac{d(2x)}{dx}$$

$$= 2(1 + 2x) \times 2x$$

$$\frac{du}{dx} = 4(1 + 2x)$$

$$\frac{dy}{du} = \frac{1}{u}$$

$$\frac{dy}{dx} = \frac{1}{u} \times 4(1 + 2x)$$

$$= \frac{4(1+2x)}{u}$$

$$= \frac{4(1+2x)}{(1+2x)^2}$$

$$= \frac{4}{1+2x}$$

13. Differentiate with respect to $x$: $5x\ln(3x^2 - 2)$

<u>Solution</u>

$y = 5x\ln(3x^2 - 2)$

By using product rule:

$u = 5x$

$v = \ln(3x^2 - 2)$

$$\frac{du}{dx} = 5$$

$$\frac{dv}{dx} = \frac{6x}{3x^2 - 2}$$

Hence, $\dfrac{dy}{dx} = u\dfrac{dv}{dx} + v\dfrac{du}{dx}$

$$= 5x\left(\frac{6x}{3x^2 - 2}\right) + \ln(3x^2 - 2)(5)$$

$$\frac{dy}{dx} = \frac{30x^2}{3x^2 - 2} + 5\ln(3x^2 - 2)$$

14. Find the derivative of $\dfrac{\ln x}{x}$

<u>Solution</u>

$y = \dfrac{\ln x}{x}$

We are going to use quotient rule as follows:

$u = \ln x$

$v = x$

$$\frac{du}{dx} = \frac{1}{x}$$

$$\frac{dv}{dx} = 1$$

Hence, $\dfrac{dy}{dx} = \dfrac{v\dfrac{du}{dx} - u\dfrac{dv}{dx}}{v^2}$

$$= \frac{x\left(\frac{1}{x}\right) - \ln x\,(1)}{x^2}$$

$$\frac{dy}{dx} = \frac{1 - \ln x}{x^2}$$

## Exercise 27

1. Find the derivative of $\log_a 7x$.

2. Find the derivative of $\log_a(x^3 + 5)$

3. Differentiate $\log_a(2x^3 - 5)^6$ with respect to $x$

4. If $y = \log_a x^{\frac{2}{3}}$, find $\dfrac{dy}{dx}$

5. Find $\dfrac{dy}{dx}$ given that $y = \log_a \dfrac{1 + x^2}{1 - x^2}$

6. If $y = \log_5(5x^3 - 1)$, find $\dfrac{dy}{dx}$.

7. Find $\dfrac{dy}{dx}$ if $y = \log_2 \dfrac{1}{x^2}$

8. Find the derivative of $\log_e(x^2 + 3)$

9. Find the derivative of $(\log_e x)^3$

10. If $y = \ln\sqrt{1 - 2x^5}$, find $\dfrac{dy}{dx}$.

11. Find $\dfrac{dy}{dx}$ if $y = x^4 \ln x^2$

12. Find $\dfrac{dy}{dx}$ given that $y = \ln(3 - 7x)^4$

13. Differentiate with respect to $x$:  $3x^2\ln(4x^3 + 1)$

14. Find the derivative of $\dfrac{\ln x^2}{x^2}$

15. Find the derivative of $\log_e(1 - 5x^2)^3$

# CHAPTER 28
## DERIVATIVE OF EXPONENTIAL FUNCTIONS

If $y = a^x$, then $\dfrac{dy}{dx} = a^x\log_e a$, where 'a' is any number.

If $y = e^x$, then $\dfrac{dy}{dx} = e^x$

**Examples**

1. Differentiate with respect to $x$: $a^{5x}$

Solution

$\quad y = a^{5x}$

Let $u = 5x$

$\quad y = a^u$

$\dfrac{du}{dx} = 5$

$\dfrac{dy}{du} = a^u\log_e a$

Hence, $\dfrac{dy}{dx} = \dfrac{dy}{du} \ \times \ \dfrac{du}{dx}$

$\quad = a^u\log_e a \ \times \ 5$

$\quad = 5a^u\log_e a$

$\dfrac{dy}{dx} = 5a^{5x}\log_e a \quad$ (since $u = 5x$)

2. Find the derivative of $a^{x^2-3x+4}$

Solution

$\quad y = a^{x^2-3x+4}$

Let us solve this example directly without using $u = x^2 - 3x + 4$

$\dfrac{dy}{dx} = \dfrac{d(x^2-3x+4)}{dx} \ \times \ a^{x^2-3x+4} \ \times \log_e a$

$\quad = (2x-3)(a^{x^2-3x+4} \ \times \ \log_e a)$

3. If $y = 3x^2 a^{5x}$, find $\dfrac{dy}{dx}$.

Solution

$\quad y = 3x^2 a^{5x}$

We apply product rule as follows:

$\quad u = 3x^2$

$\quad v = a^{5x}$

$$\frac{du}{dx} = 6x$$

$$\frac{dv}{dx} = 5a^{5x}\log_e a$$

Hence, $\dfrac{dy}{dx} = u\dfrac{dv}{dx} + v\dfrac{du}{dx}$

$$= 3x^2(5a^{5x}\log_e a) + a^{5x}(6x)$$

$$= 15x^2 a^{5x}\log_e a + 6x a^{5x}$$

Factorizing the expression above gives:

$$\frac{dy}{dx} = 3xa^{5x}(5x\log_e a + 2)$$

4. Find the derivative of $\dfrac{e^x + e^{-x}}{e^x - e^{-x}}$

Solution

$$y = \frac{e^x + e^{-x}}{e^x - e^{-x}}$$

$$u = e^x + e^{-x}$$

$$v = e^x - e^{-x}$$

$$\frac{du}{dx} = e^x - e^{-x} \quad \text{(The derivative of } e^{-x} = -1 \times e^{-x} = -e^{-x}\text{. The value } -1 \text{ is from the derivative of } -x)$$

$$\frac{dv}{dx} = e^x + e^{-x}$$

Hence, $\dfrac{dy}{dx} = \dfrac{v\dfrac{du}{dx} - u\dfrac{dv}{dx}}{v^2}$ (Quotient rule)

$$= \frac{(e^x - e^{-x})(e^x - e^{-x}) - (e^x + e^{-x})(e^x + e^{-x})}{(e^x - e^{-x})^2}$$

Expanding the numerator and denominator gives:

$$\frac{dy}{dx} = \frac{(e^x)(e^x) - (e^x)(e^{-x}) - (e^{-x})(e^x) + (e^{-x})(e^{-x}) - [(e^x)(e^x) + (e^x)(e^{-x}) + (e^{-x})(e^x) + (e^{-x})(e^{-x})]}{(e^x - e^{-x})^2}$$

$$= \frac{e^{2x} - 1 - 1 + e^{-2x} - (e^{2x} + 1 + 1 + e^{-2x})}{(e^x - e^{-x})^2} \quad \text{(Note that exponents have been added, and } e^0 = 1)$$

$$= \frac{e^{2x} - 2 + e^{-2x} - e^{2x} - 1 - 1 - e^{-2x})}{(e^x - e^{-x})^2}$$

$$\frac{dy}{dx} = \frac{-4}{(e^x - e^{-x})^2} \quad (e^{2x} \text{ and } e^{-2x} \text{ cancel out)}$$

5. If $y = e^{\sqrt{x}}$ find $\dfrac{dy}{dx}$.

Solution

$$y = e^{\sqrt{x}} = e^{x^{\frac{1}{2}}}$$

$$\frac{dy}{dx} = \frac{d(x)^{\frac{1}{2}}}{dx} \times e^{x^{\frac{1}{2}}}$$

$$= \frac{1}{2}x^{-\frac{1}{2}} \times e^{x^{\frac{1}{2}}}$$

$$= \frac{1}{2}\left(\frac{1}{x^{\frac{1}{2}}}\right) \times e^{x^{\frac{1}{2}}}$$

$$= \frac{e^{x^{\frac{1}{2}}}}{2x^{\frac{1}{2}}}$$

$$\frac{dy}{dx} = \frac{e^{\sqrt{x}}}{2\sqrt{x}}$$

6. Find $\frac{dy}{dx}$ if $y = x^{\frac{1}{2}} e^{x^{\frac{1}{2}}}$

<u>Solution</u>

$$y = x^{\frac{1}{2}} e^{x^{\frac{1}{2}}}$$

From product rule:

$$\frac{dy}{dx} = x^{\frac{1}{2}} \frac{d(e^{x^{\frac{1}{2}}})}{dx} + e^{x^{\frac{1}{2}}} \frac{d(x^{\frac{1}{2}})}{dx}$$

$$= x^{\frac{1}{2}}\left(\frac{e^{x^{\frac{1}{2}}}}{2x^{\frac{1}{2}}}\right) + e^{x^{\frac{1}{2}}}\left(\frac{1}{2}x^{-\frac{1}{2}}\right) \quad \text{(Note that the derivative of } e^{x^{\frac{1}{2}}} \text{ is } \frac{e^{x^{\frac{1}{2}}}}{2x^{\frac{1}{2}}} \text{ from example 5)}$$

$$= \frac{e^{x^{\frac{1}{2}}}}{2} + \frac{e^{x^{\frac{1}{2}}}}{2x^{\frac{1}{2}}}$$

$$= \frac{x^{\frac{1}{2}}e^{x^{\frac{1}{2}}} + e^{x^{\frac{1}{2}}}}{2x^{\frac{1}{2}}}$$

$$\frac{dy}{dx} = \frac{e^{x^{\frac{1}{2}}}(x^{\frac{1}{2}} + 1)}{2x^{\frac{1}{2}}}$$

7. Find the derivative of $a^{2x} - a^{-2x}$

<u>Solution</u>

$$y = a^{2x} - a^{-2x}$$

$$\frac{dy}{dx} = \frac{d(2x)}{dx}(a^{2x}\log_e a) - \left[\frac{d(-2x)}{dx}(a^{-2x}\log_e a)\right]$$

$$= 2a^{2x}\log_e a - (-2a^{-2x}\log_e a)$$

$$= 2a^{2x}\log_e a + 2a^{-2x}\log_e a$$

$$= 2\log_e a(a^{2x} + a^{-2x})$$

## 8. Find the derivative of $e^{2x}\log_e 3x$

<u>Solution</u>

$$y = e^{2x}\log_e 3x$$

By product rule:

$$u = e^{2x}$$

$$v = \log_e 3x \quad (\text{or } v = \ln 3x)$$

$$\frac{du}{dx} = 2e^{2x}$$

$$\frac{dv}{dx} = \frac{3}{3x}$$

$$= \frac{1}{x}$$

$$\frac{dy}{dx} = e^{2x}\left(\frac{1}{x}\right) + \log_e 3x(2e^{2x})$$

$$= \frac{e^{2x}}{x} + 2e^{2x}\log_e 3x$$

$$\frac{dy}{dx} = e^{2x}\left(\frac{1}{x} + 2\log_e 3x\right)$$

## 9. Differentiate with respect to $x$: $2\sqrt{xe^x}$

<u>Solution</u>

$$y = 2\sqrt{xe^x}$$

Or, $y = 2\sqrt{x}\ \sqrt{e^x}$

From product rule:

$$u = 2\sqrt{x} = 2x^{\frac{1}{2}}$$

$$v = \sqrt{e^x} = (e^x)^{\frac{1}{2}}$$

$$v = e^{\frac{1}{2}x} \quad (\text{When the exponents are multiplied})$$

$$\frac{du}{dx} = \frac{1}{2} \times 2x^{-\frac{1}{2}}$$

$$= x^{-\frac{1}{2}}$$

$$\frac{dv}{dx} = \frac{1}{2}e^{\frac{1}{2}x} \quad (\text{Note that } \frac{1}{2} \text{ is from the derivative of } \frac{1}{2}x)$$

$$\frac{dy}{dx} = u\frac{dv}{dx} + v\frac{du}{dx}$$

$$= 2x^{\frac{1}{2}}\left(\frac{1}{2}e^{\frac{1}{2}x}\right) + e^{\frac{1}{2}x}\left(x^{-\frac{1}{2}}\right)$$

$$= \left(\frac{1}{2} \times 2x^{\frac{1}{2}} \times e^{\frac{1}{2}x}\right) \times \frac{e^{\frac{1}{2}x}}{x^{\frac{1}{2}}}$$

$$= x^{\frac{1}{2}}e^{\frac{1}{2}x} + \frac{e^{\frac{1}{2}x}}{x^{\frac{1}{2}}}$$

$$= \frac{x \, e^{\frac{1}{2}x} + e^{\frac{1}{2}x}}{x^{\frac{1}{2}}}$$

$$= \frac{e^{\frac{1}{2}x}(x+1)}{x^{\frac{1}{2}}}$$

$$= \frac{e^{\frac{x}{2}}(x+1)}{x^{\frac{1}{2}}}$$

$$\frac{dy}{dx} = \frac{\sqrt{e^x}(x+1)}{\sqrt{x}}$$

10. Find the derivative of $2x\log_{10}x$

Solution

$y = 2x\log_{10}x$      (Note that this is similar to $2x\log_a x$)

$u = 2x$

$v = \log_{10}x$

$\frac{du}{dx} = 2$

$\frac{dv}{dx} = \frac{1}{x}\log_{10}e$      (Recall that the derivative of $\log_a x$ is $\frac{1}{x}\log_a e$)

$\frac{dy}{dx} = u\frac{dv}{dx} + v\frac{du}{dx}$

$\qquad = 2x\left(\frac{1}{x}\log_{10}e\right) + \log_{10}x(2)$

$\qquad = 2\log_{10}e + 2\log_{10}x$

$\frac{dy}{dx} = 2(\log_{10}e + \log_{10}x)$

Or, $\frac{dy}{dx} = 2\left(\frac{\log_e e}{\log_e 10} + \log_{10}x\right)$

Note that the rule of change of base in logarithm that has been applied above is given by:

If $\log_a b$ is to be converted to a new base, e, then: $\log_a b = \frac{\log_e b}{\log_e a}$

$\frac{dy}{dx} = 2\left(\frac{1}{\log_e 10} + \log_{10}x\right)$      (Recall that $\log_x x = 1$, or $\log_e e = 1$)

11. Find the derivative of $10^{2x}$

Solution

$y = 10^{2x}$      (This is similar to $a^{2x}$)

$\frac{dy}{dx} = \frac{d(2x)}{dx} \times 10^{2x}\log_e 10$

290

$$= 2 \times 10^{2x}\log_e 10$$
$$= (10^{2x})2\log_e 10$$
$$= 10^{2x}\log_e 10^2 \qquad \text{(Recall that } a\log_x y = \log_x y^a)$$
$$\frac{dy}{dx} = 10^{2x}\log_e 100$$

12. If $y = e^{\ln x}$ find $\frac{dy}{dx}$

<u>Solution</u>

$y = e^{\ln x}$     (Or $y = e^{\log_e x}$ since $\ln x = \log_e x$)

Let, $u = \ln x$

$$y = e^u$$
$$\frac{du}{dx} = \frac{1}{x}$$
$$\frac{dy}{du} = e^u$$
$$\frac{dy}{dx} = e^u \times \frac{1}{x}$$
$$= \frac{e^u}{x}$$
$$= \frac{e^{\ln x}}{x} \qquad \text{(Since } u = \ln x)$$
$$= \frac{e^{\log_e x}}{x} \qquad \text{(Since } \ln x = \log_e x)$$
$$= \frac{x}{x} \qquad \text{(Recall the identity that } m^{\log_m n} = n)$$
$$\frac{dy}{dx} = 1$$

This example shows that calculus works, since the derivative of $x$ is 1, and the derivative of $e^{\log_e x} = 1$ because $e^{\log_e x} = x$. Hence, we could have solved this example by stating that:

$$\frac{d(e^{\ln x})}{dx} = \frac{d(x)}{dx} = 1$$

13. What is the derivative of $\frac{e^{\frac{1}{x}}}{x^2}$

<u>Solution</u>

$$y = \frac{e^{\frac{1}{x}}}{x^2}$$

From quotient rule:

$$u = e^{\frac{1}{x}}$$
$$v = x^2$$

291

$$\frac{du}{dx} = \frac{d\left(\frac{1}{x}\right)}{dx} \times e^{\frac{1}{x}}$$

$$= \frac{d(x^{-1})}{dx} \times e^{\frac{1}{x}}$$

$$= -1x^{-2} \times e^{\frac{1}{x}}$$

$$= \frac{-1}{x^2} \times e^{\frac{1}{x}}$$

$$\frac{du}{dx} = \frac{-e^{\frac{1}{x}}}{x^2}$$

$$\frac{du}{dx} = 2x$$

$$\frac{dy}{dx} = \frac{v\frac{du}{dx} - u\frac{dv}{dx}}{v^2} \qquad \text{(Quotient rule)}$$

$$= \frac{x^2\left(\frac{-e^{\frac{1}{x}}}{x^2}\right) - e^{\frac{1}{x}}(2x)}{(x^2)^2}$$

$$= \frac{-e^{\frac{1}{x}} - 2x\,e^{\frac{1}{x}}}{x^4} \qquad \text{(Note that } x^2 \text{ has cancelled out)}$$

14. Find the derivative of $e^{-5x} + 5e$

Solution

$$y = e^{-5x} + 5e$$

$$\frac{dy}{dx} = \frac{d(-5x)}{dx} \times e^{-5x} + \frac{d(5e)}{dx}$$

$$= -5e^{-5x} + 0 \qquad \text{(Note that 5e is a constant and the derivative of a constant is zero)}$$

$$\frac{dy}{dx} = -5e^{-5x}$$

15. If $y = e^{(2x+3)^2}$ find $\frac{dy}{dx}$.

Solution

$$y = e^{(2x+3)^2}$$

Let $u = (2x+3)^2$

Hence, $y = e^u$

$$\frac{du}{dx} = 2(2x+3)^{2-1} \times \frac{d(2x)}{dx} \qquad \text{(Using chain rule)}$$

$$= 2(2x+3) \times 2$$

$$= 4(2x+3)$$

$$\frac{dy}{du} = e^u$$

$$\frac{dy}{dx} = \frac{dy}{du} \times \frac{du}{dx}$$
$$= e^u \times 4(2x+3)$$
$$= e^{(2x+3)^2} \times 4(2x+3) \qquad \text{(Note that } u = (2x+3)^2\text{)}$$
$$\frac{dy}{dx} = 4(2x+3)e^{(2x+3)^2}$$

16. Differentiate with respect to $\theta$, $5^{2\theta}$

Solution

$$y = 5^{2\theta} \qquad \text{(This is like } y = a^{2\theta}\text{)}$$
$$\frac{dy}{d\theta} = \frac{d(2\theta)}{d\theta} \times 5^{2\theta} \log_e 5$$
$$= 2 \times 5^{2\theta} \log_e 5$$
$$= (5^{2\theta})2\log_e 5$$
$$= 5^{2\theta} \log_e 5^2 \qquad \text{(Note that } 2\log_e 5 = \log_e 5^2\text{)}$$
$$\frac{dy}{d\theta} = 5^{2\theta} \log_e 25$$

## Exercise 28

1. Differentiate with respect to $x$:  $5a^{2x}$

2. Find the derivative of $a^{5x^3 - x}$

3. If $y = x^4 a^{3x}$, find $\dfrac{dy}{dx}$.

4. Find the derivative of $\dfrac{3 + e^{-x}}{e^x}$

5. If $y = 2e^{\sqrt{3x}}$ find $\dfrac{dy}{dx}$.

6. Find $\dfrac{dy}{dx}$ if $y = 6x^3 e^{x^2}$

7. Find the derivative of $a^{-5x} + a^{5x}$

8. Find the derivative of $3e^{4x}\log_e 2x^2$

9. Differentiate with respect to $x$:  $3x^2\sqrt{e^{5x}}$

10. Find the derivative of $x^4\log_{10}7x$

11. Find the derivative of $6^{3x}$

12. Find the derivative of $e^{x^2 - 3x}$

13. What is the derivative of $\dfrac{2e^{\frac{3}{x}}}{10x^5}$

14. Find the derivative of $1 - 2e^{-3x}$

293

15. If $y = e^{(x^2+x)^3}$ find $\dfrac{dy}{dx}$.

16. Differentiate with respect to $x$, $2^{x^2}$

17. Find the derivative of $\dfrac{a^x - a^{-x}}{a^{-x}}$

18. Find the derivative of $e^x \ln 10x^3$

19. Differentiate with respect to $x$:  $e^{2x}\sqrt{5x}$

20. Find the derivative of $2x^5 \log_{10} 5x^4$

# CHAPTER 29
## LOGARITHMIC DIFFERENTIATION

A function of $x$ could be raised to an exponent which is also a function of $x$. An example is $x^x$.

The differentiation of such a function is called logarithmic differentiation.

This type of function is differentiated by first taking the logarithm of both sides of the equation, and then differentiating implicitly before making $\frac{dy}{dx}$ the subject of the formula (i.e. solving for $\frac{dy}{dx}$). Finally, y (i.e. the function) is substituted into the final answer.

**Examples**

1. Find the derivative of $x^x$.

<u>Solution</u>

$$y = x^x$$

This is a function of $x$ raised to an exponent which is also a function of $x$.

Hence, take the logarithm to base e of both sides as follows:

$$\log_e y = \log_e x^x$$
$$\log_e y = x\log_e x \quad \text{(Note that } \log_m n^b = b\log_m n\text{)}$$

Now differentiate y implicitly and differentiate the right hand side using product rule. This gives:

$$\frac{1}{y}\frac{dy}{dx} = x\frac{1}{x} + \log_e x(1) \quad \text{(Use of product rule on the right hand side, with u = } x \text{ and v = } \log_e x\text{)}$$

$$\frac{1}{y}\frac{dy}{dx} = 1 + \log_e x$$

Multiply both sides by y to make $\frac{dy}{dx}$ the subject of the formula as follows:

$$\frac{dy}{dx} = y(1 + \log_e x)$$
$$\frac{dy}{dx} = x^x(1 + \log_e x) \quad \text{(Since y = } x^x\text{)}$$

2. Find the derivative of $x^{\ln x}$.

<u>Solution</u>

$$y = x^{\ln x}$$

Taking the logarithm of both sides gives:

$$\log_e y = \log_e x^{\ln x}$$
$$\log_e y = \ln x\log_e x$$

Differentiate y implicitly and treat the right hand side using product rule as follows:

$$\frac{1}{y}\frac{dy}{dx} = \ln x\left(\frac{1}{x}\right) + \log_e x\left(\frac{1}{x}\right)$$

$$\frac{1}{y}\frac{dy}{dx} = \left(\frac{\ln x}{x}\right) + \left(\frac{\ln x}{x}\right)$$    (Note that $\log_e x$ can also be written as $\ln x$)

$$\frac{1}{y}\frac{dy}{dx} = \frac{2\ln x}{x}$$

Multiply both sides by y. This gives:

$$\frac{dy}{dx} = y\left(\frac{2\ln x}{x}\right)$$

$$\frac{dy}{dx} = x^{\ln x}\left(\frac{2\ln x}{x}\right)$$    (Since $y = x^{\ln x}$)

Or, $$\frac{dy}{dx} = \frac{2}{x}x^{\ln x}\ln x$$

3. If $y = \ln(2 + x)^x$ find $\frac{dy}{dx}$.

Solution

$$y = \ln(2 + x)^x$$

In this case we are not going to take the logarithm of both sides since there is already logarithm on the right hand side. Hence, we proceed as follows:

$$y = \ln(2 + x)^x$$

$$y = x\log_e(2 + x)$$    [Note that $\ln(2 + x)$ can also be written as $\log_e(2 + x)$]

Using product rule gives:

$$\frac{dy}{dx} = x\left(\frac{1}{2 + x}\right) + \log_e(2 + x)(1)$$

$$\frac{dy}{dx} = \frac{x}{2 + x} + \log_e(2 + x)$$

4. Find $\frac{dy}{dx}$ if $y = (x^2 - 3)^{5x}$

Solution

$$y = (x^2 - 3)^{5x}$$

Taking the natural logarithm of both sides gives:

$$\log_e y = \log_e(x^2 - 3)^{5x}$$

$$\log_e y = 5x\log_e(x^2 - 3)$$

Using implicit differentiation for the left hand side and product rule for the right hand side gives:

$$\frac{1}{y}\frac{dy}{dx} = 5x\left(\frac{2x}{x^2 - 3}\right) + \log_e(x^2 - 3)(5)$$

$$\frac{1}{y}\frac{dy}{dx} = \frac{10x^2}{x^2 - 3} + 5\log_e(x^2 - 3)$$

Multiply both sides by y to obtain:

$$\frac{dy}{dx} = y\left[\frac{10x^2}{x^2 - 3} + 5\log_e(x^2 - 3)\right]$$

$$\frac{dy}{dx} = (x^2 - 3)^{5x}\left[\frac{10x^2}{x^2 - 3} + 5\log_e(x^2 - 3)\right] \qquad \text{[Note that } y = (x^2 - 3)^{5x}]$$

5. If $y = \dfrac{(x^2 - 5)(3x - 1)^2}{x^7(2x^3 - 3)}$

Solution

This is not a function of $x$ raised to an exponent which is also a function of $x$. This function can be differentiated by the use of product and quotient rule. However, the use of these rules will be a nightmare. Hence, we have to apply logarithmic differentiation. This is done as follows:

$$y = \frac{(x^2 - 5)(3x - 1)^2}{x^7(2x^3 - 3)}$$

Taking the logarithm of both sides gives:

$$\log_e y = \log_e\left[\frac{(x^2 - 5)(3x - 1)^2}{x^7(2x^3 - 3)}\right]$$

In order to continue, we have to recall the theory of logarithm as follows:

1. $\log_b(cd) = \log_b c + \log_b d$

2. $\log_b\left(\dfrac{c}{d}\right) = \log_b c - \log_b d$

3. $\log_b c^d = d\log_b c$

Therefore, we continue the logarithmic differentiation as follows:

$$\log_e y = \log_e\left[\frac{(x^2 - 5)(3x - 1)^2}{x^7(2x^3 - 3)}\right]$$

Applying the theory above gives:

$$\log_e y = \log_e[(x^2 - 5)(3x - 1)^2] - \log_e[x^7(2x^3 - 3)]$$
$$\log_e y = \log_e(x^2 - 5) + \log_e(3x - 1)^2 - [\log_e x^7 + \log_e(2x^3 - 3)]$$
$$\log_e y = \log_e(x^2 - 5) + 2\log_e(3x - 1) - 7\log_e x - \log_e(2x^3 - 3)$$

Differentiating each term accordingly gives:

$$\frac{1}{y}\frac{dy}{dx} = \frac{2x}{x^2 - 5} + \frac{2 \times 3}{3x - 1} - \frac{7}{x} - \frac{6x}{2x^3 - 3}$$

Note that each function of $x$ is differentiated first, and the value obtained is then divided by the original function. Hence, the expression above simplifies to gives:

$$\frac{1}{y}\frac{dy}{dx} = \frac{2x}{x^2 - 5} + \frac{6}{3x - 1} - \frac{7}{x} - \frac{6x}{2x^3 - 3}$$

Multiply both sides by y to obtain $\dfrac{dy}{dx}$ as follows:

$$\frac{dy}{dx} = y\left(\frac{2x}{x^2 - 5} + \frac{6}{3x - 1} - \frac{7}{x} - \frac{6x}{2x^3 - 3}\right)$$

Now replace y with its original value from the question as follows:

$$\frac{dy}{dx} = \left[\frac{(x^2 - 5)(3x - 1)^2}{x^7(2x^3 - 3)}\right]\left(\frac{2x}{x^2 - 5} + \frac{6}{3x - 1} - \frac{7}{x} - \frac{6x}{2x^3 - 3}\right)$$

6. If $y = \dfrac{(2x-1)(x+4)(6x-5)}{(3-x)^2}$

<u>Solution</u>

$$y = \dfrac{(2x-1)(x+4)(6x-5)}{(3-x)^2}$$

This is a complex function where product rule and quotient rule would be difficult to apply. Hence, we apply logarithmic differentiation as follows:

$$\log_e y = \log_e\left[\dfrac{(2x-1)(x+4)(6x-5)}{(3-x)^2}\right]$$

Applying the theory of logarithm gives:

$$\log_e y = \log_e[(2x-1)(x+4)(6x-5)] - \log_e(3-x)^2$$

$$\log_e y = \log_e(2x-1) + \log_e(x+4) + \log_e(6x-5) - 2\log_e(3-x)$$

Differentiating accordingly gives:

$$\frac{1}{y}\frac{dy}{dx} = \frac{2}{2x-1} + \frac{1}{x+4} + \frac{6}{6x-5} - \frac{2(-1)}{3-x}$$

$$\frac{1}{y}\frac{dy}{dx} = \frac{2}{2x-1} + \frac{1}{x+4} + \frac{6}{6x-5} + \frac{2}{3-x}$$

Multiply both sides by y. This gives:

$$\frac{dy}{dx} = y\left(\frac{2}{2x-1} + \frac{1}{x+4} + \frac{6}{6x-5} + \frac{2}{3-x}\right)$$

Finally replace y with its original expression as follows:

$$\frac{dy}{dx} = \left[\frac{(2x-1)(x+4)(6x-5)}{(3-x)^2}\right]\left(\frac{2}{2x-1} + \frac{1}{x+4} + \frac{6}{6x-5} + \frac{2}{3-x}\right)$$

7. Find the derivative of $x^{x^2}$

<u>Solution</u>

$$y = x^{x^2}$$

$$\log_e y = \log_e x^{x^2}$$

$$\log_e y = x^2\log_e x$$

$$\frac{1}{y}\frac{dy}{dx} = x^2\left(\frac{1}{x}\right) + \log_e x(2x)$$

$$\frac{1}{y}\frac{dy}{dx} = x + 2x\log_e x$$

$$\frac{1}{y}\frac{dy}{dx} = x(1 + 2\log_e x)$$

$$\frac{dy}{dx} = yx(1 + 2\log_e x)$$

$$= x^{x^2}x(1 + 2\log_e x) \qquad \text{(y has been replaced with } x^{x^2})$$

$$\frac{dy}{dx} = x^{x^2+1}(1 + 2\log_e x)$$

Note that the exponents of $x^{x^2}$ and $x$ were added to obtain $x^{x^2+1}$ since $x$ is also $x^1$

8. If $y = e^{e^x}$ find $\dfrac{dy}{dx}$.

Solution

$y = e^{e^x}$

$\log_e y = \log_e e^{e^x}$

$\log_e y = e^x \log_e e$

$\log_e y = e^x$        (Recall that logarithm of a number to the same base is 1. Hence, $\log_e e = 1$)

$\dfrac{1}{y}\dfrac{dy}{dx} = e^x$        (Note that the derivative of $e^x$ is $e^x$)

$\dfrac{dy}{dx} = ye^x$

$= e^{e^x} e^x$

$\dfrac{dy}{dx} = e^{e^x + x}$        (Their exponents have been added)

9. Find the derivative of $5^{2^x}$

Solution

$y = 5^{2^x}$

$\log_e y = \log_e 5^{2^x}$

$\log_e y = 2^x \log_e 5$

Differentiate the left hand side implicitly. On the right hand side, differentiate $2^x$ and take $\log_e 5$ as a constant. This gives:

$\dfrac{1}{y}\dfrac{dy}{dx} = 2^x \log_e 2 \log_e 5$

Note that $2^x$ is treated in a similar way as $a^x$, and recall that the derivative of $a^x$ is $a^x \log_e a$. Hence the derivative of $2^x$ is $2^x \log_e 2$, and $\log_e 5$ multiplies it since it is a constant. Therefore:

$\dfrac{dy}{dx} = y(2^x \log_e 2 \log_e 5)$

$= 5^{2^x}(2^x \log_e 2 \log_e 5)$

$\dfrac{dy}{dx} = 5^{2^x}(2^x \log_e 2)(\log_e 5)$

10. If $y = x^{\ln(2x^2 - 5)}$, find $\dfrac{dy}{dx}$.

Solution

$y = x^{\ln(2x^2 - 5)}$

$\log_e y = \log_e x^{\ln(2x^2 - 5)}$

$\log_e y = \ln(2x^2 - 5)\log_e x$

$\dfrac{1}{y}\dfrac{dy}{dx} = \ln(2x^2 - 5)\left(\dfrac{1}{x}\right) + \log_e x\left(\dfrac{4x}{2x^2 - 5}\right)$        (Use of product rule)

$$\frac{1}{y}\frac{dy}{dx} = \frac{1}{x}\ln(2x^2 - 5) + \left(\frac{4x}{2x^2 - 5}\right)\log_e x$$

$$\frac{dy}{dx} = y\left[\frac{1}{x}\ln(2x^2 - 5) + \left(\frac{4x}{2x^2 - 5}\right)\log_e x\right]$$

$$\frac{dy}{dx} = x^{\ln(2x^2 - 5)}\left[\frac{\ln(2x^2 - 5)}{x} + \left(\frac{4x}{2x^2 - 5}\right)\log_e x\right]$$

## Exercise 29

1. Find the derivative of $x^{2x}$

2. Find the derivative of $3x^{\ln 2x}$.

3. If $y = \ln(1 - 4x^2)^x$ find $\frac{dy}{dx}$.

4. Find $\frac{dy}{dx}$ if $y = (6x^2 - 5)^{3x}$

5. If $y = \frac{(2x^3 + 1)(x - 2)^3}{3x^2(x^3 - 1)^2}$

6. If $y = \frac{(2x - 1)(x^2 - 2)}{(1 - x)(x - 3)^2}$

7. Find the derivative of $5x^{3x^4}$

8. If $y = e^{e^{3x}}$ find $\frac{dy}{dx}$.

9. Find the derivative of $2^{e^x}$

10. If $y = x^{\ln(x^2 - 4x)}$, find $\frac{dy}{dx}$.

11. If $y = \ln(10 + 2x^2)^x$ find $\frac{dy}{dx}$.

12. Find $\frac{dy}{dx}$ if $y = (3x^3 - 8)^{2x}$

13. If $y = \frac{(x + 3)^2(x - 3)^2}{(x^3 - 1)}$

14. If $y = 5x^{(\ln x)^2}$, find $\frac{dy}{dx}$.

15. Find the derivative of $10x^{3x^2}$

# CHAPTER 30
## DERIVATIVE OF ONE FUNCTION WITH RESPECT TO ANOTHER

We can differentiate one function with respect to another as illustrated by the examples shown below.

**Examples**

1. Differentiate $x^{12}$ with respect to $x^7$.

Solution

Let $u = x^{12}$

And $v = x^7$

$$\frac{du}{dx} = 12x^{11}$$

$$\frac{dv}{dx} = 7x^6$$

Differentiating $x^{12}$ with respect to $x^7$ means differentiating u (i.e. $x^{12}$) with respect to v (i.e. $x^7$).

This means $\frac{du}{dv}$. This follows the parametric equation rule given by:

$$\frac{du}{dv} = \frac{\frac{du}{dx}}{\frac{dv}{dx}}$$

$$= \frac{12x^{11}}{7x^6}$$

$$= \frac{12}{7}x^5$$

2. Differentiate $2e^x$ with respect to $\ln 2x$.

Solution

The differentiation of $2e^x$ with respect to $\ln 2x$ simply means:

$$\frac{\text{The derivative of } 2e^x}{\text{The derivative of } \ln 2x}$$

Hence, $\frac{d(2e^x)}{dx} = 2e^x$

And, $\frac{d(\ln 2x)}{dx} = \frac{2}{2x} = \frac{1}{x}$

Therefore, the differentiation of $2e^x$ with respect to $\ln 2x$ is given by:

$$\frac{2e^x}{\frac{1}{x}}$$

$$= 2xe^x$$

3. Differntiate $\sin 5x$ with respect to $\cos x$.

## Solution

Hence, $\dfrac{d(\sin 5x)}{dx} = 5\cos 5x$

And, $\dfrac{d(\cos x)}{dx} = -\sin x$

Therefore, the differentiation of sin5x with respect to cosx is given by:

$$\dfrac{\text{The derivative of } \sin 5x}{\text{The derivative of } \cos x}$$

$$= \dfrac{5\cos 5x}{-\sin x}$$

$$= -\dfrac{5\cos 5x}{\sin x}$$

4. Find the derivative of $2x^2 - 5$ with respect to $4x - 1$

## Solution

This is obtained as follows:

$$\dfrac{\frac{d(2x^2 - 5)}{dx}}{\frac{d(4x - 1)}{dx}}$$

$$= \dfrac{4x}{4}$$

$$= x$$

## Exercise 30

1. Differentiate $x^3$ with respect to $x^5$.
2. Differentiate $e^{5x}$ with respect to $\ln x^2$
3. Differentiate $\cos x^2$ with respect to $\sin x$.
4. Find the derivative of $4x^3 - 5x^2$ with respect to $(x - 1)^2$
5. Differentiate $7x^4$ with respect to $3x^2$.
6. Differentiate $2a^x$ with respect to $e^x$.
7. Differentiate $\tan^2 x$ with respect to $\cos 5x$.
8. Find the derivative of $\ln(x - 1)$ with respect to $e^{x-1}$
9. Differentiate $\log_e x^5$ with respect to $e^{5x}$.
10. Differentiate $\ln(e^x - e^{-x})$ with respect to $\ln x^x$

# CHAPTER 31
## HIGHER DERIVATIVES (SUCCESSIVE DIFFERENTIATION)

If f($x$) is differentiated it gives the first derivative denoted by f'($x$) or $\dfrac{dy}{dx}$. If we differentiate f'($x$),

it gives the second derivative denoted by f''($x$) or $\dfrac{d^2y}{dx^2}$ (read as, dee two y dee $x$ squared). Other

higher derivatives such as $\dfrac{d^3y}{dx^3}, \dfrac{d^4y}{dx^4}$ etc can also be obtained depending on the function.

**Examples**

1. Find the first, second and third derivatives of $2x^5 - 3x^4 + 5x^2 - 6$

Solution

$$y = 2x^5 - 3x^4 + 5x^2 - 6$$

$$\frac{dy}{dx} = 10x^4 - 12x^3 + 10x$$

$\dfrac{d^2y}{dx^2}$ is obtained by differentiating $\dfrac{dy}{dx}$, which means to differentiate $10x^4 - 12x^3 + 10x$.

Hence, $\dfrac{d^2y}{dx^2} = 40x^3 - 36x^2 + 10$

$\dfrac{d^3y}{dx^3}$ is obtained by differentiating $40x^3 - 36x^2 + 10$ as follows:

$$\frac{d^3y}{dx^3} = 120x^2 - 72x$$

In summary, the first derivative $\left(\dfrac{dy}{dx}\right)$ is $10x^4 - 12x^3 + 10x$, the second derivative $\left(\dfrac{d^2y}{dx^2}\right)$ is $40x^3 -$

$36x^2 + 10$, while the third derivative $\left(\dfrac{d^3y}{dx^3}\right)$ is $120x^2 - 72x$.

2. If $y = \ln x^2$, find $\dfrac{d^2y}{dx^2}$.

Solution

$$y = \ln x^2$$

$$\frac{dy}{dx} = \frac{2x}{x^2}$$

$$= \frac{2}{x}$$

$$\frac{d^2y}{dx^2} = \frac{d\left(\frac{2}{x}\right)}{dx}$$

$$= \frac{d(2x^{-1})}{dx}$$

$$= -2x^{-2}$$

$$\frac{d^2y}{dx^2} = \frac{-2}{x^2}$$

3. Find $\frac{d^3y}{dx^3}$ given that $y = e^{x^3}$

Solution

$$y = e^{x^3}$$

$$\frac{dy}{dx} = 3x^2 e^{x^3} \qquad \text{(Note that } 3x^2 \text{ is from the derivative of } x^3\text{)}$$

We now use product rule to obtain $\frac{d^2y}{dx^2}$ as follows:

$$\frac{d^2y}{dx^2} = 3x^2 \left[\frac{d\left(e^{x^3}\right)}{dx}\right] + e^{x^3}\left[\frac{d\left(3x^2\right)}{dx}\right]$$

$$= 3x^2 \left(3x^2 e^{x^3}\right) + e^{x^3}(6x)$$

$$= 9x^4 e^{x^3} + 6x e^{x^3}$$

$$\frac{d^2y}{dx^2} = e^{x^3}\left(9x^4 + 6x\right)$$

Finally, let us also use product rule to obtain $\frac{d^3y}{dx^3}$ as follows:

$$\frac{d^3y}{dx^3} = e^{x^3}\left[\frac{d\left(9x^4 + 6x\right)}{dx}\right] + \left(9x^4 + 6x\right)\left[\frac{d\left(e^{x^3}\right)}{dx}\right]$$

$$= e^{x^3}\left(36x^3 + 6\right) + \left(9x^4 + 6x\right)\left(3x^2 e^{x^3}\right)$$

$$= 36x^3 e^{x^3} + 6e^{x^3} + 27x^6 e^{x^3} + 18x^3 e^{x^3}$$

$$= 27x^6 e^{x^3} + 36x^3 e^{x^3} + 18x^3 e^{x^3} + 6e^{x^3}$$

$$= 27x^6 e^{x^3} + 54x^3 e^{x^3} + 6e^{x^3}$$

$$\frac{d^3y}{dx^3} = 3e^{x^3}\left(9x^6 + 18x^3 + 2\right)$$

4. If $y = \sin 2x^3$, find the third derivative of $y$.

Solution

$$y = \sin 2x^3$$

$$\frac{dy}{dx} = \cos 2x^3 \frac{d\left(2x^3\right)}{dx} \qquad \text{(Recall that the derivative of } \sin x \text{ is } \cos x\text{)}$$

$$= \cos 2x^3 \left(6x^2\right)$$

$$\frac{dy}{dx} = 6x^2 \cos 2x^3$$

We now use product rule to obtain $\frac{d^2y}{dx^2}$ as follows:

$$\frac{d^2y}{dx^2} = 6x^2\left[\frac{d\left(\cos 2x^3\right)}{dx}\right] + \cos 2x^3\left[\frac{d\left(6x^2\right)}{dx}\right]$$

$$= 6x^2\left[-\sin 2x^3 \frac{d(2x^3)}{dx}\right] + \cos 2x^3(12x)$$

$$= 6x^2\left[-\sin 2x^3(6x^2)\right] + 12x\cos 2x^3$$

$$\frac{d^2y}{dx^2} = -36x^4\sin 2x^3 + 12x\cos 2x^3$$

Again, the use of product rule gives us $\frac{d^3y}{dx^3}$ as follows:

$$\frac{d^3y}{dx^3} = -36x^4\left[\frac{d(\sin 2x^3)}{dx}\right] + \sin 2x^3\left[\frac{d(-36x^4)}{dx}\right] + 12x\left[\frac{d(\cos 2x^3)}{dx}\right] + \cos 2x^3\left[\frac{d(12x)}{dx}\right]$$

Note that the derivative of $\sin 2x^3$ is $6x^2\cos 2x^3$ as obtained from $\frac{dy}{dx}$. Similarly, the differentiation of $\cos 2x^3$ will give us $-6x^2\sin 2x^3$ (since the derivative of $\sin x$ and $\cos x$ differs only by sign and the interchanging of sin with cos or cos with sin).

We now replace the derivative of $\sin 2x^3$ and $\cos 2x^3$ with $6x^2\cos 2x^3$ and $-6x^2\sin 2x^3$ respectively in the expression above. This gives:

$$\frac{d^3y}{dx^3} = -36x^4(6x^2\cos 2x^3) + \sin 2x^3(-144x^3) + 12x(-6x^2\sin 2x^3) + \cos 2x^3(12)$$

$$= -216x^6\cos 2x^3 - 144x^3\sin 2x^3 - 72x^3\sin 2x^3 + 12\cos 2x^3$$

$$\frac{d^3y}{dx^3} = -216x^6\cos 2x^3 - 216x^3\sin 2x^3 + 12\cos 2x^3$$

Or, $\frac{d^3y}{dx^3} = -12(18x^6\cos 2x^3 + 18x^3\sin 2x^3 - \cos 2x^3)$

5. Find $\frac{d^2y}{dx^2}$ given that $y = a^{x^2} - 5$

Solution

$$y = a^{x^2} - 5$$

$$\frac{dy}{dx} = a^{x^2}\left[\frac{d(x^2)}{dx}\right]\log_e a$$

$$= a^{x^2}(2x)\log_e a$$

$$\frac{dy}{dx} = 2x\,a^{x^2}\log_e a$$

We now use product rule to find $\frac{d^2y}{dx^2}$ as follows:

$$\frac{d^2y}{dx^2} = 2x\left[\frac{d\left(a^{x^2}\log_e a\right)}{dx}\right] + a^{x^2}\log_e a\left[\frac{d(2x)}{dx}\right]$$

$$= 2x\left[\frac{\log_e a\,d\left(a^{x^2}\right)}{dx}\right] + a^{x^2}\log_e a(2)$$

$$= 2x[\log_e a(2x a^{x^2}\log_e a)] + 2a^{x^2}\log_e a$$

$$= 4x^2 a^{x^2}(\log_e a)^2 + 2a^{x^2}\log_e a$$

$$\frac{d^2y}{dx^2} = 2a^{x^2}\log_e a(2x^2\log_e a + 1)$$

6. Find the second derivative of $y = \dfrac{\cos x}{x^3}$

<u>Solution</u>

$$y = \frac{\cos x}{x^3}$$

We use quotient rule as follows:

$$\frac{dy}{dx} = \frac{x^3\frac{d(\cos x)}{dx} - \cos x\frac{d(x^3)}{dx}}{(x^3)^2}$$

$$= \frac{x^3(-\sin x) - \cos x(3x^2)}{x^6}$$

$$\frac{dy}{dx} = \frac{-x^3\sin x - 3x^2\cos x}{x^6}$$

Divide each part by $x^6$ to separate into fractions as follows:

$$\frac{dy}{dx} = \frac{-x^3\sin x}{x^6} - \frac{3x^2\cos x}{x^6}$$

$$\frac{dy}{dx} = \frac{-\sin x}{x^3} - \frac{3\cos x}{x^4}$$

We now apply quotient rule again as follows:

$$\frac{d^2y}{dx^2} = \frac{x^3\frac{d(-\sin x)}{dx} - (-\sin x)\frac{d(x^3)}{dx}}{(x^3)^2} - \left[\frac{x^4\frac{d(3\cos x)}{dx} - 3\cos x\frac{d(x^4)}{dx}}{(x^4)^2}\right]$$

$$= \frac{x^3(-\cos x) + \sin x(3x^2)}{x^6} - \left[\frac{x^4(-3\sin x) - 3\cos x(4x^3)}{x^8}\right]$$

$$= \frac{-x^3\cos x + 3x^2\sin x}{x^6} + \frac{3x^4\sin x + 12x^3\cos x}{x^8}$$

Separating into fractions gives:

$$\frac{d^2y}{dx^2} = \frac{-x^3\cos x}{x^6} + \frac{3x^2\sin x}{x^6} + \frac{3x^4\sin x}{x^8} + \frac{12x^3\cos x}{x^8}$$

$$= \frac{-\cos x}{x^3} + \frac{3\sin x}{x^4} + \frac{3\sin x}{x^4} + \frac{12\cos x}{x^5}$$

Combining the fractions again by using $x^5$ as LCM gives:

$$\frac{d^2y}{dx^2} = \frac{-x^2\cos x + 3x\sin x + 3x\sin x + 12\cos x}{x^5}$$

$$\frac{d^2y}{dx^2} = \frac{6x\sin x + 12\cos x - x^2\cos x}{x^5}$$

7. Find the third derivative of $y = \log_e(1 + 2x)^2$

<u>Solution</u>

$$y = \log_e(1 + 2x)^2$$

$$\frac{dy}{dx} = \frac{2(1+2x)^{2-1} \; x \; \frac{d(2x)}{dx}}{(1+2x)^2}$$  [Note the use of chain rule in finding the derivative of $(1 + 2x)^2$]

$$= \frac{2(1+2x) \; x \; 2}{(1+2x)^2}$$

$$= \frac{4(1+2x)}{(1+2x)^2}$$

$$= \frac{4}{1+2x} \qquad (1 + 2x \text{ cancels out})$$

$$\frac{dy}{dx} = 4(1 + 2x)^{-1}$$

Using chain rule gives $\frac{d^2y}{dx^2}$ as follows:

$$\frac{d^2y}{dx^2} = -1 \; x \; 4(1 + 2x)^{-1-1} \; x \; \frac{d(2x)}{dx}$$

$$= -4(1 + 2x)^{-2} \; x \; 2$$

$$= -8(1 + 2x)^{-2}$$

Use of chain rule again gives $\frac{d^3y}{dx^3}$ as follows:

$$\frac{d^3y}{dx^3} = -2 \; x \; -8(1 + 2x)^{-2-1} \; x \; \frac{d(2x)}{dx}$$

$$= 16(1 + 2x)^{-3} \; x \; 2$$

$$= 32(1 + 2x)^{-3}$$

$$\frac{d^3y}{dx^3} = \frac{32}{(1+2x)^3}$$

8. If y = sec2x, find $\frac{d^3y}{dx^3}$.

Solution

   y = sec2x

$$\frac{dy}{dx} = 2\sec2x\tan2x$$

We now use product rule to obtain $\frac{d^2y}{dx^2}$ as follows:

$$\frac{d^2y}{dx^2} = 2\sec2x \; \frac{d(\tan 2x)}{dx} \; + \; \tan2x \; \frac{d(2\sec 2x)}{dx}$$

$$= 2\sec2x(2\sec^2 2x) \; + \; \tan2x(2 \; x \; 2 \; \sec2x\tan2x)$$

$$\frac{d^2y}{dx^2} = 4\sec^3 2x \; + \; 4\sec2x\tan^2 2x$$

We now use chain rule for $4\sec^3 2x$ and product rule for $4\sec2x\tan^2 2x$ to obtain $\frac{d^3y}{dx^3}$ as follows:

$$\frac{d^3y}{dx^3} = 3 \; x \; 4 \sec^2 2x \; \frac{d(\sec 2x)}{dx} \; + \; 4\sec2x \; \frac{d(\tan^2 2x)}{dx} \; + \; \tan^2 2x \; \frac{d(4\sec 2x)}{dx}$$

$$= 12\sec^2 2x(2\sec2x\tan2x) \; + \; 4\sec2x \left[2\tan2x \; \frac{d(\tan2x)}{dx}\right] \; + \; \tan^2 2x[4(2\sec2x\tan2x)]$$

307

$$= 24\sec^3 2x\tan 2x \; + \; 8\sec 2x\tan 2x(2\sec^2 2x) \; + \; 8\sec 2x\tan^3 2x$$

$$= 24\sec^3 2x\tan 2x \; + \; 16\sec^3 2x\tan 2x \; + \; 8\sec 2x\tan^3 2x$$

$$\frac{d^3 y}{dx^3} = 40\sec^3 2x\tan 2x + \; 8\sec 2x\tan^3 2x$$

9. Given that y = 2cos5$x$ + 3sin5$x$, prove that $\dfrac{d^2 y}{dx^2}$ + 25y = 0

Solution

$$y = 2\cos5x + 3\sin5x$$

$$\frac{dy}{dx} = 2 \times 5(-\sin5x) + 3 \times 5(\cos5x)$$

$$= -10\sin5x \; + \; 15\cos5x$$

$$\frac{d^2 y}{dx^2} = -10 \times 5(\cos5x) + \; 15 \times 5(-\sin5x)$$

$$= -50\cos5x - 75\sin5x$$

We now obtain 25y as follows:

$$y = 2\cos5x + 3\sin5x$$

$$25y = 25(2\cos5x + 3\sin5x)$$

$$25y = 50\cos5x + 75\sin5x$$

We now simplify $\dfrac{d^2 y}{dx^2}$ + 25y as follows:

$$\frac{d^2 y}{dx^2} + 25y = -50\cos5x - 75\sin5x + 50\cos5x + 75\sin5x$$

$$= 0 \qquad \text{(Equal terms with opposite signs cancel out)}$$

Therefore, $\dfrac{d^2 y}{dx^2}$ + 25y = 0     (As proven above)

10. If y = $x + \sqrt{4 + x^2}$, show that $(4 + x^2)\dfrac{d^2 y}{dx^2} + x\dfrac{dy}{dx} - y = 0$

Solution

$$y = x + \sqrt{4 + x^2}$$

$$y = x + (4 + x^2)^{\frac{1}{2}}$$

$$\frac{dy}{dx} = 1 + \frac{1}{2}(4 + x^2)^{\frac{1}{2} - 1} \times \frac{d(x^2)}{dx}$$

$$= 1 + \frac{1}{2}(4 + x^2)^{-\frac{1}{2}} \times 2x$$

$$\frac{dy}{dx} = 1 + x(4 + x^2)^{-\frac{1}{2}}$$

Using product rule, we obtain $\dfrac{d^2 y}{dx^2}$ as follows:

$$\frac{d^2y}{dx^2} = 0 + x\left[\frac{d(4+x^2)^{-\frac{1}{2}}}{dx}\right] + (4+x^2)^{-\frac{1}{2}}\frac{d(x)}{dx}$$

$$= x\left[-\frac{1}{2}(4+x^2)^{-\frac{1}{2}-1}\frac{d(x^2)}{dx}\right] + (4+x^2)^{-\frac{1}{2}}(1)$$

$$= x\left[-\frac{1}{2}(4+x^2)^{-\frac{3}{2}} \times 2x\right] + (4+x^2)^{-\frac{1}{2}}$$

$$= -x^2(4+x^2)^{-\frac{3}{2}} + (4+x^2)^{-\frac{1}{2}}$$

$$\frac{d^2y}{dx^2} = \frac{-x^2}{(4+x^2)^{\frac{3}{2}}} + \frac{1}{(4+x^2)^{\frac{1}{2}}}$$

Let us now simplify $(4+x^2)\dfrac{d^2y}{dx^2} + x\dfrac{dy}{dx} - y$ as follows:

$$(4+x^2)\left[\frac{-x^2}{(4+x^2)^{\frac{3}{2}}} + \frac{1}{(4+x^2)^{\frac{1}{2}}}\right] + x[1 + x(4+x^2)^{-\frac{1}{2}}] - [x + (4+x^2)^{\frac{1}{2}}]$$

Expanding the brackets gives:

$$-x^2(4+x^2)^{1-\frac{3}{2}} + (4+x^2)^{1-\frac{1}{2}} + x + x^2(4+x^2)^{-\frac{1}{2}} - x - (4+x^2)^{\frac{1}{2}}$$

$$= -x^2(4+x^2)^{-\frac{1}{2}} + (4+x^2)^{\frac{1}{2}} + x + x^2(4+x^2)^{-\frac{1}{2}} - x - (4+x^2)^{\frac{1}{2}}$$

$$= 0 \quad \text{(Note that equal terms with opposite signs cancel out one another to give zero)}$$

Therefore, $(4+x^2)\dfrac{d^2y}{dx^2} + x\dfrac{dy}{dx} - y = 0$ (As proven above)

## Exercise 31

1. Find the first, second and third derivatives of $x^4 - 2x^3 - 7x^2 + 1$

2. If $y = \ln 2x^3$, find $\dfrac{d^2y}{dx^2}$.

3. Find $\dfrac{d^2y}{dx^2}$ given that $y = e^{5x^4}$

4. If $y = \sin^2 x^5$, find the second derivative of $y$.

5. Find $\dfrac{d^3y}{dx^3}$ given that $y = \ln(4x - 10)$

6. Find the second derivative of $y = \dfrac{\sin x}{x}$

7. Find the third derivative of $y = (\ln x)^2$

8. If $y = \cot x$, find $\dfrac{d^3y}{dx^3}$.

9. Given that $y = \sin^2 x + \cos 2x$, find $\dfrac{d^2y}{dx^2} - y$

10. If $y = 2e^{2x} + 5e^{-x}$ evaluate $\dfrac{d^2y}{dx^2} - \dfrac{dy}{dx} - 2y$

11. Given that $y = \cos^2 x + 2\sin x^2$ find $\dfrac{d^2y}{dx^2} - \dfrac{dy}{dx}$

12. Find the second derivative of $y = \dfrac{\tan x}{3x}$

13. If $y = e^{-x}\cos 2x$, find the second derivative of y.

14. If $y = 2\cos 3x + 5\sin 3x$, evaluate $\dfrac{d^2y}{dx^2} + 9y$

15. Find the second derivative of $y = \dfrac{\sin 3x}{3x}$

# CHAPTER 32
## MISCELLANEOUS PROBLEMS ON DIFFERENTIAL CALCULUS

This chapter covers worked examples on general problems involving combination of topics treated in the various previous chapters. More challenging problems would also be covered here.

**Examples**

1. If $y = \dfrac{1}{(2x^3 - 5)^4}$ find $\dfrac{dy}{dx}$.

Solution

$$y = \frac{1}{(2x^3 - 5)^4}$$

This can be written as:

$$y = (2x^3 - 5)^{-4}$$

We now use chain rule to differentiate it as follows:

$$\frac{dy}{dx} = -4(2x^3 - 5)^{-4-1} \times \frac{d(2x^3 - 5)}{dx}$$
$$= -4(2x^3 - 5)^{-5} \times 6x^2$$
$$= -24x^2(2x^3 - 5)^{-5}$$
$$\frac{dy}{dx} = \frac{-24x^2}{(2x^3 - 5)^5}$$

2. Given that $f(x) = 5x^4 - 3x^3 - 7x^2 + 9$, find $f'(2)$

Solution

$$f(x) = 5x^4 - 3x^3 - 7x^2 + 9$$

We differentiate $f(x)$ to obtain $f'(x)$ as follows:

$$f'(x) = 20x^3 - 9x^2 - 14x$$

In order to find $f'(2)$, we simply substitute 2 for $x$ in $f'(x)$ as follows:

$$f'(x) = 20x^3 - 9x^2 - 14x$$
$$f'(2) = 20(2^3) - 9(2^2) - 14(2)$$
$$= 20(8) - 9(4) - 28$$
$$= 160 - 36 - 28$$
$$f'(2) = 96$$

3. A function $f(x)$ is given by $f(x) = \dfrac{\sqrt{4 + 3x^2}}{x^3}$. Find:

(a) the derivative of $f(x)$

(b) the gradient of $f(x)$ at the point $(2, \dfrac{1}{2})$

<u>Solution</u>

(a) $f(x) = \dfrac{\sqrt{4 + 3x^2}}{x^3}$

We apply quotient rule to differentiate $f(x)$ as follows:

$$f'(x) = \frac{x^3\left[\dfrac{d(\sqrt{4 + 3x^2})}{dx}\right] - \sqrt{4 + 3x^2}\left[\dfrac{d(x^3)}{dx}\right]}{(x^3)^2}$$

$$= \frac{x^3\left[\dfrac{d(4 + 3x^2)^{\frac{1}{2}}}{dx}\right] - (4 + 3x^2)^{\frac{1}{2}}(3x^2)}{x^6}$$

$$= \frac{x^3\left[\dfrac{1}{2}(4 + 3x^2)^{\frac{1}{2}-1} \times \dfrac{d(3x^2)}{dx}\right] - 3x^2(4 + 3x^2)^{\frac{1}{2}}}{x^6}$$

$$= \frac{x^3\left[\dfrac{1}{2}(4 + 3x^2)^{-\frac{1}{2}} \times 6x\right] - 3x^2(4 + 3x^2)^{\frac{1}{2}}}{x^6}$$

$$= \frac{x^3\left[3x (4 + 3x^2)^{-\frac{1}{2}}\right] - 3x^2(4 + 3x^2)^{\frac{1}{2}}}{x^6}$$

$$= \frac{3x^4(4 + 3x^2)^{-\frac{1}{2}} - 3x^2(4 + 3x^2)^{\frac{1}{2}}}{x^6}$$

Let us factorize the expression above by taking $3x^2(4 + 3x^2)^{-\frac{1}{2}}$ as the common factor. This gives:

$$f'(x) = \frac{3x^2(4 + 3x^2)^{-\frac{1}{2}}[x^2 - (4 + 3x^2)]}{x^6}$$

Note that $(4 + 3x^2)^{\frac{1}{2}} \div (4 + 3x^2)^{-\frac{1}{2}} = (4 + 3x^2)^{\frac{1}{2}-(-\frac{1}{2})} = 4 + 3x^2$ as the exponent becomes 1. Recall that exponents are subtracted during division.

$$= \frac{3x^2(4 + 3x^2)^{-\frac{1}{2}}(x^2 - 4 - 3x^2)}{x^6}$$

$$= \frac{3x^2(4 + 3x^2)^{-\frac{1}{2}}(-2x^2 - 4)}{x^6}$$

$$= \frac{-3x^2(4 + 3x^2)^{-\frac{1}{2}}(2x^2 + 4)}{x^6}$$

$$= \frac{-3(4 + 3x^2)^{-\frac{1}{2}}(2x^2 + 4)}{x^4} \qquad \text{(Note that } x^2 \text{ has cancelled out of } x^6\text{)}$$

$$f'(x) = \frac{-3(2x^2 + 4)}{x^4(4 + 3x^2)^{\frac{1}{2}}}$$

(b) The gradient of f(x) at the point $(2, \frac{1}{2})$ is obtained by substituting 2 for $x$ in f'(x). Note that the point $(2, \frac{1}{2})$ is at $x = 2$ and $y = \frac{1}{2}$. We ignore $y = \frac{1}{2}$ since $y$ is not in the expression for f'(x).

Hence, $f'(x) = \dfrac{-3(2x^2 + 4)}{x^4(4 + 3x^2)^{\frac{1}{2}}}$

At $(2, \frac{1}{2})$, $f'(2) = \dfrac{-3(2(2)^2 + 4)}{(2)^4(4 + 3(2)^2)^{\frac{1}{2}}}$

$= \dfrac{-3(8 + 4)}{16(4 + 12)^{\frac{1}{2}}}$

$= \dfrac{-3(12)}{16(16)^{\frac{1}{2}}}$

$= \dfrac{-36}{16 \times 4}$ \qquad (Note that $(16)^{\frac{1}{2}} = \sqrt{16} = 4$)

$f'(2) = \dfrac{-9}{16}$ \qquad (In its lowest term)

Therefore the gradient of f(x) at the point $(2, \frac{1}{2})$ is $\dfrac{-9}{16}$

4. If $x^2y^2 - 3xy + 4xy^3 = 4$, find:

(a) the derivative of the expression

(b) the gradient at $(-1, 2)$

Solution

(a) $\quad x^2y^2 - 3xy + 4xy^3 = 4$

The use of implicit differentiation combined with product rule gives us the derivative as follows:

$$x^2\frac{d(y^2)}{dx} + y^2\frac{d(x^2)}{dx} - \left[3x\frac{d(y)}{dx} + y\frac{d(3x)}{dx}\right] + 4x\frac{d(y^3)}{dx} + y^3(4) = \frac{d(4)}{dx}$$

$$x^2\left(2y\frac{dy}{dx}\right) + y^2(2x) - \left[3x\frac{dy}{dx} + y(3)\right] + 4x\left(3y^2\frac{dy}{dx}\right) + 4y^3 = 0$$

$$2x^2y\frac{dy}{dx} + 2xy^2 - 3x\frac{dy}{dx} - 3y + 12xy^2\frac{dy}{dx} + 4y^3 = 0$$

Collect terms in $\dfrac{dy}{dx}$ on one side of the equation. This gives:

$$2x^2y\frac{dy}{dx} - 3x\frac{dy}{dx} + 12xy^2\frac{dy}{dx} = 3y - 2xy^2 - 4y^3$$

Factorizing the left hand side gives:

$$\frac{dy}{dx}(2x^2y - 3x + 12xy^2) = 3y - 2xy^2 - 4y^3$$

Divide both sides by the terms in the bracket. This gives:

$$\frac{dy}{dx} = \frac{3y - 2xy^2 - 4y^3}{12x^2y - 3x + 12xy^2}$$

(b) At the point (−1, 2) the gradient of the expression is obtained by simply substituting −1 for $x$ and 2 for y in the expression for the derivative. This is done as follows:

$$\frac{dy}{dx} = \frac{3y - 2xy^2 - 4y^3}{12x^2y - 3x + 12xy^2}$$

$$= \frac{3(2) - 2(-1)(2)^2 - 4(2)^3}{12(-1)^2(2) - 3(-1) + 12(-1)(2)^2}$$

$$= \frac{6 + 8 - 32}{24 + 3 - 48}$$

$$= \frac{-18}{-21}$$

$$= \frac{6}{7}$$

5. Find the derivative of $(x + 3y^2)^3 = 7$

Solution

$(x + 3y^2)^3 = 7$

We differentiate implicitly and apply chain rule as follows:

$$3(x + 3y^2)^{3-1} \times \frac{d(x+3y^2)}{dx} = 0$$

$$3(x + 3y^2)^2\left(1 + 6y\frac{dy}{dx}\right) = 0$$

Dividing both sides by $3(x + 3y^2)^2$ gives:

$$1 + 6y\frac{dy}{dx} = 0 \qquad \left(\text{Note that } \frac{0}{3(x + 3y^2)^2} = 0\right)$$

$$6y\frac{dy}{dx} = -1$$

$$\frac{dy}{dx} = -\frac{1}{6y}$$

6. If y = $(3x^2 - 2x + 5)(2x - 3)$, find $\frac{dy}{dx}$.

Solution

y = $(3x^2 - 2x + 5)(2x - 3)$

We differentiate the expression by applying product rule as follows:

$$\frac{dy}{dx} = (3x^2 - 2x + 5)(2) + (2x - 3)(6x - 2)$$

$$= 6x^2 - 4x + 10 + 12x^2 - 4x - 18x + 6$$

$$\frac{dy}{dx} = 18x^2 - 26x + 16$$

7. Find the derivative of $a^{1 + \tan x}$

Solution

y = $a^{1 + \tan x}$

315

Let u = 1 + tan$x$

Hence, y = a$^u$

$$\frac{du}{dx} = \sec^2 x$$

$$\frac{dy}{du} = a^u \log_e a$$

Hence, $\frac{dy}{dx} = \frac{dy}{du} \times \frac{du}{dx}$

$$= a^u \log_e a \times \sec^2 x$$

$$\frac{dy}{dx} = a^{1+\tan x} \sec^2 x \log_e a$$

8. Find the derivative of $\dfrac{\log_e x}{1+\cos x}$

<u>Solution</u>

$$y = \frac{\log_e x}{1+\cos x}$$

We apply product rule as follows:

$$\frac{dy}{dx} = \frac{(1+\cos x)\frac{d(\log_e x)}{dx} - \log_e x \frac{d(1+\cos x)}{dx}}{(1+\cos x)^2}$$

$$= \frac{(1+\cos x)\frac{1}{x} - \log_e x(-\sin x)}{(1+\cos x)^2}$$

$$= \frac{\frac{1+\cos x}{x} + \sin x \log_e x}{(1+\cos x)^2}$$

$$= \frac{\frac{1+\cos x + x\sin x \log_e x}{x}}{(1+\cos x)^2}$$

$$\frac{dy}{dx} = \frac{1+\cos x + x\sin x \log_e x}{x(1+\cos x)^2}$$

9. Differentiate with respect to $x$:  ln(cos$x$ + sin$x$)

<u>Solution</u>

y = ln(cos$x$ + sin$x$)

$$\frac{dy}{dx} = \frac{\frac{d(\cos x + \sin x)}{dx}}{\cos x + \sin x}$$

$$= \frac{-\sin x + \cos x}{\cos x + \sin x}$$

$$\frac{dy}{dx} = \frac{\cos x - \sin x}{\cos x + \sin x}$$

10. If $y = e^{\sin 2x + \cos x}$ find $\frac{dy}{dx}$.

Solution

$y = e^{\sin 2x + \cos x}$

$$\frac{dy}{dx} = \frac{d(\sin 2x + \cos x)}{dx} \times e^{\sin 2x + \cos x}$$

$$= (2\cos 2x - \sin x)(e^{\sin 2x + \cos x})$$

11. Given that $y = e^x - e^{-x}$ show that $\frac{d^3y}{dx^3} + \frac{d^2y}{dx^2} + \frac{dy}{dx} + y = 4e^x$

Solution

$y = e^x - e^{-x}$

$\frac{dy}{dx} = e^x - (-e^{-x})$    (Note that $\frac{d(e^{-x})}{dx} = \frac{d(-x)}{dx} \times e^{-x} = -1 \times e^{-x} = -e^{-x}$)

$\quad = e^x + e^{-x}$

$\frac{d^2y}{dx^2} = e^x - e^{-x}$

$\frac{d^3y}{dx^3} = e^x + e^{-x}$

Let us now substitute corresponding term into $\frac{d^3y}{dx^3} + \frac{d^2y}{dx^2} + \frac{dy}{dx} + y$ as follows:

$(e^x + e^{-x}) + (e^x - e^{-x}) + (e^x + e^{-x}) + (e^x - e^{-x})$

Collecting like terms together gives:

$e^x + e^x + e^x + e^x + e^{-x} - e^{-x} + e^{-x} - e^{-x}$

$= 4e^x$    (Note that $e^{-x}$ cancels out each other)

Hence, $\frac{d^3y}{dx^3} + \frac{d^2y}{dx^2} + \frac{dy}{dx} + y = 4e^x$    (As proven above)

12. Differentiate with respect to $x$:    $\ln\left(\frac{1-3x^2}{1+3x^2}\right)^{\frac{1}{2}}$

Solution

$$y = \ln\left(\frac{1-3x^2}{1+3x^2}\right)^{\frac{1}{2}}$$

$$= \log_e\left(\frac{1-3x^2}{1+3x^2}\right)^{\frac{1}{2}}    \text{(Note that "In" is } \log_e)$$

$$= \log_e\left[\frac{(1-3x^2)^{\frac{1}{2}}}{(1+3x^2)^{\frac{1}{2}}}\right]$$

$$= \log_e(1 - 3x^2)^{\frac{1}{2}} - \log_e(1 + 3x^2)^{\frac{1}{2}}$$

$$y = \frac{1}{2}\log_e(1 - 3x^2) - \frac{1}{2}\log_e(1 + 3x^2)$$

$$\frac{dy}{dx} = \frac{1}{2}\frac{\frac{d(1-3x^2)}{dx}}{1-3x^2} - \frac{1}{2}\frac{\frac{d(1+3x^2)}{dx}}{1+3x^2}$$

$$= \frac{1}{2}\frac{-6x}{1-3x^2} - \frac{1}{2}\frac{6x}{1+3x^2}$$

$$= \frac{-3x}{1-3x^2} - \frac{3x}{1+3x^2} \qquad \text{(Note that } \frac{1}{2} \text{ reduces } 6x \text{ to } 3x)$$

$$= \frac{-3x(1+3x^2) - 3x(1-3x^2)}{(1-3x^2)(1+3x^2)}$$

$$= \frac{-3x - 9x^3 - 3x + 9x^3)}{(1-3x^2)(1+3x^2)}$$

$$= \frac{-6x}{1 + 3x^2 - 3x^2 - 9x^4}$$

$$\frac{dy}{dx} = \frac{-6x}{1 - 9x^4}$$

13. If $y = \dfrac{x}{\sqrt{9 - x^2}}$ show that: $(9 - x^2)\dfrac{d^2y}{dx^2} = 3x\dfrac{dy}{dx}$

Solution

$$y = \frac{x}{\sqrt{9 - x^2}}$$

Applying quotient rule gives $\dfrac{dy}{dx}$ as follows:

$$\frac{dy}{dx} = \frac{\sqrt{9 - x^2}(1) - x\frac{d(\sqrt{9 - x^2})}{dx}}{(\sqrt{9 - x^2})^2}$$

$$= \frac{(9 - x^2)^{\frac{1}{2}}(1) - x\frac{d[(9 - x^2)^{\frac{1}{2}}]}{dx}}{[(9 - x^2)^{\frac{1}{2}}]^2}$$

$$= \frac{(9 - x^2)^{\frac{1}{2}} - x\left[\frac{1}{2}(9 - x^2)^{\frac{1}{2} - 1}\right] \times (-2x)}{9 - x^2} \qquad \text{(Note that } -2x \text{ is from the derivative of } -x^2)$$

$$= \frac{(9 - x^2)^{\frac{1}{2}} - x\left[-x(9 - x^2)^{-\frac{1}{2}}\right]}{9 - x^2}$$

$$= \frac{(9 - x^2)^{\frac{1}{2}} + x^2(9 - x^2)^{-\frac{1}{2}}}{9 - x^2}$$

$$= \frac{(9 - x^2)^{\frac{1}{2}} + \frac{x^2}{(9 - x^2)^{\frac{1}{2}}}}{9 - x^2}$$

$$= \frac{\frac{9 - x^2 + x^2}{(9 - x^2)^{\frac{1}{2}}}}{9 - x^2}$$

$$= \frac{9}{(9 - x^2)^{\frac{1}{2}}(9 - x^2)}$$

$$= \frac{9}{(9 - x^2)^{\frac{3}{2}}}$$

$$\frac{dy}{dx} = 9(9 - x^2)^{-\frac{3}{2}}$$

Applying chain rule gives $\frac{d^2y}{dx^2}$ as follows:

$$\frac{d^2y}{dx^2} = \frac{-3}{2} \times 9(9 - x^2)^{-\frac{3}{2} - 1} \times -2x \qquad \text{(Note that } -2x \text{ is from the derivative of } -x^2\text{)}$$

$$= \frac{-27}{2} \times -2x(9 - x^2)^{-\frac{5}{2}}$$

$$= 27x(9 - x^2)^{-\frac{5}{2}}$$

$$\frac{d^2y}{dx^2} = \frac{27x}{(9 - x^2)^{\frac{5}{2}}}$$

Let us now show that $(9 - x^2)\frac{d^2y}{dx^2} = 3x\frac{dy}{dx}$

We simplify $(9 - x^2)\frac{d^2y}{dx^2}$ as follows:

$$(9 - x^2)\frac{27x}{(9 - x^2)^{\frac{5}{2}}}$$

$$= \frac{27x(9 - x^2)}{(9 - x^2)^{\frac{5}{2}}}$$

$$= 27x(9 - x^2)^{1 - \frac{5}{2}}$$

$$= 27x(9 - x^2)^{-\frac{3}{2}}$$

$$= \frac{27x}{(9 - x^2)^{\frac{3}{2}}}$$

Hence $(9 - x^2)\frac{d^2y}{dx^2}$ gives us $\dfrac{27x}{(9 - x^2)^{\frac{3}{2}}}$

Let us now simplify $3x\frac{dy}{dx}$ as follows:

$$3x[9(9 - x^2)^{-\frac{3}{2}}] \qquad \text{(Note that } \frac{dy}{dx} = 9(9 - x^2)^{-\frac{3}{2}} \text{ as obtained above)}$$

$$= 27x(9 - x^2)^{-\frac{3}{2}}$$

$$= \frac{27x}{(9 - x^2)^{\frac{3}{2}}}$$

319

Hence $3x\dfrac{dy}{dx}$ also gives us $\dfrac{27x}{(9-x^2)^{\frac{3}{2}}}$

Therefore, $(9-x^2)\dfrac{d^2y}{dx^2} = 3x\dfrac{dy}{dx}$ as both sides give $\dfrac{27x}{(9-x^2)^{\frac{3}{2}}}$

14. Given that $y = \dfrac{x}{x-1}$, show that: $(x-1)\dfrac{d^2y}{dx^2} + 2\dfrac{dy}{dx} = 0$

Solution

$$y = \dfrac{x}{x-1}$$

Using quotient rule gives $\dfrac{dy}{dx}$ as follows:

$$\dfrac{dy}{dx} = \dfrac{(x-1)(1) - x(1)}{(x-1)^2}$$

$$= \dfrac{x-1-x}{(x-1)^2}$$

$$\dfrac{dy}{dx} = \dfrac{-1}{(x-1)^2}$$

Using chain rule gives $\dfrac{d^2y}{dx^2}$ as follows:

$$\dfrac{dy}{dx} = \dfrac{-1}{(x-1)^2}$$

$$= -1(x-1)^{-2}$$

$$\dfrac{d^2y}{dx^2} = -2 \times -1(x-1)^{-2-1} \times \dfrac{d(x)}{dx}$$

$$= 2(x-1)^{-3} \times 1$$

$$\dfrac{d^2y}{dx^2} = \dfrac{2}{(x-1)^3}$$

From the question, let us now simplify $(x-1)\dfrac{d^2y}{dx^2} + 2\dfrac{dy}{dx}$ as follows:

$$(x-1)\dfrac{d^2y}{dx^2} + 2\dfrac{dy}{dx}$$

$$= (x-1)\dfrac{2}{(x-1)^3} + (2)\dfrac{-1}{(x-1)^2}$$

$$= \dfrac{2}{(x-1)^2} - \dfrac{2}{(x-1)^2}$$

$$= 0$$

This shows that $(x-1)\dfrac{d^2y}{dx^2} + 2\dfrac{dy}{dx}$ is equal to zero.

15. Find, with respect to $x$, the derivative of $\left(x - \dfrac{5}{x}\right)^3$

## Solution

$$y = \left(x - \frac{5}{x}\right)^3$$

This can also be written as:

$$y = (x - 5x^{-1})^3$$

$$\frac{dy}{dx} = 3(x - 5x^{-1})^{3-1} \times \frac{d(x - 5x^{-1})}{dx} \qquad \text{(By use of chain rule)}$$

$$= 3(x - 5x^{-1})^2 \times 1 - (-1 \times 5x^{-1-1})$$

$$= 3(x - 5x^{-1})^2 \times 1 - (-5x^{-2})$$

$$= 3(x - 5x^{-1})^2 \times (1 + 5x^{-2})$$

$$= 3(x - 5x^{-1})^2(1 + 5x^{-2})$$

$$\frac{dy}{dx} = 3\left(x - \frac{5}{x}\right)^2\left(1 + \frac{5}{x^2}\right)$$

16. Given that $\exp(2x^2 + 2y^2 - 16) = x + y$, find:

(a) $\dfrac{dy}{dx}$

(b) $\dfrac{dy}{dx}$ at $\left(\dfrac{1}{2}, \dfrac{1}{2}\right)$

## Solution

$$\exp(2x^2 + 2y^2 - 16) = x + y$$

This can also be written as

$$e^{2x^2 + 2y^2 - 16} = x + y \qquad \text{(Note that } \exp x = e^x\text{)}$$

We now differentiate implicitly as follows:

$$\frac{d(2x^2 + 2y^2 - 16)}{dx} \times (e^{2x^2 + 2y^2 - 16}) = \frac{d(x)}{dx} + \frac{d(y)}{dx}$$

$$\left(4x + 4y\frac{dy}{dx}\right)(e^{2x^2 + 2y^2 - 16}) = 1 + \frac{dy}{dx}$$

Expanding the bracket gives:

$$4x(e^{2x^2 + 2y^2 - 16}) + 4y\frac{dy}{dx}(e^{2x^2 + 2y^2 - 16}) = 1 + \frac{dy}{dx}$$

Collecting terms in $\dfrac{dy}{dx}$ on the left hand side gives:

$$4y\frac{dy}{dx}(e^{2x^2 + 2y^2 - 16}) - \frac{dy}{dx} = 1 - 4x(e^{2x^2 + 2y^2 - 16})$$

Factorizing the left hand side gives:

$$\frac{dy}{dx}[4y(e^{2x^2 + 2y^2 - 16}) - 1] = 1 - 4x(e^{2x^2 + 2y^2 - 16})$$

Hence, $\dfrac{dy}{dx} = \dfrac{1 - 4x(e^{2x^2 + 2y^2 - 16})}{4y(e^{2x^2 + 2y^2 - 16}) - 1}$

Or, $\dfrac{dy}{dx} = \dfrac{-[4x(e^{2x^2 + 2y^2 - 16}) - 1]}{4y(e^{2x^2 + 2y^2 - 16}) - 1}$

(b) In order to find $\dfrac{dy}{dx}$ at $\left(\dfrac{1}{2}\ \dfrac{1}{2}\right)$, we simply substitute $x = \dfrac{1}{2}$ and $y = \dfrac{1}{2}$ into the expression for $\dfrac{dy}{dx}$ as follows:

$$\frac{dy}{dx} = \frac{-[4x(e^{2x^2+2y^2-16}) - 1]}{4y(e^{2x^2+2y^2-16}) - 1}$$

Since $x = \dfrac{1}{2}$ and $y = \dfrac{1}{2}$, this simplifies to give:

$$\frac{dy}{dx} \text{ at } \left(\frac{1}{2}\ \frac{1}{2}\right) = \frac{-[4\left(\frac{1}{2}\right)(e^{2x^2+2y^2-16}) - 1]}{4\left(\frac{1}{2}\right)(e^{2x^2+2y^2-16}) - 1}$$

$$= \frac{-[2(e^{2x^2+2y^2-16}) - 1]}{2(e^{2x^2+2y^2-16}) - 1}$$

The numerator cancels out the denominator to give:

$$\frac{dy}{dx} = \frac{-1}{1}$$

$$\frac{dy}{dx} = -1$$

17. If $y = x^3 - 2x^2 + 5$, show that: $x\dfrac{dy}{dx} - 3y - 2x^2 + 15 = 0$

<u>Solution</u>

$$y = x^3 - 2x^2 + 5$$

$$\frac{dy}{dx} = 3x^2 - 4x$$

Let us now simplify $x\dfrac{dy}{dx} - 3y - 2x^2 + 15$ as follows:

$x\dfrac{dy}{dx} - 3y - 2x^2 + 15$

$= x(3x^2 - 4x) - 3y - 2x^2 + 15$

$= x(3x^2 - 4x) - 3(x^3 - 2x^2 + 5) - 2x^2 + 15$    (Note that $x^3 - 2x^2 + 5$ has been substituted for y)

$= 3x^3 - 4x^2 - 3x^3 + 6x^2 - 15 - 2x^2 + 15$

$= 0$

Therefore, $x\dfrac{dy}{dx} - 3y - 2x^2 + 15 = 0$ as proven above.

18. Determine $\dfrac{d^2}{dx^2}\left(x\sin\dfrac{1}{x}\right)$

<u>Solution</u>

This means the second derivative of $x\sin\dfrac{1}{x}$

Let $y = x\sin\dfrac{1}{x}$

Or, $y = x\sin x^{-1}$

We now apply product rule to obtain $\dfrac{dy}{dx}$ as follows:

$$\frac{dy}{dx} = x\left[\frac{d(x^{-1})}{dx}\cos x^{-1}\right] + \sin x^{-1}\left[\frac{d(x)}{dx}\right]$$

$$= x(-x^{-2}\cos x^{-1}) + \sin x^{-1}(1)$$

$$= x\left(\frac{-1}{x^2}\cos x^{-1}\right) + \sin x^{-1}$$

$$= \frac{-x}{x^2}\cos x^{-1} + \sin x^{-1}$$

$$= \frac{-\cos x^{-1}}{x} + \sin x^{-1}$$

$$\frac{dy}{dx} = -\frac{1}{x}\cos\frac{1}{x} + \sin\frac{1}{x}$$

Or, $\dfrac{dy}{dx} = -x^{-1}\cos x^{-1} + \sin x^{-1}$

We now obtain $\dfrac{d^2y}{dx^2}$ as follows:

$$\frac{d^2y}{dx^2} = -x^{-1}\left[\frac{d(\cos x^{-1})}{dx}\right] + \cos x^{-1}\left[\frac{d(-x^{-1})}{dx}\right] + \frac{d(\sin x^{-1})}{dx}$$

$$= -x^{-1}\left[\frac{d(x^{-1})}{dx}(-\sin x^{-1})\right] + \cos x^{-1}(x^{-2}) + \left[\frac{d(x^{-1})}{dx}\cos x^{-1}\right]$$

$$= -x^{-1}[-x^{-2}(-\sin x^{-1})] + \cos x^{-1}(x^{-2}) - x^{-2}(\cos x^{-1})$$

$$= -x^{-1}(x^{-2}\sin x^{-1}) + x^{-2}\cos x^{-1} - x^{-2}\cos x^{-1}$$

$$= -x^{-1}(x^{-2}\sin x^{-1}) \quad \text{(Note that } x^{-2}\cos x^{-1} \text{ cancels out)}$$

$$= -x^{-3}\sin x^{-1}$$

$$\frac{d^2y}{dx^2} = -\frac{1}{x^3}\sin\frac{1}{x}$$

19. Given that $y = \dfrac{e^x + e^{-x}}{e^x - e^{-x}}$

(a) find $\dfrac{d^2y}{dx^2}$

(b) show that $\dfrac{d^2y}{dx^2} + 2y\dfrac{dy}{dx} = 0$

Solution

(a) $\quad y = \dfrac{e^x + e^{-x}}{e^x - e^{-x}}$

By using quotient rule we obtain $\dfrac{dy}{dx}$ as follows:

$$\frac{dy}{dx} = \frac{e^x - e^{-x}\left[\frac{d(e^x + e^{-x})}{dx}\right] - \left(e^x + e^{-x}\left[\frac{d(e^x - e^{-x})}{dx}\right]\right)}{(e^x - e^{-x})^2}$$

$$= \frac{(e^x - e^{-x})(e^x - e^{-x}) - (e^x + e^{-x})(e^x + e^{-x})}{(e^x - e^{-x})^2}$$

$$= \frac{(e^x - e^{-x})^2 - (e^x + e^{-x})^2}{(e^x - e^{-x})^2}$$

Separating into fractions gives:

$$= \frac{(e^x - e^{-x})^2}{(e^x - e^{-x})^2} - \frac{(e^x + e^{-x})^2}{(e^x - e^{-x})^2}$$

$$\frac{dy}{dx} = 1 - \frac{(e^x + e^{-x})^2}{(e^x - e^{-x})^2}$$

We now apply quotient and chain rules to obtain $\dfrac{d^2y}{dx^2}$ as follows:

$$\frac{d^2y}{dx^2} = 0 - \frac{(e^x - e^{-x})^2[2(e^x + e^{-x})(e^x - e^{-x})] - (e^x + e^{-x})^2[2(e^x - e^{-x})(e^x + e^{-x})]}{[(e^x - e^{-x})^2]^2}$$

$$= -\left[\frac{2(e^x - e^{-x})^3(e^x + e^{-x}) - 2(e^x + e^{-x})^3(e^x - e^{-x})}{(e^x - e^{-x})^4}\right]$$

$$= \frac{-2(e^x - e^{-x})^3(e^x + e^{-x}) + 2(e^x + e^{-x})^3(e^x - e^{-x})}{(e^x - e^{-x})^4}$$

$$= \frac{2(e^x + e^{-x})^3(e^x - e^{-x}) - 2(e^x - e^{-x})^3(e^x + e^{-x})}{(e^x - e^{-x})^4} \qquad \text{(After rearranging the numerator)}$$

Separating into fractions gives:

$$= \frac{2(e^x + e^{-x})^3(e^x - e^{-x})}{(e^x - e^{-x})^4} - \frac{2(e^x - e^{-x})^3(e^x + e^{-x})}{(e^x - e^{-x})^4}$$

$$\frac{d^2y}{dx^2} = \frac{2(e^x + e^{-x})^3}{(e^x - e^{-x})^3} - \frac{2(e^x + e^{-x})}{e^x - e^{-x}}$$

(b) Let us simplify $\dfrac{d^2y}{dx^2} + 2y\dfrac{dy}{dx}$ as follows:

$$\frac{d^2y}{dx^2} + 2y\frac{dy}{dx}$$

$$= \frac{2(e^x + e^{-x})^3}{(e^x - e^{-x})^3} - \frac{2(e^x + e^{-x})}{e^x - e^{-x}} + 2\left(\frac{e^x + e^{-x}}{e^x - e^{-x}}\right)\left[1 - \frac{(e^x + e^{-x})^2}{(e^x - e^{-x})^2}\right]$$

Expanding bracket gives:

$$= \frac{2(e^x + e^{-x})^3}{(e^x - e^{-x})^3} - \frac{2(e^x + e^{-x})}{e^x - e^{-x}} + 2\left(\frac{e^x + e^{-x}}{e^x - e^{-x}}\right) - \frac{2(e^x + e^{-x})^3}{(e^x - e^{-x})^3}$$

$$= 0 \qquad \text{(Since equal terms with opposite signs cancel out each other)}$$

Therefore, $\dfrac{d^2y}{dx^2} + 2y\dfrac{dy}{dx}$ gives zero as proven above.

Or, $\dfrac{d^2y}{dx^2} + 2y\dfrac{dy}{dx} = 0$

20. Find the derivative of In(tan2x)

Solution

$y = \text{In(tan}2x)$

$$\frac{dy}{dx} = \frac{\frac{d(\tan 2x)}{dx}}{\tan 2x}$$

$$= \frac{2\sec^2 2x}{\tan 2x}$$

Or, $\dfrac{dy}{dx} = \dfrac{2\left(\frac{1}{\cos 2x}\right)\left(\frac{1}{\cos 2x}\right)}{\left(\frac{\sin 2x}{\cos 2x}\right)}$  (Note that $\sec 2x = \dfrac{1}{\cos 2x}$ and $\tan 2x = \dfrac{\sin 2x}{\cos 2x}$)

$$= \left(\frac{2}{\cos 2x}\right)\left(\frac{1}{\cos 2x}\right) \times \frac{\cos 2x}{\sin 2x}$$

$$\frac{dy}{dx} = \frac{2}{\sin 2x \cos 2x}$$  (Since cos2x cancels cos2x)

21. Given that $y = x^3 + 3x^2$, determine $2\dfrac{dy}{dx} - x\dfrac{d^2y}{dx^2}$

Solution

$$y = x^3 + 3x^2$$

$$\frac{dy}{dx} = 3x^2 + 6x$$

$$\frac{d^2y}{dx^2} = 6x + 6$$

Hence, $2\dfrac{dy}{dx} - x\dfrac{d^2y}{dx^2}$ is simplified as follows:

$$2(3x^2 + 6x) - x(6x + 6)$$
$$= 6x^2 + 12x - 6x^2 - 6x$$
$$= 6x$$

Therefore, $2\dfrac{dy}{dx} - x\dfrac{d^2y}{dx^2} = 6x$

22. Find $\dfrac{dy}{dx}$ if $y = 3(3x + \sqrt{x})^2$

Solution

$$y = 3(3x + \sqrt{x})^2$$

Applying chain rule gives:

$$\frac{dy}{dx} = 3 \times 2\,(3x + x^{\frac{1}{2}})^{2-1} \times \frac{d(3x + x^{\frac{1}{2}})}{dx}$$  (Note that $\sqrt{x} = x^{\frac{1}{2}}$)

$$= 6(3x + x^{\frac{1}{2}}) \times (3 + \frac{1}{2}x^{-\frac{1}{2}})$$

$$= 6(3x + x^{\frac{1}{2}})\left(3 + \frac{1}{2x^{\frac{1}{2}}}\right)$$

$$= 6(3x + \sqrt{x})\left(3 + \frac{1}{2\sqrt{x}}\right)$$

Expanding the bracket gives:

$$= 6\left(9x + \frac{3x}{2\sqrt{x}} + 3\sqrt{x} + \frac{1}{2}\right)$$

$$= 6\left(9x + \frac{3}{2}\sqrt{x} + 3\sqrt{x} + \frac{1}{2}\right) \qquad \left(\text{Note that } \frac{3x}{2\sqrt{x}} = \frac{3}{2}x^{1-\frac{1}{2}} = \frac{3}{2}x^{\frac{1}{2}} = 3\sqrt{x}\right)$$

$$= 6\left(9x + \frac{9}{2}\sqrt{x} + \frac{1}{2}\right)$$

$$\frac{dy}{dx} = 54x + 27\sqrt{x} + 3$$

23. If $y = \dfrac{5x^2 + 7}{x^4}$,

(a) find $\dfrac{d^2y}{dx^2}$

(b) show that $x^2\dfrac{d^2y}{dx^2} + 7x\dfrac{dy}{dx} + 8y = 0$

Solution

(a) $y = \dfrac{5x^2 + 7}{x^4}$

$$= \frac{5x^2}{x^4} + \frac{7}{x^4}$$

$$= \frac{5}{x^2} + \frac{7}{x^4}$$

$$y = 5x^{-2} + 7x^{-4}$$

$$\frac{dy}{dx} = -10x^{-3} - 28x^{-5}$$

Similarly,

$$\frac{d^2y}{dx^2} = 30x^{-4} + 140x^{-6}$$

(b) Let us now simplify $x^2\dfrac{d^2y}{dx^2} + 7x\dfrac{dy}{dx} + 8y$ by substituting appropriately as follows:

$$x^2(30x^{-4} + 140x^{-6}) + 7x(-10x^{-3} - 28x^{-5}) + 8(5x^{-2} + 7x^{-4})$$

$$= 30x^{-2} + 140x^{-4} - 70x^{-2} - 196x^{-4} + 40x^{-2} + 56x^{-4}$$

$$= 30x^{-2} + 40x^{-2} - 70x^{-2} + 140x^{-4} + 56x^{-4} - 196x^{-4}$$

$$= 0$$

Therefore $x^2\dfrac{d^2y}{dx^2} + 7x\dfrac{dy}{dx} + 8y$ gives zero as obtained above.

Or, $x^2\dfrac{d^2y}{dx^2} + 7x\dfrac{dy}{dx} + 8y = 0$

24. If $y = (2x + 5)^4 + \dfrac{x - 1}{2x - 1}$ find $\dfrac{dy}{dx}$.

Solution

$$(2x + 5)^4 + \frac{x - 1}{2x - 1}$$

We now use chain rule and quotient rule as follows:

$$\frac{dy}{dx} = 4(2x+5)^{4-1} \times \frac{d(2x+5)}{dx} + \frac{(2x-1)\frac{d(x-1)}{dx} - \left[(x-1)\frac{d(2x-1)}{dx}\right]}{(2x-1)^2}$$

$$= 4(2x+5)^3(2) + \frac{(2x-1)(1) - [(x-1)(2)]}{(2x-1)^2}$$

$$= 8(2x+5)^3 + \frac{(2x-1) - 2(x-1)}{(2x-1)^2}$$

$$= 8(2x+5)^3 + \frac{2x - 1 - 2x + 2}{(2x-1)^2}$$

$$\frac{dy}{dx} = 8(2x+5)^3 + \frac{1}{(2x-1)^2}$$

## Exercise 32

1. If $y = x^2(3x^4 - 5)^3$ find $\frac{dy}{dx}$.

2. Given that $f(x) = 2x^5 - x^4 + 2x^3 + x^2 - 3x + 4$ find $f'(-2)$

3. A function $f(x)$ is given by $f(x) = \frac{\sqrt{(3x^2 - 1)^3}}{x^2}$. Find:

(a) the derivative of $f(x)$

(b) the gradient of $f(x)$ at the point $(1, -2)$

4. If $3xy^3 - y^2 - 4x^3 = 5y$, find:

(a) the derivative of the expression

(b) the gradient at $(-1, 1)$

5. Find the derivative of $(2x^2 + y^2)^2 = 0$

6. If $y = (x^2 + 5)^3(x^3 - 1)$, find $\frac{dy}{dx}$.

7. Find the derivative of $e^{\sin x + \tan x}$

8. Find the derivative of $\frac{x^2}{\sin^2 x}$

9. Differentiate with respect to $x$:  $\ln(\sin^2 x + \cos 3x)$

10. If $y = a^{\cos 5x}$ find $\frac{dy}{dx}$.

11. Given that $y = x^2 e^{-3x}$ evaluate $\frac{d^2y}{dx^2} + \frac{dy}{dx} - y$

12. Differentiate with respect to $x$:  $\sin\left(\frac{1 - 2x^3}{x^2}\right)$

13. If $y = \frac{2x - 1}{x^2}$, find $x^4 \frac{d^2y}{dx^2} - 3(2x - 1)\frac{dy}{dx}$

327

14. Given that $y = 5e^{-2x} + 3e^x$, evaluate $\dfrac{d^2y}{dx^2} + \dfrac{dy}{dx} - 2y$

15. Find the derivative of $\left(\dfrac{1}{x^2} - \dfrac{2}{x}\right)^5$

16. Given that $e^{x^3 - y^3} = xy$, find:

(a) $\dfrac{dy}{dx}$

(b) $\dfrac{dy}{dx}$ at $(1, -1)$

17. If $y = \sin(\sin x)$, evaluate $\dfrac{d^2y}{dx^2} + \tan x \dfrac{dy}{dx} + y\cos^2 x$

18. Determine $\dfrac{d^2}{dx^2}\left(x^2\cos\dfrac{1}{x^2}\right)$

19. Given that $y = \dfrac{1 + e^{-x}}{1 - e^{-x}}$

(a) find $\dfrac{d^2y}{dx^2}$

(b) evaluate $\dfrac{d^2y}{dx^2} - \dfrac{dy}{dx} + y$

20. Find the derivative of $\ln(\sec^2 x)$

21. Given that $y = x + \tan x$, determine $\cos^2 x \dfrac{d^2y}{dx^2} - 2y + 2x$

22. Find $\dfrac{dy}{dx}$ if $y = \left(3x + \dfrac{\sqrt{5x}}{2}\right)^3$

23. If $y = \dfrac{x^3 + 2x^2 - 5x - 3}{x^2}$, find $\dfrac{d^2y}{dx^2}$

24. If $y = (x^3 + 1)^3 + \dfrac{5x - 2}{x^2 + 1}$ find $\dfrac{dy}{dx}$.

25. Given that $y = 2\cos 5x + 7\sin 5x$, determine $\dfrac{d^2y}{dx^2} + 25y$

# ANSWERS TO EXERCISES

## Exercise 1

1. (a) -15    (b) -11    (c) -5    (d) -12    (e) 19    (f) 6    (g) -82

2. (a) -10    (b) -22    (c) 36    (d) 30    (e) -4    (f) -6    (g) 8    (h) 4

3. (a) 12    (b) 24    (c) $\frac{71}{3}$ or $23\frac{2}{3}$    (d) $\frac{19}{8}$ or $2\frac{3}{8}$    (e) -1910

## Exercise 2

1. 1000y metres    2. (38 + m)years, (38 + m + y)years    3. (b - c)years

4. (a) 100y cents    (b) $\frac{n}{60}$ minutes, $\frac{n}{60 \times 60}$    5. $(22x - x^2)$cm$^2$    6. $(x - 1)(x + 10)$cm$^2$

7. $10x - 5y$    8. $8x - 3y - 5$    9. $2a - 3b$    10. $11x^2 + 15x$    11. $2a^2 - 7ab - 15b^2$

12. $15x^2 - 13xy + 2y^2$    13. $16a^2 - 40a + 25$    14. 24    15. 3    16. (a) $lmn$    (b) 6cde

(c) $30a^3x^2y^2$    (d) 20mn    (e) $(x + y)(3x - y)$    17. (a) $\frac{3x}{5}$    (b) $\frac{4x + 3}{9x^2}$    (c) $\frac{3ax - 14b}{30bx}$

(d) $\frac{10bx + 25b + 8ab - 24a}{40ab}$    (e) $\frac{25a^3 - 15a^2 - 15a^2b - 3a + 15b}{15a^2}$    18. (a) $\frac{-15x + 8}{30}$    (b) $\frac{3}{2(3a + 7)}$

(c) $\frac{-7a(a + 2) + 6}{(3a - 1)(a - 2)}$    19. (a) 5    (b) 2    (c) $6a^2b^2$    (d) $ab^2c^2$    (e) $12x^2yz^2$    20. $2(2m - 4n)$

21. $4(3y - 2)$    22. $-10y(2 + 5z)$    23. $m^2(5m^2 - 10 - m)$    24. $(2 + 5m)(x + 3)$    25. $(a - 1)(3b + 7)$

26. $(5b + n)(x - y)$    27. $(5 + y)(3 - x)$    28. $(10 + m)(10 - m)$    29. $(5x + 3y)(5x - 3y)$

30. $7(1 + b)(1 - b)$    (31) $(6p + 2)(6p - 2)$    32. 2600    33. 8400    34. $(x + 3)(x + 3)$

35. $(b - 5)(b + 4)$    36. $(n - 8)(n - 6)$    37. $(5a + 4)(a - 3)$    38. (a) -6    (b) -5    39. (a) 23    (b) -78

(c) $\frac{6}{7}$    40. -12    41. (a) 0    (b) -24

## Exercise 3

1. (a) $-3t^4e^{12}$    (b) $64a^3b^9$    (c) $-(a^4)$    (d) $g^{20}$    (e) $-m$    2. (a) $\frac{1}{3a}$    (b) $\frac{1}{a}$    (c) $7x^{\frac{3}{2}}$    (d) $9x$

3. (a) $x = \frac{1}{25}$    (b) $a = \frac{1}{3}$    (c) $x = 3\frac{1}{2}$    (d) $x = 4$

## Exercise 4

1. (a) $x = \frac{4}{3}$    (b) $a = \frac{5}{16}$    (c) $b = -\frac{4}{7}$    2. (a) $x = \frac{5}{11}$    (b) $x = 1\frac{11}{20}$    (c) $\frac{113}{62}$    3. $p = \frac{md^2}{3(md - t)}$

4. $C = \sqrt{\frac{P^2}{I^2V^2} - \frac{E^2}{I^2}}$    or    $C = \frac{1}{I}\sqrt{\frac{P^2}{V^2} - E^2}$    5. $x = \frac{R^2T - S}{b - aR^2}$    6. $m = 0$    7. $Y = \sqrt[3]{\frac{TC^2}{L} - B}$

or    $Y = \sqrt[3]{\frac{TC^2 + BL}{L}}$    8. $x = \sqrt{\frac{M}{2Q^2}}$    or    $x = \frac{1}{Q}\sqrt{\frac{M}{2}}$

## Exercise 5

1. 8    2. 4    3. Man = 100years, daughter = 20years    4. 10years    5. 150km/hr    6. 150km/hr

7. Eighteen \$10 notes and six \$1 notes    8. 15days    9. 1050litres    10. 2250km    11. 53.3km

12. 12    13. 13, 15 and 17    14. 5 and 24    15. 12    16. 15 and 25    17. 32years    18. $-\frac{5}{4}$

**Exercise 6**

1. $x = -1$, $y = 1$  2. $x = 6$  3. $m = 2$, $n = 2$  4. $a = 10$, $b = 5$  5. $P = 4$, $q = -3$  6. $x = -2$, $y = 2$

7. $x = 3$, $y = 6$  8. $x = \frac{31}{23}$, $y = \frac{3}{23}$  9. $m = 1$, $n = -1$  10. $p = -\frac{11}{23}$, $q = -\frac{3}{23}$  11. $c = 3\frac{1}{2}$, $d = -1\frac{1}{2}$

12. $x = -\frac{60}{53}$, $y = \frac{250}{53}$

**Exercise 7**

1. Book = \$10, pen = \$5  2. Mathematics = 30, Biology = 70  3. 18.75 and 56.25  4. m = 5hours,

n = 3hours  5. $x = -2$, $y = 3$. Sides are 14cm, 18cm and 18cm  6. (a) $x = 6$, $y = 7.5$  (b) 30cm each

7. 47  8. 92  9. 35  10. $\frac{3}{8}$

**Exercise 8**

1. (a) P: (Closed, True)  (b) Q: (Open, Impossible to say)  (c) R: (Neither, not applicable)

2. (a) Stanley is not neat  (b) Jane is a doctor  (c) $m \neq -5$  3. (a) $Y \Rightarrow X$  (b) $\sim Z \Rightarrow Y$

4. (a) If r is negative then $r^3$ is negative  (b) If $y^3$ is equal to -64 then y is equal to -4

5. Abel is intelligent $\Rightarrow$ Abel is clever  6. (a) If you do not travel by road, it will not take you 11 hours.

(b) Those students who do not offer mathematics will not score above 80

7.

| p | q | ~p | ~q | ~p ∧ ~q | ~(~p ∧ ~q) | p ∨ q |
|---|---|----|----|---------|------------|-------|
| T | T | F | F | F | T | T |
| T | F | F | T | F | T | T |
| F | T | T | F | F | T | T |
| F | F | T | T | T | F | F |

8. Valid  9. Valid  10. Not valid  11. Not valid

**Exercise 9**

1. (a) a(7a + 9)  (b) (m + 4)(m - 4)  (c) (3p + 11)(3p - 11)  (d) 8(2b + 5)(2b - 5)  2. (a) (3$x$ + 5)($x$ - 3)

(b) ($x$ + 9)($x$ - 8)  (c) (5$x$ + 2)($x$ + 4)  (d) (10$x$ + 6)($x$ - 1)  3. (a) Yes  (b) Yes  (c) No  (d) No

(e) No  4. (a) a = 0, or -4  (b) b = -6 or 6  (c) $y = \frac{7}{2}$ or $-\frac{7}{2}$  (d) $x = 4$ or 1  5. (a) $x = \frac{4}{3}$ or -1

(b) y = 2 or -3  (c) y = 2 or 18  (d) $m = \frac{1}{2}$ or $-\frac{4}{5}$  (e) $x = \frac{6}{7}$ or -1  6. (a) $x^2 - 3x - 40 = 0$

(b) $4x^2 + 27x - 7 = 0$  (c) $5x^2 + 32x + 12 = 0$  7. (a) $\frac{15}{2}$  (b) -4  8. (a) $-\frac{4}{3}$  (b) $-\frac{11}{3}$

9. M = 5. The other root is $-\frac{9}{2}$    10. K = -8. The other root is $-\frac{11}{3}$    11. K = -6, P = -27

12. A = 9, B = 30    13. K = 12 or -12. The two possible equations are $m^2 - 12m + 27 = 0$ and $m^2 +$

$12m + 27 = 0$    14. 64    15. $\frac{169}{4}$    16. $a^2 - 6a + 9$    17. $36m^2 + 60m + 26$    18. $x - 4$    19. $x + 0.6$

20. (a) y = -11 or 2    (b) $x = 3$ or 8    (c) $x = 4$ or $-\frac{3}{2}$    (d) $x = 1$ or $\frac{7}{3}$    (e) y = -1 or $-\frac{6}{5}$    (f) m = 2

or $-\frac{7}{2}$

(g) $x = 5.56$ or 1.44    21. (a) $x = 3$ or -4    (b) $x = 1$ or $\frac{10}{7}$    (c) $x = -\frac{1}{2}$ or 4    (d) $x = 2.76$ or -1.09

(e) $x = 0.52$ or -1.92    (f) n = $\frac{3}{2}$ twice

## Exercise 10

1. -4 and -18 or 4 and 18    2. -3 and 7 or -7 and 3    3. 11years and 18years    4. 2.89cm    5. 2m
6. 11years    7. 4 years    8. -16 and -18 or 16 and 18    9. 10    10. 12years and 36years

## Exercise 11

1. (a) $x = \frac{5y}{4}$    (b) $x = 12.5$    (c) y = 12.8    2. (a) h $= \frac{5}{3}\sqrt{P}$    (b) p = 144    (c) h $= \frac{25}{6}$    3. (a) p $= \frac{36}{q}$

(b) q $= \frac{9}{5}$    (c) p $= 7\frac{1}{5}$    4. (a) m $= \frac{15}{\sqrt[3]{n}}$    (b) m $= 7\frac{1}{2}$    (c) n = 24,146    5. (a) a $= \frac{8b^2}{3c}$    (b) a $= \frac{20}{3}$

(c) b = 1.225    6. (a) w $= \sqrt{\frac{9h}{5L}}$    (b) w = 2.32    (c) L = 0.576    7. (a) $x = 2 + 3y$    (b) $x = 26$

8. (a) E = 5 + 10F    (b) E = 30    (c) F $= 3\frac{1}{2}$

## Exercise 12

1. (a) (-12, 73) or (2, 3)    (b) (-$\frac{23}{9}$, -$\frac{10}{9}$) or (-1, 2)    (c) (-$\frac{13}{2}$, -$\frac{19}{2}$) or (-2, 1)    (d) (-$\frac{3}{5}$, -$\frac{5}{6}$) or ($\frac{1}{2}$, 1)

(e) ($\frac{23}{5}$, -$\frac{3}{5}$) or (3, -1)    (f) (-5, 3)    (g) (2, 1) or ($\frac{14}{17}$, -$\frac{33}{17}$)    2. Boy = 30years old, Father = 60 years old

3. ($\frac{47}{4}$, -$\frac{13}{2}$) or (5, 7)    4. 6years and 11years

## Exercise 13

1. $x > 2$

2. $x < -1$

3. $x < 2\frac{32}{43}$

4. $x \leq -4$

5. -1, -2 and -3    6. 1, 2 and 3    7. 65, 66 and 67    8. 3, 4 and 5

9. $-1 \leq y \leq 2$

10. $-2 < y < 3$

11. $y > -1$ or $y \leq -7$

12. $4 < m < 5$    13. (a) $x \geq 5$, $y \geq 2$, $x + y \leq 10$ and $2x + y \leq 15$, where x = drivers and y = supervisors

(b) 5 supervisors    (c) 5 drivers and 5 supervisors    14. (a) 50 items consisting of 40 bicycles and 10 tricycles    (b) 11 tricycles and 33 bicycles    15. (a) 7 shirts and 5 trousers    (b) $10
16. (a) $320f + 80c \geq 1680$, $8f + 6c \geq 62$ and $f + c \leq 20$, where f = fux tin and c = clusion tin
(b) 4 fux tins and 5 clusion tins    17. (a) $4500    (b) 5 manual and 3 electric mixing machines
18. (a) A = 62units, B = 55 units    (b) $3440    19. 57.45    20. 94.93

## Exercise 14

1. $-5 < x < 6$    2. $x \geq \frac{2}{3}$ or $x \leq -4$    3. $\frac{4}{5} \leq x \leq 1$    4. $-\frac{3}{5} < x < \frac{3}{2}$    5. $-1.95 \leq x \leq 2.95$    6. $-2 < x < \frac{9}{2}$

7. $x \geq \frac{6}{7}$ or $x \leq -1$    8. $-2 \leq x \leq -\frac{7}{5}$    9. $-3 < x < 5$    10. $-1.5 \leq x \leq 3.7$

## Exercise 15

1. 10 units    2. $x = 12$    3. (a) $\begin{pmatrix} -13 \\ 27 \end{pmatrix}$    (b) $\begin{pmatrix} -1 \\ 1 \end{pmatrix}$    (c) $\begin{pmatrix} -2\frac{1}{2} \\ 8 \end{pmatrix}$    (d) 59.1    4. (a) $p = -\frac{41}{23}$, $q = -\frac{7}{23}$

(b) 37.6    5. (a) $\begin{pmatrix} -11 \\ 38 \end{pmatrix}$    (b) $(\sqrt{50}, 81.9°)$    6. $\theta = 33.7°$, $(7.2, 33.7°)$    7. (a) $\begin{pmatrix} -2 \\ 4 \end{pmatrix}$, $\sqrt{20}$    (b) $\begin{pmatrix} -6 \\ -2 \end{pmatrix}$, $\sqrt{40}$

(c) $\begin{pmatrix} 6 \\ -3 \end{pmatrix}$, 6.7    (d) $\begin{pmatrix} -4 \\ -6 \end{pmatrix}$, 7.2    8. (a) Q = (-1, 6), R = (8, 9)    (b) $\begin{pmatrix} 1\frac{1}{2} \\ 2\frac{1}{2} \end{pmatrix}$    9. 29.4units    10. 14.3

11. (a) $\begin{pmatrix} 12 \\ -8 \end{pmatrix}$    (b) $\begin{pmatrix} -7 \\ 9 \end{pmatrix}$    (c) $\begin{pmatrix} -5 \\ 8\frac{3}{4} \end{pmatrix}$    (d) 42.2    12. (a) $p = \frac{22}{7}$, $q = \frac{26}{21}$    (b) 19.6

13. (a) PR = $\begin{pmatrix} -4 \\ 5 \end{pmatrix}$, |PR| = $\sqrt{41}$    (b) RQ = $\begin{pmatrix} 4 \\ 2 \end{pmatrix}$, |RQ| = $\sqrt{20}$    (c) -5QS = $\begin{pmatrix} 15 \\ -10 \end{pmatrix}$, |-5QS| = $\sqrt{325}$
(d) 2PQ = $\begin{pmatrix} 16 \\ -6 \end{pmatrix}$, |2PQ| = $\sqrt{292}$ or 17.09

## Exercise 16

1. 9     2. $-\dfrac{8}{5}$     3. 9     4. 14     5. −14     6. $\dfrac{2}{5}$     7. $\dfrac{1}{3}$     8. −1     9. −15

10. 10     11. $\dfrac{1}{4}$     12. 27     13. 12     14. $\dfrac{1}{10}$     15. $\dfrac{3}{7}$     16. Continuous

17. Continuous     18. Discontinuous     19. Not continuous (Discontinuous)

20. Not continuous     21. Continuous     22. Not continuous     23. Not continuous

24(a) 8     (b) 5     26. $\dfrac{7}{2}$     27. $\dfrac{1}{3}$     28. 1     29. $-\dfrac{1}{2}$     30. $2\dfrac{1}{4}$

## Exercise 17

1. 2     2. $2x$     3. $-3x^{-4}$ or $-\dfrac{3}{x^4}$     4(a) $10x + 5h$     (b) $10x$

5(a) $27x^2 + 27xh + 9h^2$     (b) $27x^2$     6(a) $\dfrac{2x^3 + 3x^2\Delta x + x(\Delta x)^2 + 2}{x(x + \Delta x)}$     (b) $2x + \dfrac{2}{x^2}$

7. $6x - 10$     8. $1 - \dfrac{3}{x^2}$     9. $5 - 6x$     10. $2 + \dfrac{1}{5} = 2\dfrac{1}{5}$

## Exercise 18

1(a) $40x^4$     (b) $2x^4$     (c) $\dfrac{1}{3x^{\frac{2}{3}}}$ or $\dfrac{1}{3\sqrt[3]{x^2}}$     (d) $\dfrac{1}{x^{\frac{6}{7}}}$ or $\dfrac{1}{\sqrt[7]{x^6}}$     (e) $-\dfrac{5}{8x^{\frac{13}{8}}}$ or $-\dfrac{5}{8\sqrt[8]{x^{13}}}$

(f) $-\dfrac{5}{x^{\frac{7}{2}}}$ or $-\dfrac{5}{\sqrt{x^7}}$     2. (a) $10x^4 - 12x^3 - 12x^2 + 10x - 6$     (b) $7x^6 + 8x^3 + \dfrac{3}{x^2}$

(c) $18x^5 - 4x^3 - 5 - \dfrac{2}{x^2} + \dfrac{3}{x^4}$     (d) $\dfrac{5}{4\sqrt[4]{x^3}} - \dfrac{5}{3\sqrt[3]{2x^4}}$     3. $8(2x - 5)^3$     4. $\dfrac{-6}{(x^3 - 7)^3}$

5. $\dfrac{6x^2 + 7}{2(2x^3 + 7x)^{\frac{1}{2}}}$     6. $5(21x^2 - 2x)(7x^3 - x^2 + 3)^4$     7. $\left(9 + \dfrac{2}{x^2}\right)\left(3x - \dfrac{2}{3x}\right)^2$

8. $1 + \dfrac{9(6x - 1)}{(3x^2 - x - 10)^2}$     9. $\dfrac{-4x^3}{\sqrt{1 - 2x^4}}$     10. $\dfrac{-5x^2}{\sqrt[3]{(5x^3 - 1)^4}}$

## Exercise 19

1. $12x - 1$     2. $12x^3 + 50x$     3. $\dfrac{15x + 30}{2\sqrt{3 + x}}$     4. $(7x^2 + 30 - 7)(x^2 - 7)^2$

5. $48x^3 + 24x^2 - 6x - 20$     6. $\dfrac{\sqrt{2}\,[(3x^2 - 1)^3 + 36x^2(9x^4 - 6x^2 + 1)]}{2\sqrt{x}}$

7. $\dfrac{15 - 7x}{4(x + 3)^{\frac{1}{4}}}$ or $\dfrac{15 - 7x}{4\sqrt[4]{x + 3}}$     8. $12x^3 + 9x^2 - 46x - 5$     9. $15x^4 - 12x^3 - 3x^2 + 6x - 2$

10. $\dfrac{12x + 1}{2}$     11. $\dfrac{9x^{\frac{7}{2}}}{2}$ or $\dfrac{9\sqrt{x^7}}{2}$     12. $\dfrac{4x^3(7x - 33)}{3\sqrt[3]{2x - 11}}$

13. $-30x^5 - 175x^4 + 20x^3 + 102x^2 - 14x + 1$     14. $75x^4 + 28x^3 + 3x^2 + 28x - 7$

15. $-\dfrac{2}{x^2} + \dfrac{12}{x^5} - \dfrac{25}{x^6}$

## Exercise 20

1. $\dfrac{x^2 - 2x + 4}{(x-1)^2}$    2. $\dfrac{8x^2 + 24x - 3}{(2x+3)^2}$    3. $\dfrac{42x}{(3x^2+1)^2}$    4. $\dfrac{1}{(1-x)^{\frac{3}{2}}\sqrt{x+1}}$

5. $\dfrac{4(x^3-2)(x^3+1)}{x^3}$    6. $\dfrac{-6(x^2+1)}{x^4\sqrt{3x^2+2}}$    7. $\dfrac{-8x^2+12x+24}{3(2x+1)^3(x^2-x-4)^{\frac{2}{3}}}$    8. $\dfrac{10x^3-2x^2-x+4}{2(2-x)^{\frac{3}{2}}}$

9. $\dfrac{-6x}{(1-3x^2)^2}$    10. $\dfrac{4}{(x-2)^2}$    or    $\dfrac{4}{(2-x)^2}$

## Exercise 21

1. $\dfrac{3t}{2}$    2. $\dfrac{2(4-t^3)^2}{3t}$    3. $\dfrac{2r}{3(l+2r)}$    4. $\dfrac{2t-1}{10t-1}$    5. $\dfrac{-2m(v-u)}{(u+v)t^2}$    6. $\dfrac{5(t^4+1)^2}{3(t^2-1)^2}$

7. $\dfrac{1-12t^2}{2t}$    8. $\dfrac{15t^2}{2}$    9. $\dfrac{8}{r}$    10. $\dfrac{2s^3}{3}$

## Exercise 22

1. $\dfrac{-15x^2}{3-2y}$    or    $\dfrac{15x^2}{2y-3}$    2. $\dfrac{-4x}{9y^2}$    3. $\dfrac{3y^2+12x^2-y}{x-6xy}$    4. $\dfrac{3x^2-y^2-4x}{2xy}$    5. $\dfrac{x^2}{y}$

6. $\dfrac{4xy}{2x^2+5}$    7. $\dfrac{-2xy}{x^2+6y^2-1}$    8. $\dfrac{4y}{10y-4x-1}$    9. $\dfrac{2xy^2-1}{2y(x^2-1)}$    10. $\dfrac{-5x^4y^2}{3}$

## Exercise 23

1. $2\sec^2 2x$    2. $-\dfrac{1}{5}\sin\dfrac{1}{5}x$    3. $-50\sin 5x$    4. $3\sin^2 x\cos x$    5. $2\sec^2 x\tan x$

6. $60x^4\sin^3 3x^5\cos 3x^5$    7. $-12x\cot 6x^2\csc 6x^2$    8. $10x^4\sec 2x^5\tan 2x^5$    9. $6x\sec^2 3x^2$

10. $-x(3x\sin 3x - 2\cos 3x)$    11. $\dfrac{6x\cos 2x - 9\sin 2x}{x^4}$    12. $48x^3\sec x^4\tan x^4$

13. $\dfrac{-5\tan x\cos 5x - \sec^2 x(2 - \sin 5x)}{\tan^2 x}$    14. $\dfrac{\sin^2 x(3x\cos x - \sin x)}{2x^2}$

15. $\dfrac{-(2x+6)\csc^2 2x - \cot 2x)}{(x+3)^2}$    16. $\dfrac{-\sin\sqrt[3]{3}}{5x^2}$    17. $\dfrac{2[x-(x^2-3)\tan 2x]}{\sec 2x}$

18. $6x\cos 3x^2 - 18x^3\sin 3x^2$    19. $-\cos x(2\cot x\sin x + \cos x\csc^2 x)$

20. $\dfrac{2(\sin^2 x + \sin^2 x)}{(\sin x + \cos x)^2}$    or    $\dfrac{2}{(\sin x + \cos x)^2}$    21. $\dfrac{4\sec 4x\sin(4x-2) + \tan 4x\cos(4x-2)}{\cos^2(4x-2)}$

22. $-3\sin 3x - x^2\sec x\tan x - 2x\sec x$    23. $27x^2\cos x^3\sin^8 x^3$    24. $\dfrac{3\cos 6\sqrt{x}}{\sqrt{x}}$

25. $3\cos x^3 - \dfrac{2\sin x^3}{x^3}$    26. $-30x^4\cos^2 2x^5\sin 2x^2$    27. $-10\sin 2x\sin 10x - 2\cos 2x\cos 10x$

28. $6x^2\cos x^4 - \dfrac{3\sin x^4}{2x^2}$    29. $\dfrac{(10x-5)\sec^2 5x - 2\tan 5x}{(2x-1)^2}$    30. $\dfrac{\sec^2 x\tan x}{x} - \dfrac{\tan^2 x}{2x^2}$

334

## Exercise 24

1. $\dfrac{1}{3\sqrt[3]{7y^2}}$  2. $\dfrac{3y^2}{2\sqrt{y^3-1}}$  3. $\dfrac{1}{5\sqrt[5]{2(y+3)^4}}$  4. $\dfrac{1}{\sqrt[3]{2(3y+27)^2}}$  5. $\dfrac{-1}{(y-2)^2}$

6. $\dfrac{-3}{2y^2\sqrt{\frac{3}{y}+5}}$  7. $2y^3$  8. $5y^4$  9. $\dfrac{-5}{(y-4)^2}$  10. $\dfrac{-1}{3y^2\sqrt[3]{(\frac{1}{y}+8)^2}}$

## Exercise 25

1. $\dfrac{2}{\sqrt{1-4x^2}}$  2. $\dfrac{-1}{\sqrt{1-x^2}}$  3. $\dfrac{-6x}{9x^4+1}$  4. $\dfrac{-4x^3}{x^8+1}$  5. $3\sec^{-1}x+\dfrac{3}{\sqrt{x^2-1}}$

6. $2x+\dfrac{3}{x^2+1}$  7. $\dfrac{1}{(x+5)^2+1}$ or $\dfrac{1}{x^2+10x+26}$  8. $\dfrac{1}{x^2\sqrt{1-\frac{1}{x^2}}}$  9. $\dfrac{-1}{5\sqrt{1-y^2}}$

10. $\dfrac{1}{2y\sqrt{y-1}}$  11. $5\tan^{-1}3x+\dfrac{15x}{9x^2+1}$  12. $2x+\dfrac{20x^4}{\sqrt{1-x^{10}}}$  13. $\dfrac{-3}{x\sqrt{x^6-1}}$

14. $\dfrac{1}{2(y^2+1)}$  15. $\dfrac{1}{15y^{\frac{2}{3}}\sqrt{1-y^{\frac{2}{3}}}}$

## Exercise 26

1. $3\cosh3x-2\sinh x$  2. $-5x\,\text{sech}x(x\tanh x-2)$  3. $\dfrac{2(x\cosh 2x-\sinh 2x)}{3x^3}$

4. $6\cosh3x\sinh3x$  5. $60x^4\cosh^24x^5\sinh4x^5$  6. $-10x^4\text{cosech}2x^5$

7. $-12x\,\text{sech}^32x^2\tanh2x^2$  8. $x(5x\sinh5x+2\cosh5x)$

9. $\dfrac{3\text{cosech}^23x\tanh 5x+5\coth 3x\,\text{sech}^25x}{\coth^23x}$  10. $\dfrac{2\cosh^2x(3x\sinh x-\cosh x)}{3x^2}$

## Exercise 27

1. $\dfrac{1}{x}\log_a e$  2. $\dfrac{3x^2}{x^3+5}\log_a e$  3. $\dfrac{36x^2}{2x^3-5}\log_a e$  4. $\dfrac{2}{3x}\log_a e$  5. $\dfrac{4x}{(x^2+1)(1-x^2)}\log_a e$

or $\dfrac{-4x}{(x^2+1)(x^2-1)}\log_a e$  or  $\dfrac{-4x}{(x^4-1)}\log_a e$  6. $\dfrac{15x^2}{5x^3-1}\log_5 e$  7. $\dfrac{-2}{x}\log_2 e$  8. $\dfrac{2x}{x^2+3}$

9. $\dfrac{3\ln^2x}{x}$  10. $\dfrac{5x^4}{2x^5-1}$  11. $2x^3(4\ln x+1)$  12. $\dfrac{-28}{3-7x}$ or $\dfrac{28}{7x-3}$

13. $6x\ln(4x^3+1)+\dfrac{36x^4}{4x^3+1}$  14. $\dfrac{2-4\ln x}{x^3}$  15. $\dfrac{-30x}{1-5x^2}$ or $\dfrac{30x}{5x^2-1}$

## Exercise 28

1. $10a^{2x}\log_e a$  2. $a^{5x^2-x}\ln a(10x-1)$  3. $a^{3x}x^3(3x\ln a+4)$

4. $-2e^{-2x}-3e^{-x}$ or $-e^{-2x}(2+3e^x)$  5. $\dfrac{\sqrt{3}\,e^{\sqrt{3x}}}{\sqrt{x}}$  6. $6x^2e^{x^2}(2x^2+3)$

7. $\dfrac{5\ln a(a^{10x}-1)}{a^{5x}}$  8. $\dfrac{6e^{4x}(2x\ln 2x^2+1)}{x}$  9. $\dfrac{3x\sqrt{e^{5x}}(5x+4)}{2a}$  10. $x^3(\log_{10}e+4\log_{10}7x)$

11. $3\ln 6(6^{3x})$  12. $2x - 3(e^{x^2-3x})$  13. $\dfrac{-(5x+3)e^{\frac{3}{x}}}{5x^7}$  14. $6e^{-3x}$

15. $3(2x+1)(x^2+x)^2 e^{(x^2+x)^3}$  16. $2^{x^2}2x\ln 2$  17. $2a^{2x}\ln a$  18. $\dfrac{e^x(x\ln 10x^3 + 3)}{x}$

19. $\dfrac{\sqrt{5}(4x+1)e^{2x}}{2\sqrt{x}}$  20. $2x^4(4\log_{10}e + 5\log_{10}5x^4)$

## Exercise 29

1. $x^{2x}(2 + 2\ln x)$  2. $\dfrac{3x^{\ln 2x}(\ln 2x + \ln x)}{x}$  3. $\ln(1 - 4x^2) - \dfrac{8x^2}{1 - 4x^2}$

4. $(6x^x - 5)^{3x}\left[3\ln(6x^2 - 5) + \dfrac{36x^2}{6x^2 - 5}\right]$  5. $\dfrac{(2x^3+1)(x-2)^3}{3x^2(x^3-1)^2}\left(\dfrac{6x^2}{2x^3+1} + \dfrac{3}{x-2} - \dfrac{2}{x} - \dfrac{6x^2}{x^3-1}\right)$

6. $\dfrac{(2x-1)(x^2-2)}{(1-x)(x-3)^2}\left(\dfrac{2}{2x-1} + \dfrac{2x}{x^2-2} + \dfrac{1}{1-x} - \dfrac{2}{x-3}\right)$

7. $15x^3x^{3x^4}(4\ln x + 1)$  or  $15x^{3x^4+3}(4\ln x + 1)$  8. $3e^{3x}(e^{e^{3x}})$  or  $3e^{e^{3x}+3x}$

9. $2e^x e^x \ln 2$  10. $\dfrac{x^{\ln(x^2-4x)}[(x-4)\ln(x^2-4x) + (2x-4)\ln x]}{x(x-4)}$  11. $\ln(2x^2 + 10) + \dfrac{4x^2}{2x^2+10}$

12. $(3x^2 - 8)^{2x}\left[2\ln(3x^3 - 8) + \dfrac{18x^3}{3x^3-8}\right]$  13. $\dfrac{(x+3)^2(x-3)^2}{x^3-1}\left(\dfrac{2}{x+3} + \dfrac{2}{x-3} - \dfrac{3x^2}{x^3-1}\right)$

14. $15x^{(\ln x)^2}(\ln x)^2$  15. $30x^{3x^2}x(2\ln x + 1)$  or  $30x^{3x^2+1}(2\ln x + 1)$

## Exercise 30

1. $\dfrac{3}{5x^2}$  2. $\dfrac{5xe^{5x}}{2}$  3. $\dfrac{-2x\sin x^2}{\cos x}$  4. $\dfrac{x(6x-5)}{x-1}$  5. $\dfrac{14x^2}{3}$  6. $\dfrac{2a^x\ln a}{e^x}$

7. $\dfrac{-2\tan x \sec^2 x}{5\sin 5x}$  8. $\dfrac{1}{(x-1)(e^{x-1})}$  9. $\dfrac{1}{xe^{5x}}$  10. $\dfrac{e^x + e^{-x}}{(e^x - e^{-x})(1 + \ln x)}$

## Exercise 31

1. $\dfrac{dy}{dx} = 2x(2x^2 - 3x - 7)$, $\dfrac{d^2y}{dx^2} = 2(6x^2 - 6x - 7)$, $\dfrac{d^3y}{dx^3} = 2(12x - 6)$  2. $-\dfrac{3}{x^2}$

3. $20x^2e^{5x^4}(20x^4 + 3)$  4. $-50x^8\sin^2 x^5 + 40x^3\cos x^5\sin x^5 + 50x^8\cos x^5$  5. $\dfrac{128}{(4x - 10)^2}$

6. $\dfrac{-(x^2-2)\sin x - 2x\cos x}{x^3}$  7. $\dfrac{4\ln x - 6}{x^3}$  8. $-2\operatorname{cosec}^2 x(\operatorname{cosec}^2 x + 2\cot^2 x)$

9. $2\cos^2 x - 3\sin^2 x - 5\cos 2x$  10. 0

11. $-8x^2\sin x^2 + 4\cos x^2 + 2\sin^2 x - 2\cos^2 x - 4x\cos x^2 + 2\cos x\sin x$

12. $\dfrac{2[(x^2\sec^2 x + 1)\tan x - x\sec^2 x]}{3x^3}$  13. $e^{-x}(4\sin 2x - 3\cos 2x)$  14. 0

15. $\dfrac{-(9x^2-2)\sin 3x - 6x\cos 3x}{3x^3}$

**Exercise 32**

1. $2x(3x^4-5)^3 + 36x(3x^4-5)^3$  or  $2x(3x^4-5)^2(21x^4-5)$    2. 209

3(a) $\dfrac{\sqrt{3x^2-1}\,(3x^2+2)}{x^3}$    (b) $5\sqrt{2}$    4(a) $\dfrac{12x^2-3y^3}{9xy^2-2y-5}$    (b) $-\dfrac{9}{16}$    5. $-\dfrac{2}{y}$

6. $3x(x^2+5)^2(3x^3+5x-2)$    7. $e^{\sin x + \tan x}(\sec^2 x + \cos x)$    8. $\dfrac{2x(\sin x - x\cos x)}{\sin^3 x}$

9. $\dfrac{2\cos x \sin x - 3\sin 3x}{\sin^2 x + \cos 3x}$    10. $-5a^{\cos 5x}\sin 5x \ln a$    11. $5x^2\,e^{-3x} - 10xe^{-3x} + 2e^{-3x}$

12. $-(2x^3+2)\cos\left(\dfrac{1-2x^3}{x^2}\right)$    13. $\dfrac{4x^4-6x^3+12x^2-18x+6}{x^3}$    14. 0

15. $5\left(\dfrac{2}{x^2} - \dfrac{2}{x^3}\right)\left(\dfrac{1}{x^2} - \dfrac{2}{x}\right)^4$  or  $\dfrac{10(x-1)(1-2x)^4}{x^{11}}$    16(a) $\dfrac{3x^2 e^{x^3-y^3} - y}{3y^2 e^{x^3-y^3} + x}$    (b) 1

17. 0    18. $\dfrac{2\sin\frac{1}{x^2}}{x^2} - \dfrac{4\cos\frac{1}{x^2}}{x^4} + 2\cos\dfrac{1}{x^2}$    19(a) $\dfrac{2e^x(e^x+1)}{(e^x-1)^3}$

(b) $\dfrac{4(e^{2x}-e^{-x}) + (1+e^{-x})(e^x-1)^3}{(e^x-1)^3(1-e^{-x})}$    20. $2\tan x$    21. 0

22. $3\left(\dfrac{\sqrt{5}}{4\sqrt{x}} + 3\right)\left(3x + \dfrac{\sqrt{5x}}{2}\right)^2$    23. $\dfrac{-10x-18}{x^4}$  or  $\dfrac{-2(5x+9)}{x^4}$

24. $9x^2(x^3+1)^2 + \dfrac{5+4x-10x^2}{(x^2+1)^2}$    25. 0

Please if you found this book well simplified enough for easier understanding, kindly give it a five star rating on amazon so as to encourage people to buy this book, thereby helping them improve on their skills on algebra and differential calculus. Thank you.

If you want to see other books written by the author, just simply search for the author's name, Kingsley Augustine on amazon.com

If you have any enquiry, suggestion or information concerning this book, please contact the author through the email below.

KINGSLEY AUGUSTINE

kingzohb2@yahoo.com